D0142026

PHYSICS FOR CAREER EDUCATION

PHYSICS FOR CAREER EDUCATION

DALE EWEN
Parkland Community College

RONALD J. NELSON
Parkland Community College

NEILL SCHURTER
Rantoul Township High School

LaVERNE M. McFADDEN
Parkland Community College

PRENTICE-HALL, INC., Englewood Cliffs, New Jersey

Library of Congress Cataloging in Publication Data
Main entry under title:

Physics for career education.

1. Physics. I. Ewen, Dale
QC23.P58 530 74–1244
ISBN 0–13–672303–9

Drawings by George Morris

Printed in the United States of America

10 9 8 7 6

PRENTICE-HALL INTERNATIONAL, INC., London
PRENTICE-HALL OF AUSTRALIA, PTY. LTD., Sydney
PRENTICE-HALL OF CANADA, LTD., Toronto
PRENTICE-HALL OF INDIA PRIVATE LIMITED, New Delhi
PRENTICE-HALL OF JAPAN, INC., Tokyo

CONTENTS

2 AREA AND VOLUME MEASUREMENT

3 MOTION

4 EQUATIONS

5 PROBLEM SOLVING

6 FORCES IN ONE DIMENSION

7 VECTORS AND TRIGONOMETRY

8 CONCURRENT FORCES

9 WORK AND ENERGY

10 SIMPLE MACHINES

11 ROTATIONAL MOTION

12 GEARS AND PULLEYS

13 *NON-CONCURRENT FORCES*

14 *MATTER*

15 *FLUIDS*

16 *TEMPERATURE AND HEAT*

17 THERMAL EXPANSION OF SOLIDS AND LIQUIDS

18 GAS LAWS

19 CHANGE OF STATE

20 STATIC ELECTRICITY

21 *DIRECT CURRENT ELECTRICITY*

22 *OHM'S LAW AND DC CIRCUITS*

23 *DC SOURCES*

24 MAGNETISM

25 GENERATORS

26 MOTORS

27 ALTERNATING CURRENT ELECTRICITY

TRANSFORMERS 28

AC CIRCUITS 29

LIGHT 30

appendix *SLIDE RULE*

TABLES

PREFACE

Physics for Career Education has been written as a preparation for students considering a vocational-technical career. It is designed to emphasize physical concepts as applied to the industrial-technical fields and to use these applications to improve the physics and mathematics competence of the student.

This text is written at a language level and at a mathematics level that is cognizant of and beneficial to *most* students in vo-tech programs which do not require a high level of mathematics rigor and sophistication. The authors have assumed that the student has successfully completed one year of high school algebra or its equivalent. Simple equations and formulas are reviewed and any mathematics beyond this level is developed in the text. The manner in which the mathematics topics are integrated into the text displays the need for mathematics in technology. For the better prepared mathematics student, the mathematics sections may be omitted with no loss in continuity. The sections are short and deal with only one concept. The need for the investigation of a physical principle is developed before undertaking its study and many diagrams are used to aid the student in visualizing the concept. A large number of examples is included. The problems at the end of each section allow the student to check his mastery of one concept before moving on to another.

This text is designed to be used in a vocational-technical program in a community college, a technical institute or a high school class for students who plan to pursue a vocational-technical career. The topics were chosen with the assistance of technicians and management of several industries and teaching consultants in various vo-tech areas.

The text is divided into four major areas: mechanics, matter and heat, electricity and magnetism, and light. The emphasis on mechanics is a result of our belief that it is basic to all technical programs.

The chapters on measurement introduce the student to basic units and some mathematics skills such as scientific notation and basic geometry. The need for vectors is developed in the chapter on motion and graphical addition of vectors is used. A separate chapter introduces the student to a problem solving method that is used in the rest of the text. One-dimensional dynamics is then discussed and the need for trigonometry in more complex problems is developed. The chapter on trigonometry includes right triangle trig and the component method for addition of vectors. A more thorough treatment of dynamics and other standard topics follow. Chapters on simple machines and gear systems are also included.

The section on matter includes a discussion of the three states of matter, density, fluids, pressure, and Pascal's principle. The section on heat includes temperature, specific heat, thermal expansion, gas laws, and change of state.

The section on electricity and magnetism begins with a brief discussion of static electricity followed by an extensive treatment of DC circuits including sources, Ohm's law, and series and parallel circuits. The chapters on magnetism, generators and motors are largely descriptive in nature, but allow for a more in-depth study if desired. AC circuits and transformers are given an extensive treatment.

The chapter on light briefly discusses the wave and particle nature of light, but deals mostly with illumination.

The appendix includes a section on the slide rule with a large number of examples and exercises, followed by an extensive array of easy-to-read tables.

The authors wish to thank the Research and Development Unit, Vocational and Technical Education Division, Illinois Board of Vocational Education and Rehabilitation, and Parkland College for their support in the initial version of this text and to the following pilot Illinois high schools: ABL (Allerton-Broadlands-Longview), Bethany, Dallas City, Mahomet-Seymour, Melvin-Sibley, Rantoul Township, St. Joseph-Ogden, and Villa Grove for using and testing these materials. We are grateful to Albert W. Lemmon for the graphic art work of the pilot edition. The typing for the manuscript by Miss Rhonda Saathoff and Mrs. Marcia Olson has been especially appreciated and our special thanks is given to them. The special efforts of Clifton H. Matz, Gayle W. Wright, Sidney E. Barnes, William O. Smith, Walter Miller, Paul N. Thompson, and John Costello of Parkland College were appreciated for serving as consultants.

Champaign, Illinois

DALE EWEN
RONALD J. NELSON
NEILL SCHURTER
LAVERNE M. MCFADDEN

PHYSICS FOR CAREER EDUCATION

1 MEASUREMENT—AN INTRODUCTION TO TECHNICAL PHYSICS

1-1 *WHAT AND HOW DO WE MEASURE?*

In every industry the technician is faced with problems to be solved. In this age, much work can be done by computers and other machines, but man's ability to think and reason makes him a necessary part of our modern industrial world. Machines are the result of man's effort to reduce his own labor and time spent working. In this book we will explore many ways in which people have attempted to make their work easier or of better quality. Let's look at the building of a bridge across a large body of water. To build from only one end would take twice the amount of construction time that building from both ends and meeting in the middle would require. But can you see any problems that might arise if construction is started on both ends?

Examples like this show how very important communication between the people building the bridge is and how important it is that construction on both ends be based on the same system and standards of measurement. The use of a standard system makes possible the employment of any person trained for a particular job.

Another industry where many complex measurements of lengths, volumes, and areas are required is the automobile industry. Let's consider the manufacture of an automobile engine. As you probably know, the many parts are mass-produced and then assembled into the final product. If the various parts are not the correct size and do not fit together properly, the engine will not work. Again, the importance of a uniform system of measurement is shown.

1-2 *LENGTH MEASUREMENT*

Length is one of the most basic measurements. Historically, man used his hands or feet or whatever was handy to make measurements. As civilization progressed, however, the need for everyone to use the same standards increased. Two systems are most widely used today. The English system has been historically used in the English speaking countries and its basic units are the foot, pound, and second. The foot is the basic unit of length in the English system and it may be divided into twelve equal parts or *inches*.

The metric system is used in most all other countries and is used by scientists all over the world. Its basic unit of length is the *meter* which is equal to 39.37 inches.

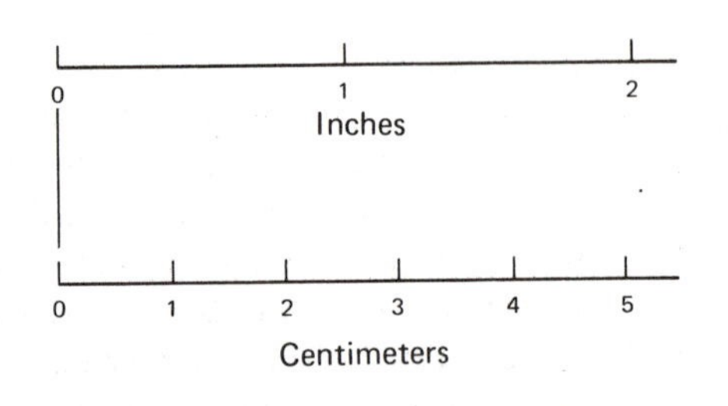

12 inches (in.) = 1 foot (ft)
100 centimeters (cm) = 1 meter (m)
1 in. = 2.54 cm
39.37 in. = 1 meter (m)
3 ft = 1 yard (yd)

The advantage of the metric system is that it is based on multiples of ten (10) of the fundamental units, as in our decimal number system. For example, from Table B in the Appendix:

$$1 \text{ centimeter} = 0.01 \text{ meter}$$
$$1 \text{ kilometer} = 1000 \text{ meters} = 10^3 \text{ meters}$$

Turn to Table C in the Appendix for a listing of the English and metric unit conversions. Lines or distances may be measured in either system directly or indirectly.

To convert from one unit to another we use the fact that, when anything is multiplied by 1, its value is not changed. The correct conversion factor is a fraction whose numerator is equal to the denominator, that is, the fraction equals 1. The numerator should be expressed in the new units and the denominator should be expressed in the old units so that the old units divide (cancel out).

EXAMPLE 1: Express 10 inches in centimeters.

$$1 \text{ in.} = 2.54 \text{ cm} \quad \text{so} \quad 10 \cancel{\text{in.}} \times \left(\frac{2.54 \text{ cm}}{1 \cancel{\text{in.}}}\right) = 25.4 \text{ cm}$$

It is very important that you use the correct units on all quantities and divide (cancel) units where possible. The answer will then appear in the desired units. This practice will aid you greatly in solving problems in later chapters.

The conversion factors you will need are to be found in the Appendix Tables. The examples below will show you how to use these tables.

EXAMPLE 2: Convert 15 miles to kilometers. From the length conversion table we find 1 mile listed in the left-hand column. Moving over to the fourth column, under the heading km, we see that 1 mile = 1.61 km. Then we have:

$$15 \cancel{\text{mile}} \times \left(\frac{1.61 \text{ km}}{1 \cancel{\text{mile}}}\right) = 24.15 \text{ km}$$

EXAMPLE 3: Convert 220 centimeters to inches. Find 1 centimeter in the left-hand column and move to the fifth column under the heading in. We find that 1 centimeter = 0.394 in. Then:

$$220 \cancel{\text{cm}} \times \left(\frac{0.394 \text{ in.}}{1 \cancel{\text{cm}}}\right) = 86.68 \text{ in.}$$

EXAMPLE 4: Convert 3 feet to centimeters. Since there is no direct conversion from feet to centimeters in the tables, we must first change feet to inches and then inches to centimeters:

$$3 \cancel{\text{ft}} \times \left(\frac{12 \cancel{\text{in.}}}{1 \cancel{\text{ft}}}\right) \times \left(\frac{2.54 \text{ cm}}{1 \cancel{\text{in.}}}\right) = 91.44 \text{ cm}$$

PROBLEMS

Use the technique of multiplying by 1 to do the following:

1. The wheel base of a certain automobile is 108 inches. What is it in yards? in feet? in centimeters?
2. The length of a connecting rod is 7 inches. What is its length in centimeters?
3. The distance between two cities is 256 miles. What is the distance in kilometers?

4. Convert the following lengths to meters:
   (a) 5 ft
   (b) 2.7 mi
   (c) 10.6 ft
   (d) 7 yd
5. Convert the following lengths as indicated:
   (a) 5.94 m to feet
   (b) 7.1 cm to inches
   (c) 1.2 in. to centimeters
6. The turning radius of an auto is 20 feet. What is this in meters?
7. Would a wrench with an opening of 25 mm be larger or smaller than a 1-in. wrench?
8. What is the width in inches of 8-mm movie film?
9. How many reamers each 20 cm long can be cut from a bar 6 ft long, allowing 3 mm for each saw cut?
10. If 214 pieces each 47 cm long are ordered to be turned from $\frac{1}{4}$-in. round steel stock and $\frac{1}{8}$ in. waste on each piece is allowed, what length of stock is required?

1-3 *TIME MEASUREMENT*

Airlines and other transportation systems run on time schedules which would be meaningless if we did not have a unit for time measurement. All the common units for time measurement are the same in both the English and metric systems. These units are based on the motion of the earth and the moon. The year is approximately the time required for one complete revolution of the earth about the sun. The month is approximately the time for one complete revolution of the moon about the earth. The day is the time for one rotation of the earth about its axis.

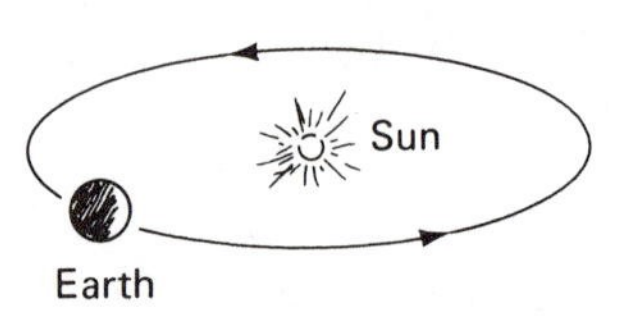

Revolution of Earth

Rotation about Axis

The basic time unit is the *second* (s) which is defined to be 1/86,400 day. The second is not always a conveniently sized unit and others are necessary. The *minute* is 60 seconds, the *hour* is 60 minutes, and the *day* is 24 hours. The *year* is 365 days, except for every fourth year when it is 366 days long. This difference is necessary to keep the seasons at the same time each year since one revolution of the earth about the sun is $365\frac{1}{4}$ days.

Common devices for time measurement are the electric clock and mechanical watches. The accuracy of an electric clock depends on how accurately the 60-Hz (Hertz=cycles per second) line voltage is controlled. In the United States this is controlled very accurately. Most mechanical watches have a balance wheel that oscillates near a given frequency, usually 18,000–36,000 vibrations per hour, and drives the hands of the watch. The accuracy of the watch depends on how well the frequency of oscillation is controlled.

PROBLEMS

Use conversion factors to:

1. Convert 12 minutes to seconds.
2. Convert 7800 seconds to minutes.
3. Convert 8 hours to seconds.
4. Convert 2 days to minutes.
5. Convert 3 years to hours.
6. How many minutes does it take to turn out 1500 mainsprings if it takes 15 seconds to turn out one mainspring?
7. An arbor requires $1\frac{1}{2}$ hours to machine. How many arbors can a department of 4 machinists turn out in a 5-day week if the workday is 8 hours?

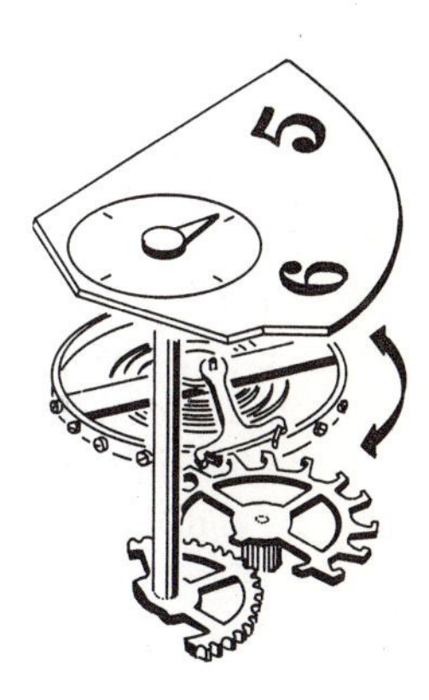

Oscillation of Balance Wheel

1-4 *WEIGHT MEASUREMENT*

The third fundamental unit is that of weight. *Weight* is a measure of gravitational force, or pull, exerted on a body by the earth. Weight is important to the consumer, who must know how much of a certain item he is purchasing, and also to anyone working in industry. The automotive engineer must know the weight of certain parts of an automobile so that weight distribution can be well planned. The structural engineer must be able to design a bridge from the knowledge of the weight it needs to support.

The *pound* (lb) is the basic unit of weight in the English system. It is defined as the pull of the earth on a cylinder of a platinum-iridium alloy which is stored in a vault at the U. S. Bureau of Standards.

The *ounce* (oz) is another common unit of weight in the English system. The relationship between ounces and pounds is:

$$1 \text{ lb} = 16 \text{ oz}$$

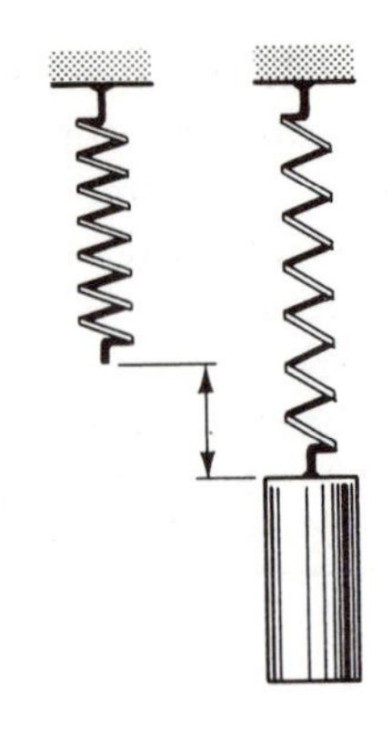

A corresponding weight unit in the metric system is the *newton* (N). The following relationships can be used for conversion between systems of units:

$$1 \text{ N} = 0.225 \text{ lb} \quad \text{or}$$
$$1 \text{ lb} = 4.45 \text{ N}$$

A common method of measuring weights is based on the spring balance. The basis of this method is the fact that the distance a spring stretches when a body is supported by it is proportional to the weight of the body. A pointer can be attached to the spring and a calibrated scale added so that the device will read directly in pounds or newtons. The common bathroom scale uses this principle to measure weights.

The other common device for measuring weights is the equal arm balance. Two platforms are connected by a horizontal rod which balances on a knife edge. This device compares the pull of gravity on objects which are on the two platforms. The platforms are at the same height only when the unknown weight of the object on the left is equal to the known weights placed on the right. It is also possible to use one platform and a weight which slides along a calibrated scale. Variations of this basic design are used in scales designed for high accuracy such as those found in meat departments of supermarkets and highway truck scales.

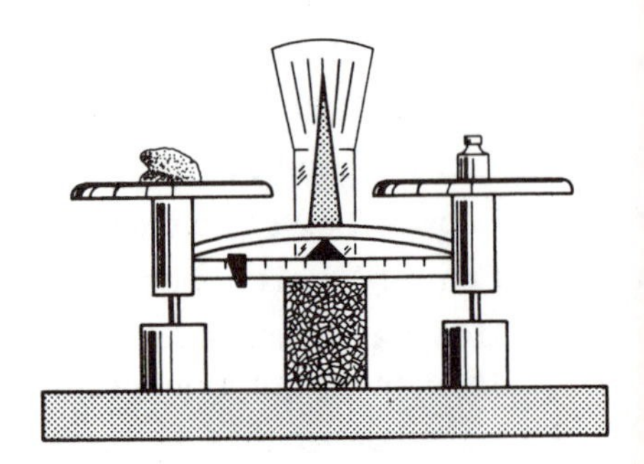

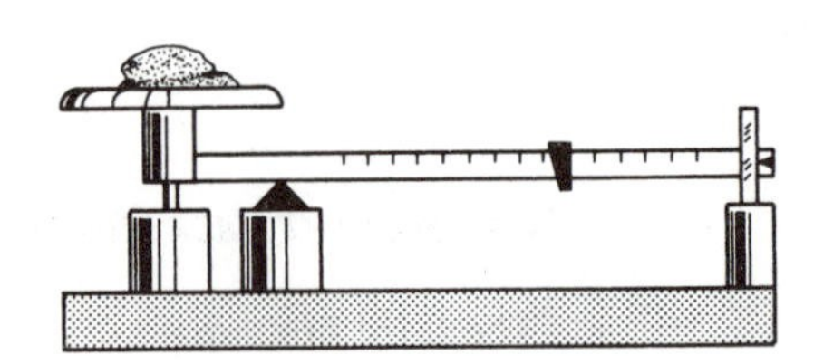

EXAMPLE: The weight of the intake valve of an auto engine is 0.157 lb. What is the weight in ounces and in newtons?

To find the weight in ounces, we simply use the correct conversion factors as follows:

$$\text{Weight} = 0.157\,\cancel{\text{lb}}\left(\frac{16 \text{ oz}}{1\,\cancel{\text{lb}}}\right) = 2.512 \text{ oz}$$

To find weight in newtons, we again use the correct conversion factor.

$$\text{Weight} = 0.157\,\cancel{\text{lb}}\left(\frac{4.45 \text{ N}}{1\ \cancel{\text{lb}}}\right) = 0.69865 \text{ N}$$

PROBLEMS

1. If the weight of a car is 3500 lb, what is its weight in newtons?
2. A certain bridge is designed to support 150,000 lb. What is the maximum supportable weight in newtons?
3. A man's weight is 200 lb. What is this in newtons?
4. Convert the following:
   (a) 75 pounds to newtons
   (b) 2000 newtons to pounds
   (c) 2000 pounds to newtons
5. Find the metric weight of a 94-lb bag of cement.
6. What is the weight in newtons of 500 blocks if each block weighs 3 lb?

1-5 *ELECTRICAL UNITS*

Later in this book we will study electricity and magnetism. *Electricity* is the flow of energy by charge transported through wires. The importance of electricity to our industrialized society cannot be underestimated. This is evidenced by the dependence of every household and every industry on it for the energy necessary to run appliances and large machinery.

For this study of electricity we will need to define one more basic unit in addition to the three units already defined. This unit is the *coulomb* which is a measure of the amount of electrical charge. All the units to be used in electricity are combinations of the metric units and the coulomb. Some of these combined units with which you are probably familiar are the *volt* (a measure of electrical energy), the *ampere* (a measure of flow of electrical charge), the *watt* (a measure of power), and the *kilowatt-hour* (a measure of work or electrical energy used).

1-6 *SIGNED NUMBERS*

Signed numbers have wide application from temperature measurement to determining tolerances in machinery. Since we will be doing these things, it is necessary for us to use signed numbers.

The fundamental rules for working with signed numbers follow:

Adding Signed Numbers

To add two negative numbers, add their absolute values and place a negative sign before the sum.

EXAMPLE: $(-4) + (-16) = -(4 + 16) = -20$

To add a negative number and a positive number, find the difference of their absolute values and place the sign of the integer whose absolute value is larger before the result.

EXAMPLES: (a) $-11 + 7 = -4$
(b) $-7 + 10 = +3$

To add three or more signed numbers:

1. add the positive numbers;
2. add the negative numbers;
3. add the sums from steps 1 and 2 according to the rules for addition of signed numbers.

EXAMPLE: $(-2) + 4 + (-6) + 10 + (-7)$

$$\begin{array}{rrr} -2 & & \\ -6 & +4 & -15 \\ \underline{-7} & \underline{+10} & \underline{+14} \\ -15 & +14 & -1 \end{array}$$

Therefore: $(-2) + 4 + (-6) + 10 + (-7) = -1$

Subtracting Signed Numbers

To subtract signed numbers, change the sign of the *second number* and *add* according to the rules for addition.

EXAMPLES: (a) $85 - (-15) = 100$
(b) $-70 - 12 = -82$

Multiplying Signed Numbers

To multiply two signed numbers:

(a) If the signs of the numbers are both positive or both negative, find the product of their absolute values.
(b) If the signs of the numbers are unlike, find the product of their absolute values and place a negative sign before the result.

To multiply more than two signed numbers, first multiply the absolute values of the numbers. If there is an odd number of negative factors, place a negative sign before the result. If there is an even number of factors, the product is positive. *Note*: An *even* number is divisible by 2. Any other number is *odd*.

EXAMPLES: (a) $(-4)(-8) = 32$
(b) $(-16)(4) = -64$
(c) $(-2)(4)(-6)(-3) = -144$
(d) $(-4)(-1)(-2)(-3) = 24$

Dividing Signed Numbers

The rules for dividing signed numbers are similar to those for multiplying signed numbers.

(a) When a positive number is divided by a negative number, the quotient is negative.
(b) When a negative number is divided by a positive number, the quotient is negative.
(c) If both numbers have the same sign, the quotient is positive.

EXAMPLES: (a) $\frac{-56}{8} = -7$
(b) $\frac{72}{-8} = -9$
(c) $\frac{-14}{-7} = 2$

PROBLEMS

Perform the indicated operations.

1. $(-10) + (-8)$
2. $(-14) + (-7) + (-8)$
3. $(-4) + (-5) + (-20)$
4. $7 + (-3)$
5. $10 + (-7)$
6. $(-16) + 8$
7. $(-5) + (-1) + 5 + (-3)$
8. $(-42) + (-73) + (-35) + 46$
9. $-62 + 3 - 10 + 51$
10. $-53 + 5 - 8 + 9$
11. $(-9) - (-7)$
12. $(-11)(-8)$
13. $(-7)(-5)$
14. $(5)(-12)$
15. $(-5)(-4)(-3)$
16. $(-7)(-8)(-3)(-2)$
17. $\frac{-40}{10}$
18. $\frac{140}{-20}$
19. $\frac{-15}{-5}$
20. $\frac{(-36)(-4)}{(-9)}$

Scientists and technicians often use very large or very small numbers which cannot be conveniently written as fractions or decimal fractions. For example, the thickness of an oil film on water is about 0.0000001 m. A more useful method of expressing such very small (or very large) numbers is known as scientific notation or exponential notation. Expressed this way, the thickness of the film is 1×10^{-7} or 10^{-7}. For example:

$$0.1 = 1 \times 10^{-1} \text{ or } 10^{-1}$$

$$10000 = 1 \times 10^{4} \text{ or } 10^{4}$$

$$0.001 = 1 \times 10^{-3} \text{ or } 10^{-3}$$

To write a number in scientific notation, write it as a product of a number between 1 and 10 and a power of 10. General form: $M \times 10^n$, *where:*

M = a number between 1 and 10

n = the exponent or power of the base number 10

EXAMPLE 1: Write 325 in scientific notation.

$325 = 3.25 \times 10^2$. (Remember that 10^2 is a short way of writing $10 \times 10 = 100$, and since multiplying 3.25 by 100 gives 325, all that has been done is to move the decimal point two places.)

EXAMPLE 2: Write 65,800 in scientific notation.

$$65{,}800 = 6.58 \times 10{,}000 = 6.58 \times (10 \times 10 \times 10 \times 10) = 6.58 \times 10^4$$

The following procedure should help you to write any decimal number in scientific notation.

1. Place a decimal point after the first nonzero digit reading from left to right.
2. Place a caret (Λ) at the position of the original decimal point.
3. If the decimal point is to the left of the caret, the exponent of 10 is the number of places from the caret to the decimal point.
   Example: $83{,}662 = 8.\underset{4}{\underline{3662}}{}_{\wedge} \times 10^{\underline{4}}$
4. If the decimal point is to the right of the caret, the exponent of 10 is the negative of the number of places from the caret to the decimal point.
   Example: $0.00683 = {}_{\wedge}\underset{3}{\underline{006}}.83 \times 10^{\underline{-3}}$
5. If the decimal point and the caret coincide, the exponent of 10 is zero.
   Example: $5.12 = 5.12 \times 10^0$

EXAMPLE 3: Write 0.0000002486 in scientific notation.

$$0.0000002486 = {}_{\wedge}\underset{7}{\underline{0000002}}.486 \times 10^{\underline{-7}} = 2.486 \times 10^{-7}$$

PROBLEMS

Write the following numbers in scientific notation.

1. 326
2. 798
3. 826.4
4. 0.00413
5. 6.432
6. 482,300
7. 0.00224
8. 777,380,000
9. 0.000299
10. 540,000
11. 732,000,000,000,000,000
12. 0.000000000000000615

Converting from scientific notation to decimal form requires the ability to multiply by powers of 10. To multiply by positive powers of 10, move the decimal point to the right the same number of places as is indicated by the exponent of 10. Supply zeros as needed. To multiply by negative powers of 10, move the decimal point to the left the same number of places as the exponent of 10 indicates, and supply zeros as needed.

EXAMPLE 4: Write 7.62×10^2 in decimal form.

$7.62 \times 10^2 = 762$ (Move the decimal point two places to the right.)

EXAMPLE 5: Write 3.15×10^{-4} in decimal form.

Move the decimal point four places to the left and insert three zeros:

$3.15 \times 10^{-4} = 0.000315$

PROBLEMS

Write the following numbers in decimal form.

1. 8.62×10^0
2. 8.67×10^2
3. 6.31×10^{-4}
4. 5.41×10^3
5. 7.68×10^{-1}
6. 9.94×10^1
7. 7.77×10^8
8. 4.19×10^{-6}
9. 6.93×10^1
10. 3.78×10^{-2}
11. 9.61×10^4
12. 7.33×10^3

1-8 *POWERS OF TEN*

The operations of multiplication and division can sometimes be performed more quickly using scientific notation, if you know the rules of exponents.

EXAMPLE 1: Find 6300×200.

$$6300 = 6.3 \times 10^3 \quad \text{and} \quad 200 = 2 \times 10^2$$

$$\begin{aligned}(6300)(200) &= (6.3 \times 10^3)(2 \times 10^2)\\ &= (6.3)(2)(10^3)(10^2)\\ &= 12.6(10^3)(10^2)\\ &= 12.6 \times 10^{(3+2)}\\ &= 12.6 \times 10^5\\ &= 1{,}260{,}000\end{aligned}$$

Notice that, when multiplying the powers of 10, we add the exponents. Then we multiply the parts involving the numbers between 1 and 10 and change back to decimal form.

EXAMPLE 2: Find $54{,}000 \times 0.004$.

$$54{,}000 = 5.4 \times 10^4 \quad \text{and} \quad 0.004 = 4 \times 10^{-3}$$

so

$$\begin{aligned}(54{,}000)(0.004) &= (5.4 \times 10^4)(4 \times 10^{-3})\\ &= (5.4)(4)(10^4)(10^{-3})\\ &= 21.6 \times 10^{[4+(-3)]}\\ &= 21.6 \times 10^1\\ &= 216\end{aligned}$$

RULE

When multiplying *two numbers having the same base,* add *the exponents, that is,* $10^a \times 10^b = 10^{a+b}$.

EXAMPLE 3:

(a) $(10^6)(10^3) = 10^{6+3} = 10^9$
(b) $(10^4)(10^2) = 10^{4+2} = 10^6$
(c) $(10^1)(10^{-3}) = 10^{1+(-3)} = 10^{-2}$
(d) $(10^{-2})(10^{-5}) = 10^{[-2+(-5)]} = 10^{-7}$

PROBLEMS

Find the following products using the rule of scientific notation.

1. $(10^{-4})(10^3)$
2. $(10^{10})(10^{-6})$
3. $(10^7)(10^{13})$
4. $42{,}000 \times 0.006$
5. $(0.050)(0.0006)$
6. $(160)(20{,}000)$
7. $(520)(0.002)$
8. $(62{,}000)(0.005)$

As you might expect, when dividing numbers involving powers of ten, you subtract exponents.

EXAMPLE 4: $10^7 \div 10^4 = 10^{7-4} = 10^3$

$10^4 \div 10^7 = 10^{4-7} = 10^{-3}$

EXAMPLE 5: Find the quotient.

$$\begin{aligned} 830 \div 0.0002 &= (8.3 \times 10^2) \div (2 \times 10^{-4}) \\ &= (8.3 \div 2) \times (10^2 \div 10^{-4}) \\ &= 4.15 \times 10^{[2-(-4)]} \\ &= 4.15 \times 10^6 \\ &= 4{,}150{,}000 \end{aligned}$$

RULE

When dividing two numbers having the same base, subtract the exponent of the divisor from the exponent of the other number, that is, $10^a \div 10^b = 10^{a-b}$.

PROBLEMS

Find the quotients for the following using the rule and scientific notation.

1. $10^4 \div 10^3$
2. $10^6 \div 10^{-3}$
3. $10^{-6} \div 10^{-6}$
4. $10^{-4} \div 10^5$
5. $160 \div 4000$
6. $0.52 \div 0.005$
7. $63 \div 0.0009$
8. $800 \div 60{,}000$

1-9 *SIGNIFICANT DIGITS*

The emphasis in this book is on problem solving and the fundamental physical concepts that students need for success in becoming technicians. Emphasis will not be given to accuracy and precision of calculations. Slide rule, which has accuracy of three significant digits, will be the tool used for doing the calculations. We will assume that all measurements have three significant digits or, if they have more, the student will round to three significant digits before calculating. Therefore, a sound knowledge of the rules for finding the number of significant digits in a number is needed. Also then, beginning with Chapter 2, the answers will be expressed with three significant digits.

To find the number of significant digits use the following rules:

1. All nonzero digits are significant.
2. Zeros are significant when they:
   (a) are between significant digits;

(b) follow the decimal point and a significant digit; or
(c) are in a whole number and a bar is placed over the zero.

EXAMPLES: (a) 376.52 m has five significant digits.
(b) 30.5 mi has three significant digits.
(c) 4007 kg has four significant digits.
(d) 35.00 cm has four significant digits.
(e) 0.0060 in. has two significant digits.
(f) $35\bar{0}$ lb has three significant digits.
(g) $40\bar{0}0$ km has three significant digits.

PROBLEMS

Give the number of significant digits in each of the following measurements.

1. 536 V
2. 307.3 mi
3. 5007 m
4. 5.00 cm
5. 0.0070 in.
6. 6.010 cm
7. $84\bar{0}0$ km
8. $30\bar{0}0$ ft
9. 187.40 m
10. $5\bar{0}0$ g
11. 0.00700 in.
12. 10.30 cm

AREA AND VOLUME MEASUREMENT 2

Note: Beginning with Chapter 2, it will be assumed that the student is familiar with the use of the slide rule. A student who has not learned the basic slide rule operations should proceed at this time to the Appendix, Slide Rule, before studying Chapter 2. Because the slide rule will be used throughout the following chapters, three significant digits will be used wherever practical.

2-1 *AREA*

If you wish to measure a surface area of an object, you must first decide upon some standard unit of area. Standard units of area are based on the square and are called square inches, square centimeters, square miles, or some square unit of measure. An area of 1 square centimeter is the amount of area found within a square 1 cm on each side. An area of 1 square inch is the amount of area found within a square of 1 in. on each side.

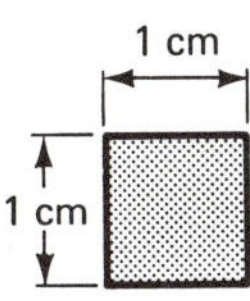

1 Square Centimeter

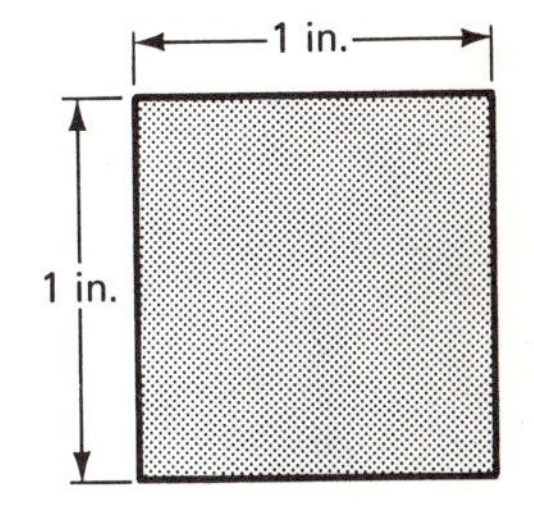

1 Square Inch

The area of a figure is the number of square units which are contained in the figure. In general, when multiplying measurements of like units, multiply the numbers and then multiply the units as follows:

$$3.00 \text{ in.} \times 5.00 \text{ in.} = (3.00 \times 5.00)(\text{in.} \times \text{in.}) = 15.0 \text{ sq in. or } 15.0 \text{ in}^2$$

$$2.00 \text{ ft} \times 4.00 \text{ ft} = (2.00 \times 4.00)(\text{ft} \times \text{ft}) = 8.00 \text{ sq ft or } 8.00 \text{ ft}^2$$

$$1.40 \text{ m} \times 6.70 \text{ m} = (1.40 \times 6.70)(\text{m} \times \text{m}) = 9.38 \text{ sq m or } 9.38 \text{ m}^2$$

EXAMPLE 1: Find the area of a rectangle which is 6.00 in. long and 4.00 in. wide.

Each square is 1.00 in². To find the area of the rectangle, simply count the number of squares in the rectangle. Therefore, you find the area = 24.0 in² or by formula $A = lw$.

$$\begin{aligned} A &= lw \\ &= (6.00 \text{ in.})(4.00 \text{ in.}) \\ &= 24.0 \text{ in}^2 \end{aligned}$$

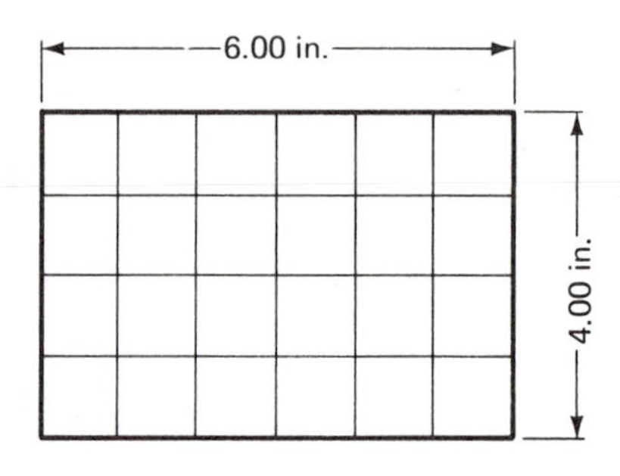

EXAMPLE 2: What is the area of the metal plate below?

To find the area of the metal plate, find the area of each of the two rectangles and then find the difference of their areas. The large rectangle is 10.0 cm long and 8.00 cm wide. The small rectangle is 6.00 cm long and 4.00 cm wide. The two areas are 80.0 cm² and 24.0 cm². Therefore, the area of the metal plate is 80.0 cm² − 24.0 cm² = 56.0 cm².

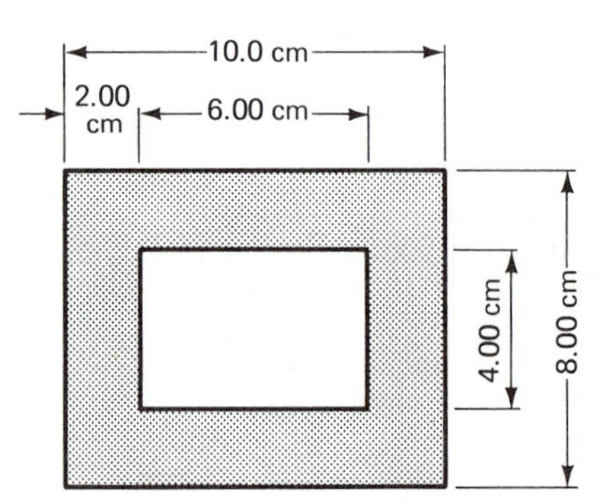

To find the area of a triangle, use the formula $A = \frac{1}{2}bh$, where b is the base and h is the height. The height of a triangle is a line from one vertex of the triangle that forms a right angle with the opposite side (base) or the opposite side extended.

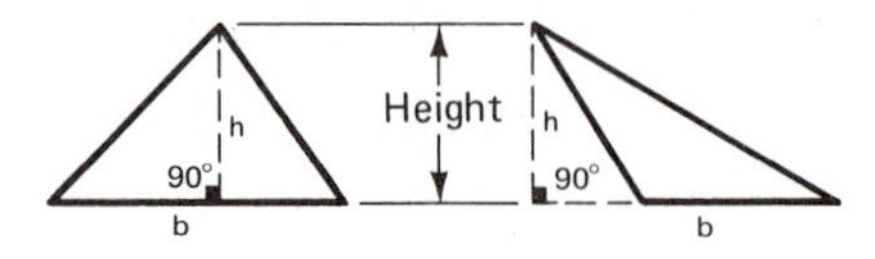

EXAMPLE 3: Find the area of the following right triangle.

$$A = \tfrac{1}{2}\,bh$$
$$= \tfrac{1}{2}(6.00 \text{ in.})(4.00 \text{ in.})$$
$$= 12.0 \text{ in}^2$$

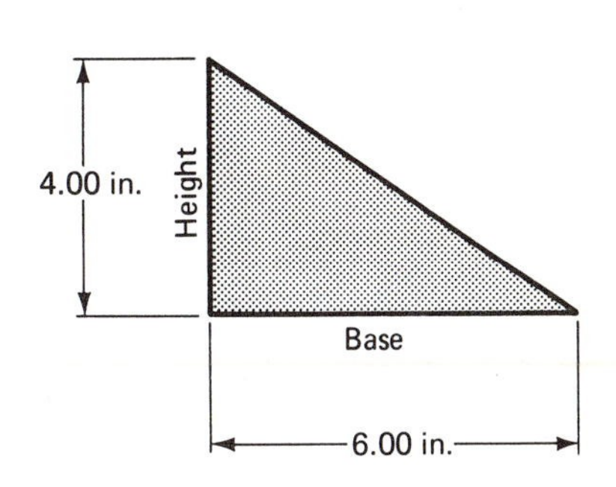

The area of a circle is πr^2, where r is the radius of the circle.

EXAMPLE 4: Find the area of the circle at the right.

$$\text{Area} = \pi r^2 \qquad (\pi = 3.14)$$
$$= (3.14)(6.00 \text{ ft})^2$$
$$= 113 \text{ ft}^2$$

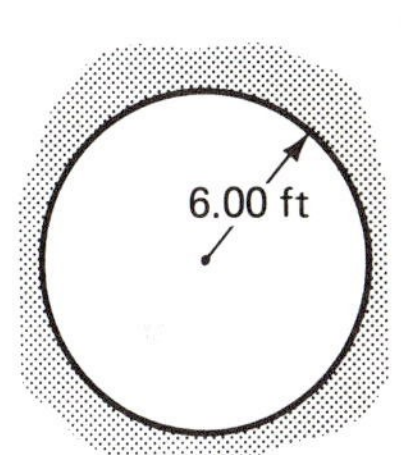

The formulas for finding the areas of other plane figures can be found in Table R in the Appendix.

2-2 CROSS-SECTIONAL AREA

The surface that would be seen by cutting a geometric solid with a thin plate represents the idea of a cross section of a solid. To find a cross-sectional area of a solid, use the correct area formula.

EXAMPLE 1: Find a cross-sectional area of the following box.

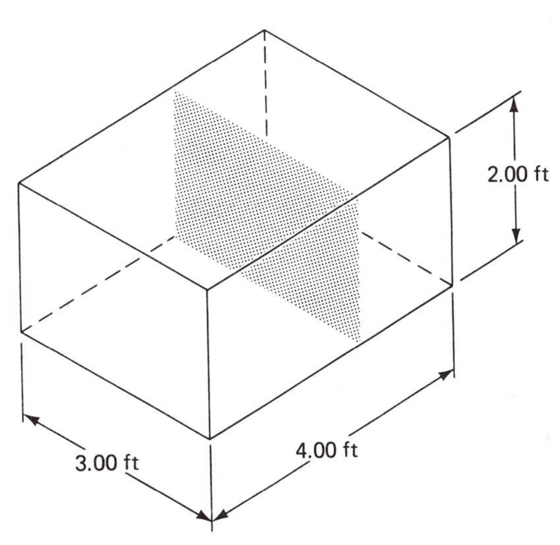

The indicated cross section of this box is a rectangle 3.00 ft long and 2.00 ft wide.

$$
\begin{aligned}
A &= lw \\
&= (3.00 \text{ ft})(2.00 \text{ ft}) \\
&= 6.00 \text{ ft}^2
\end{aligned}
$$

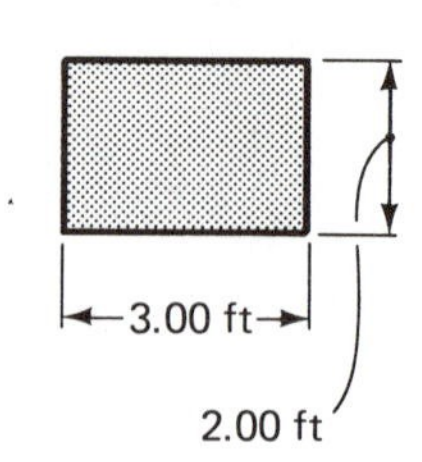

The area of this rectangle is 6.00 ft² which represents a cross-sectional area.

EXAMPLE 2: Find the cross-sectional area of a pipe whose outer diameter is 3.50 cm and inner diameter is 3.20 cm.

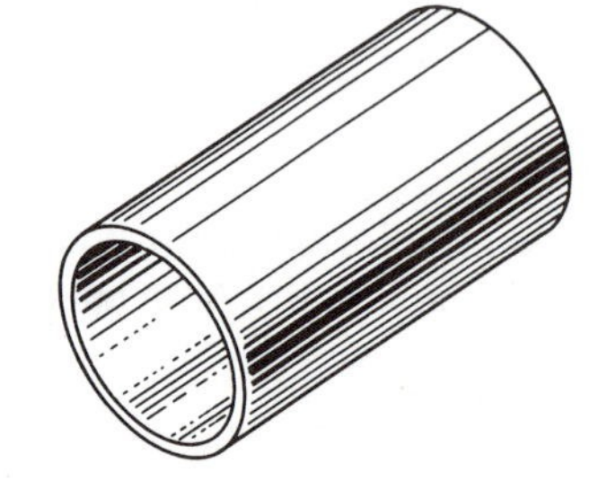

3.50 cm dia.

3.20 cm dia.

A cross section of the pipe is shown at the left.

To find the cross-sectional area, find the area of the larger circle and subtract the area of the smaller circle.

Area of a circle $= \pi r^2$

$$
\begin{aligned}
\text{Cross-sectional area of the pipe} &= \pi(1.75 \text{ cm})^2 - \pi(1.60 \text{ cm})^2 \\
&= 9.62 \text{ cm}^2 - 8.04 \text{ cm}^2 \\
&= 1.58 \text{ cm}^2
\end{aligned}
$$

Note that there are many figures and surfaces that do not contain a whole number of square units.

EXAMPLE 3: Find a cross-sectional area of the dovetail slide on page 19.

A cross section looks like the following figure. To find the area of this cross section, find the area of the large rectangle and subtract the area of the trapezoid.

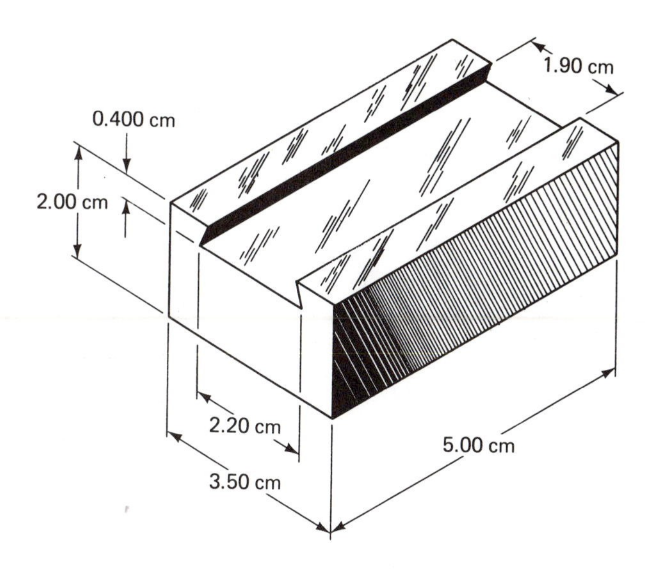

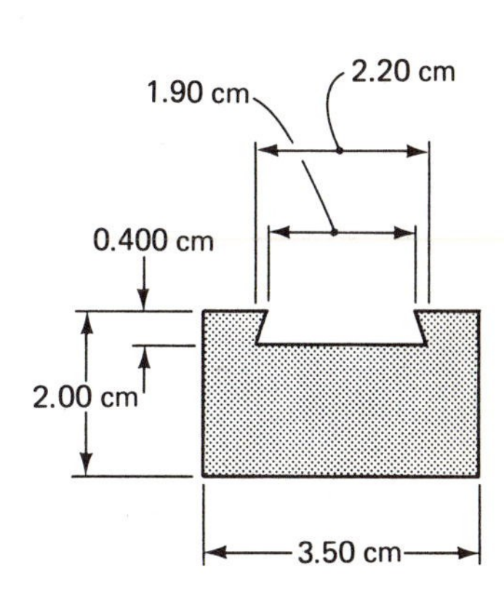

Area of rectangle = (2.00 cm)(3.50 cm) = 7.00 cm^2

Area of trapezoid = $\frac{1}{2}$(2.20 cm + 1.90 cm)(0.400 cm) = 0.820 cm^2

Area of cross section is 7.00 cm^2 − 0.820 cm^2 = 6.18 cm^2

PROBLEMS

Find the area of each of the following.

1. 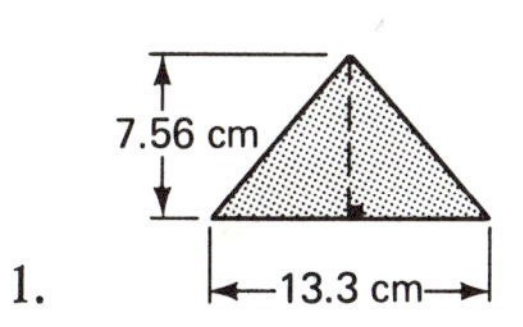

2. 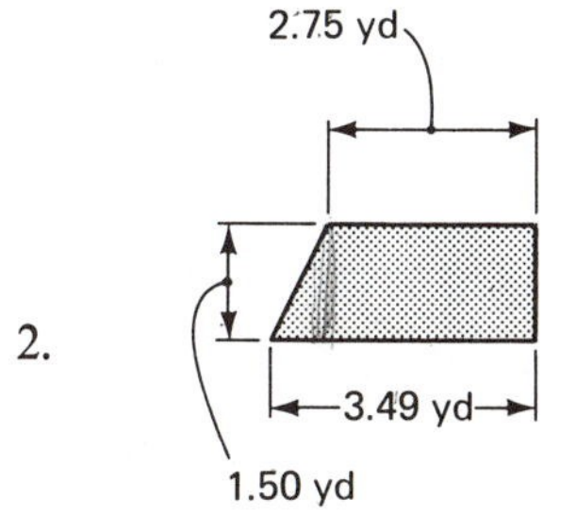

3. 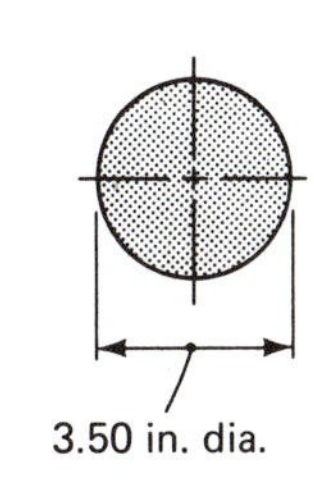

4. 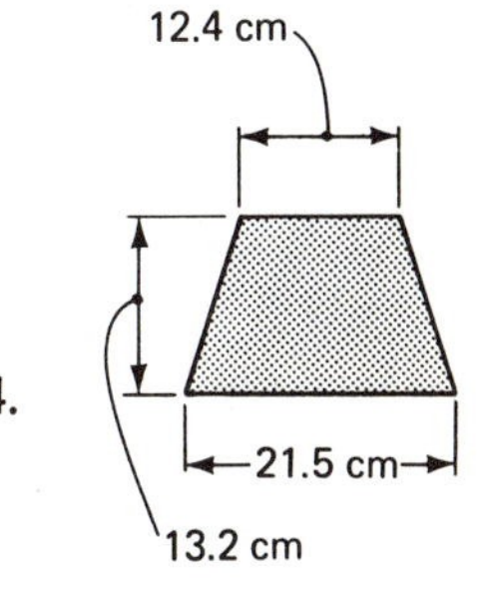

5. Find the cross-sectional area of the concrete dam shown at the right.

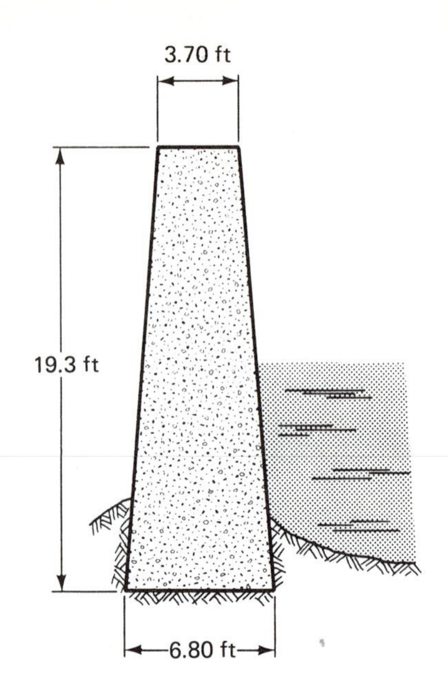

6. Find the area of a piston head with a diameter of 3.25 in.

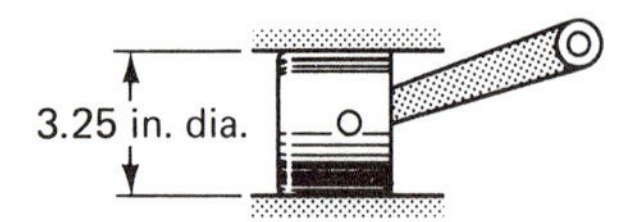

7. A large plot of ground is to be divided into four smaller lots. What will be the area of each smaller lot? What is the area of the large lot? (The large lot and each small lot are shaped like a parallelogram).

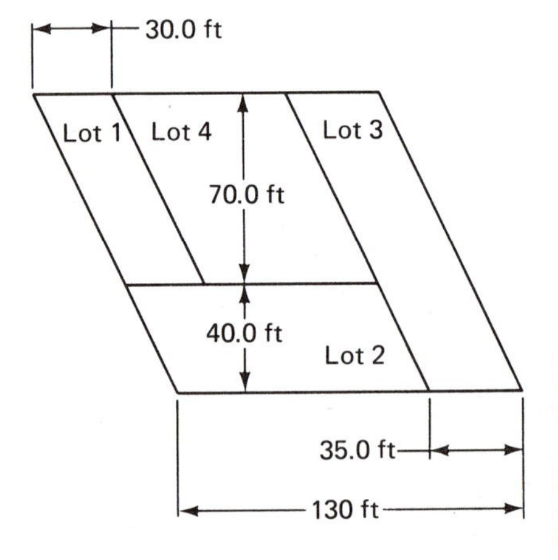

8. Find the area of the template shown at the right.

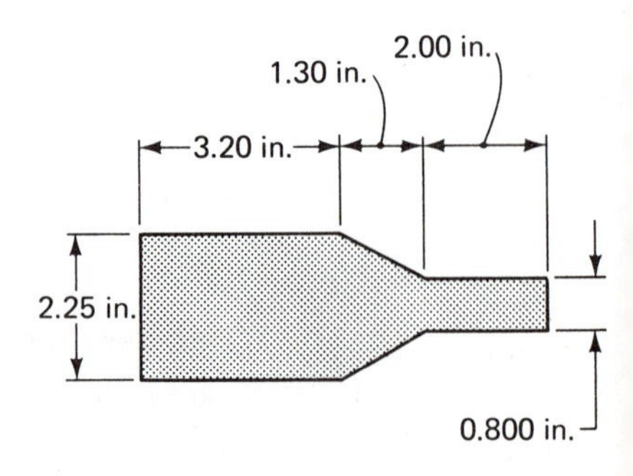

9. The diameter of the large end of a taper plug is 3.70 cm. The diameter of the small end is 2.10 cm. Find the area of each end.

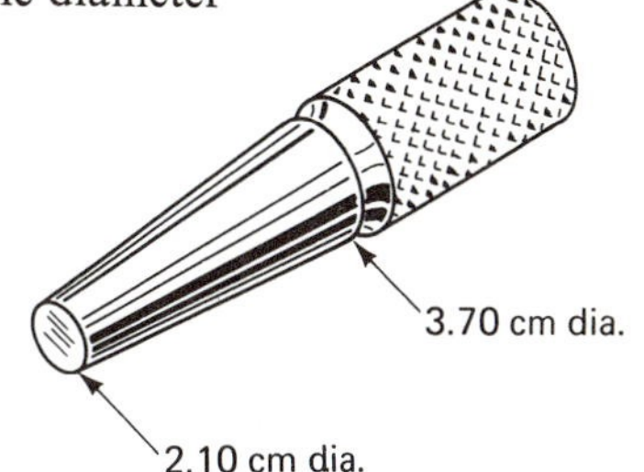

10. Find the area of the gable end (shaded end) of the building shown below.

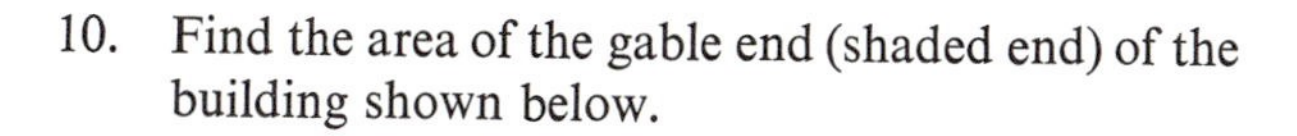

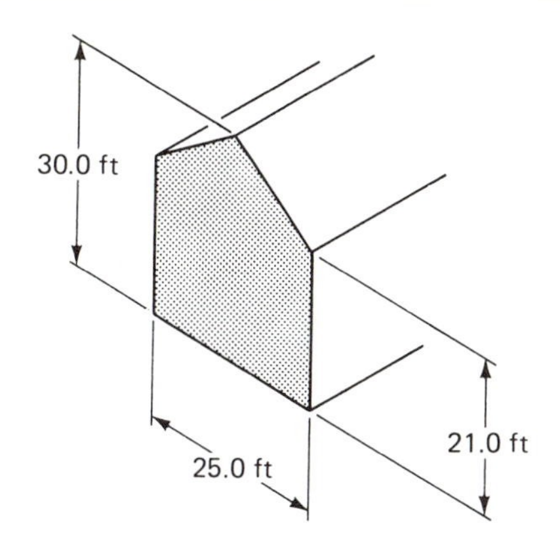

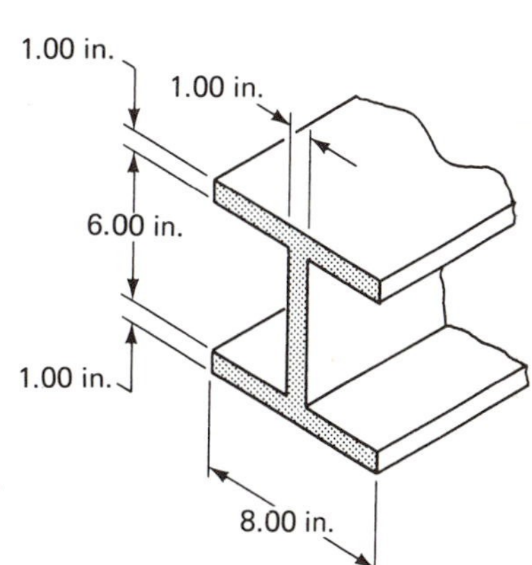

11. Find the cross-sectional area of the I beam at the right.

12. Find the cross-sectional area of a cylindrical steel shaft whose diameter is 8.13 in.

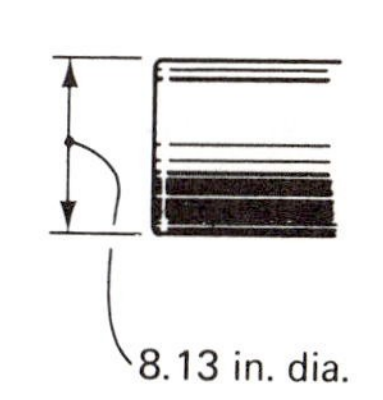

2-3 *CONVERSION OF SQUARE UNITS*

Recall the rule for finding a conversion factor. The correct conversion factor will be in fractional form and equal to 1, with the numerator expressed in the units you wish to convert to and the denominator expressed in the units given.

EXAMPLE 1: Convert 324 in^2 to yd^2.

$$324\,\not{in}^2 \times \frac{1\ yd^2}{1296\ \not{in}^2} = \frac{324}{1296}\,yd^2 = 0.250\ yd^2$$

EXAMPLE 2: Convert 258 cm^2 to m^2.

$$258\ \not{cm}^2 \times \frac{1\ m^2}{10{,}000\ \not{cm}^2} = 0.0258\ m^2$$

EXAMPLE 3: Convert 28.5 m^2 to in^2.

$$28.5\ \not{m}^2 \times \frac{1550\ in^2}{1\ \not{m}^2} = 44200\ in^2$$

PROBLEMS

1. Convert 5.00 yd^2 to m^2.
2. Convert 15.0 cm^2 to mm^2.
3. How many m^2 are in 15.0 ft^2?
4. Convert 15.0 ft^2 to cm^2.
5. How many ft^2 are there in a rectangle 15.0 m long and 12.0 m wide?
6. Convert 72.0 cm^2 to m^2.
7. Convert 99.0 in^2 to ft^2.
8. How many cm^2 are in 37,000 mm^2?
9. How many in^2 are in 51.0 cm^2?
10. How many in^2 are there in a square 11.0 yd on a side?
11. How many cm^2 are there in the cross section of a piece of metal stock which has an area of $1\frac{1}{2}$ in^2?
12. How many m^2 are there in door stoop which is 15.0 ft^2?
13. How many cm^2 in a face plate which is 4.00 in^2?
14. What is the area in cm^2 of a cross section of a rod which is 993 mm^2?

2-4 *VOLUME*

Standard units of volume are based on the cube and are called cubic inches, cubic centimeters, cubic yards, or some other cubic unit of measure. A volume of one cubic centimeter is the same as the amount of volume contained in a cube 1 cm on each side. One cubic inch is the volume contained in a cube 1 in. on each side.

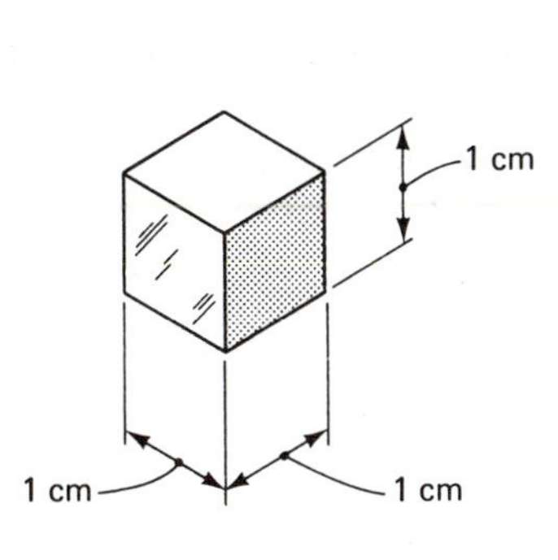

1 Cubic Centimeter

1 Cubic Inch

The volume of a figure is the number of cubic units that are contained in the figure. *Note*: When multiplying measurements of like units, multiply the numbers and then multiply the units as follows:

$$3.00 \text{ in.} \times 5.00 \text{ in.} \times 4.00 \text{ in.} = (3.00 \times 5.00 \times 4.00)(\text{in.} \times \text{in.} \times \text{in.})$$
$$= 60.0 \text{ cubic in. or } 60.0 \text{ in}^3$$

$$2.00 \text{ yd} \times 4.00 \text{ yd} \times 1.00 \text{ yd} = (2.00 \times 4.00 \times 1.00)(\text{yd} \times \text{yd} \times \text{yd})$$
$$= 8.00 \text{ cu yd or } 8.00 \text{ yd}^3$$

$$1.40 \text{ ft} \times 8.70 \text{ ft} \times 6.00 \text{ ft} = (1.40 \times 8.70 \times 6.00)(\text{ft} \times \text{ft} \times \text{ft})$$
$$= 73.1 \text{ cu ft or } 73.1 \text{ ft}^3$$

EXAMPLE 1: Find the volume of a rectangular prism which is 6.00 in. long, 4.00 in. wide, and 5.00 in. high.

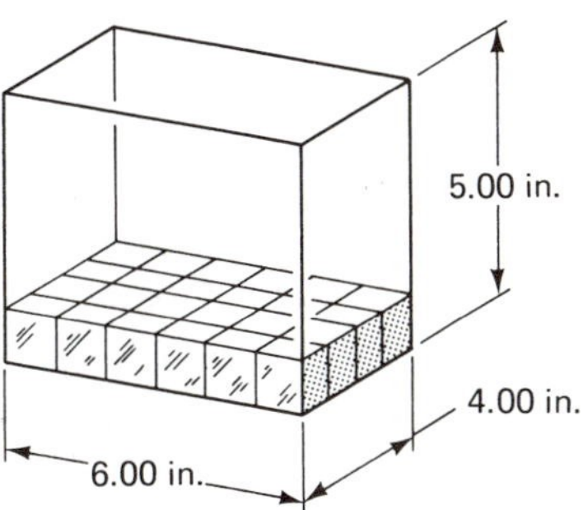

Each cube is 1 in^3. To find the volume of the rectangular solid, count the number of cubes in the bottom layer of the rectangular solid and then multiply that number by the number of layers that the solid can hold. Therefore, there are 5.00 layers of 24.0 cubes, which is 120 cubes or 120 cubic inches.

Or, by formula, $V = Bh$, where B is the area of the base and h is the height. However, the area of the base is found by lw, where l is the length and w is the width of the rectangle. Therefore, the volume of a rectangular solid can be found by the formula:

$$\begin{aligned} V &= lwh \\ &= (6.00 \text{ in.})(4.00 \text{ in.})(5.00 \text{ in.}) \\ &= 120 \text{ in}^3 \end{aligned}$$

EXAMPLE 2: Find the volume of the figure at the right.

$$\begin{aligned} V &= lwh \\ &= (8.00 \text{ cm})(4.00 \text{ cm})(5.00 \text{ cm}) \\ &= 160 \text{ cm}^3 \end{aligned}$$

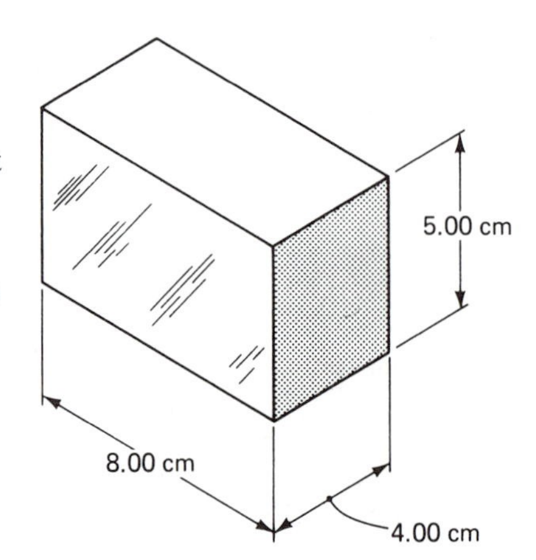

To find the volume of a cylinder, the same formula is used. However, the area of the base, which is a circle, is πr^2. Therefore, the volume of a cylinder is $\pi r^2 h$.

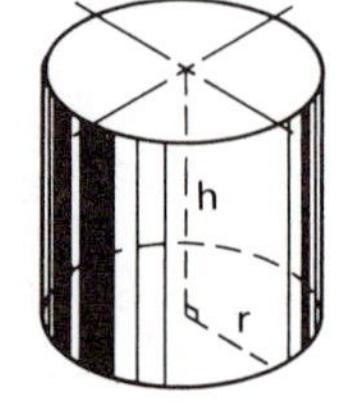

EXAMPLE 3: Find the volume of a cylinder whose height is 9.00 ft if the radius of the base is 6.00 ft. (The height is the shortest distance between the circular bases.)

$$\begin{aligned} V &= \pi r^2 h \\ &= (3.14)(6.00 \text{ ft})^2(9.00 \text{ ft}) \\ &= 1020 \text{ ft}^3 \end{aligned}$$

The formulas for finding the volumes of other solids can be found in Table R.

2-5 *SURFACE AREA*

The lateral (side) surface area of any geometric solid is the area of all the lateral faces.

The total surface area of any geometric solid is the lateral surface area plus the area of the bases.

EXAMPLE 1: Find the lateral surface area of the prism at the right.

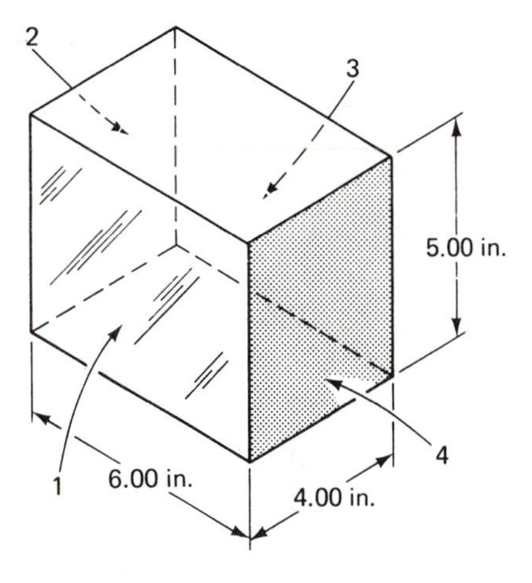

Area of lateral face 1 = (6.00 in.)(5.00 in.) = 30.0 in²
Area of lateral face 2 = (5.00 in.)(4.00 in.) = 20.0 in²
Area of lateral face 3 = (6.00 in.)(5.00 in.) = 30.0 in²
Area of lateral face 4 = (5.00 in.)(4.00 in.) = 20.0 in²

Lateral surface area = 100.0 in²

Total surface area = lateral surface area + area of the bases
Area of base = (6.00 in.)(4.00 in.) = 24.0 in²
Area of both bases = 2(24.0 in²) = 48.0 in²
Total surface area = 100.0 in² + 48.0 in² = 148.0 in²

EXAMPLE 2: Find the lateral surface area of the cylinder at the right.

Lateral surface area of a cylinder can be found by the formula $A = 2\pi rh$ where r is the radius of the circular base and h is the height. Therefore:

$A = 2\pi rh$

$A = 2(3.14)(6.00 \text{ ft})(9.00 \text{ ft})$
$= 339 \text{ ft}^2$

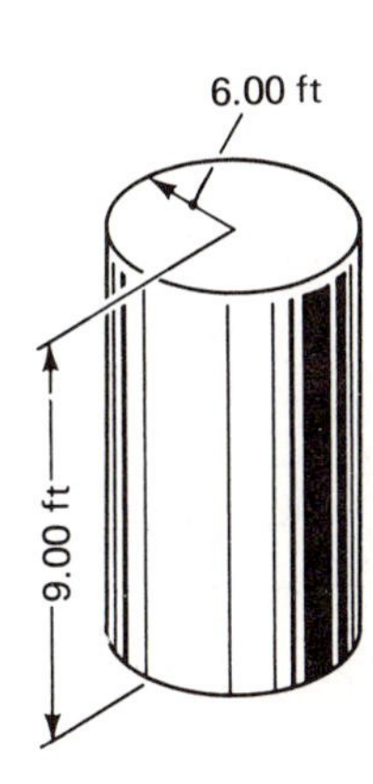

PROBLEMS

Area formulas, volume formulas, and lateral surface area formulas can be found in Table R.

1. Find the lateral surface area, total surface area, and volume of the prism below.

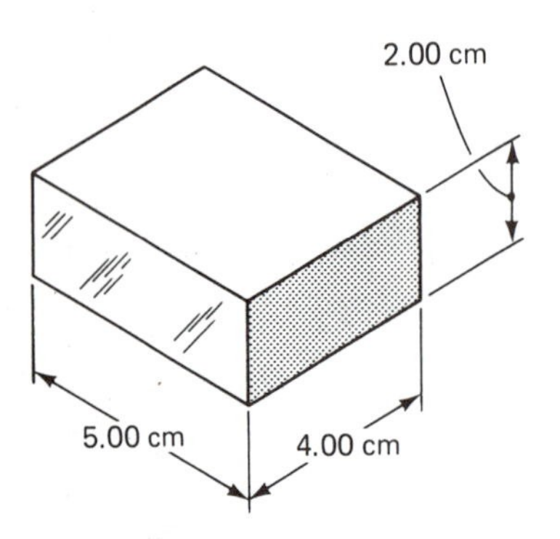

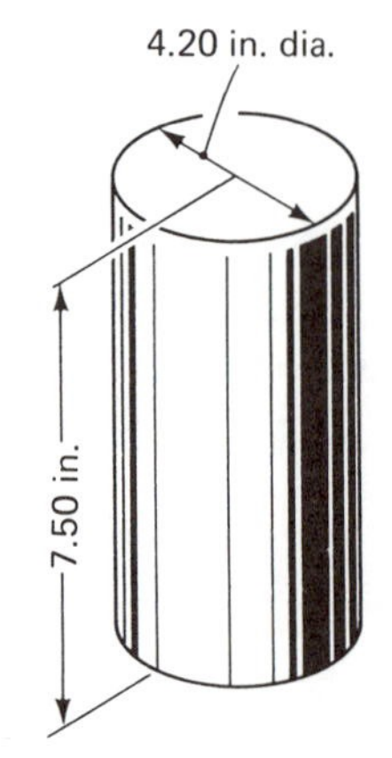

2. Find the volume of a cylinder whose height is 7.50 in. and diameter is 4.20 in.

3. Find the volume of a cone whose height is 9.30 cm if the radius of the base is 5.40 cm.

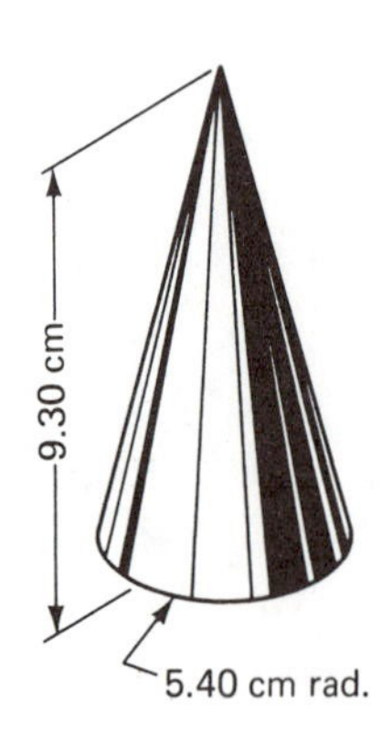

4. Find the volume of the figure below.

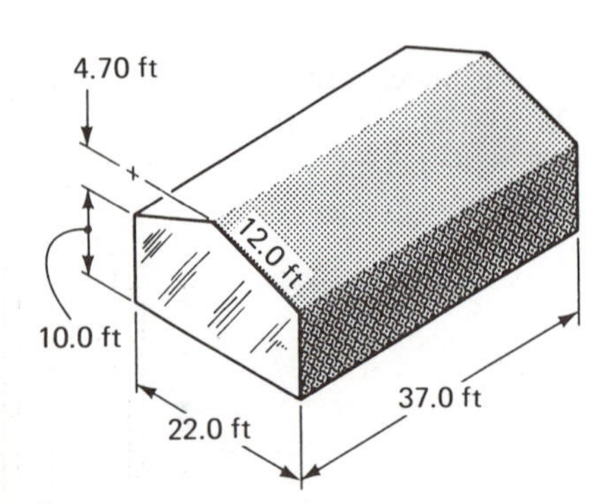

5. What is the total surface area of the figure in problem 4?
6. What is the volume of a cylinder of an engine whose height is 7.00 in. and diameter is 3.50 in.?
7. Find the lateral surface area of the cylinder opposite.
8. Find the surface area of a ball bearing whose diameter is 2.80 cm.

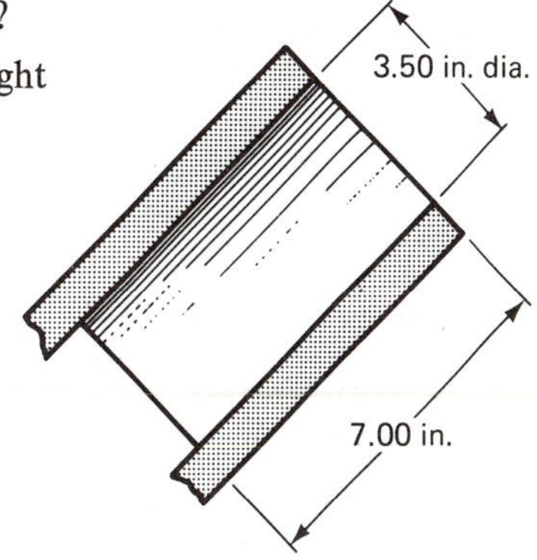

ENGINE PROBLEM:

The cylinder in an engine of a road grader is 4.50 in. in diameter and 9.00 in. high.

9. Find the volume of the cylinder.
10. Find the cross-sectional area of the cylinder.
11. Find the lateral surface area of the cylinder.

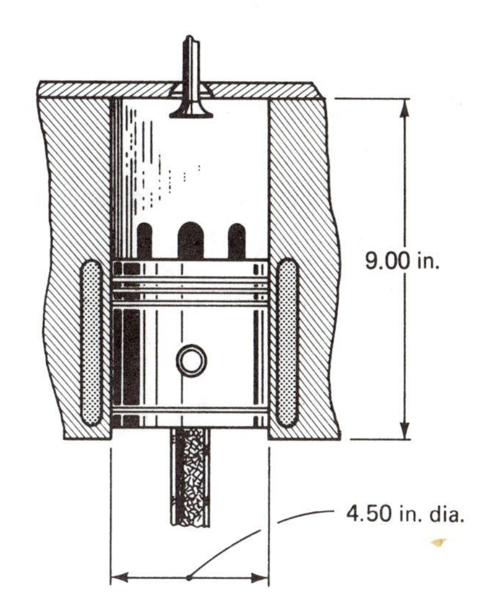

2-6 *CONVERSION OF CUBIC UNITS*

Use the same conversion process that was applied in converting square units.

EXAMPLE 1: Convert 24.0 ft³ to in³.

$$24.0\ \cancel{\text{ft}}^3 \times \frac{1728\ \text{in}^3}{1\ \cancel{\text{ft}}^3} = 41{,}500\ \text{in}^3$$

EXAMPLE 2: Convert 56.0 in³ to cm³.

$$56.0\ \cancel{\text{in}}^3 \times \frac{16.4\ \text{cm}^3}{1\ \cancel{\text{in}}^3} = 918\ \text{cm}^3$$

EXAMPLE 3: Convert 28.5 m^3 to ft^3.

$$28.5\ \text{m}^3 \times \frac{35.3\ \text{ft}^3}{1\ \text{m}^3} = 1010\ \text{ft}^3$$

PROBLEMS

1. Convert 19.0 yd^3 to ft^3.
2. Convert 5440 cm^3 to m^3.
3. How many in^3 are there in 29.0 cm^3?
4. How many yd^3 are there in 23.0 m^3?
5. How many cm^3 are in 88.0 in^3?
6. How many cm^3 are there in 27.0 m^3?
7. Convert 84.0 in^3 to ft^3.
8. Convert 79.0 ft^3 to m^3.
9. Convert 9.00 ft^3 to cm^3.
10. How many in^3 are in 12.0 m^3?
11. The volume of a casting is 38.0 in^3. What is its volume in cm^3?
12. How many castings of volume 14.0 cm^3 could be made from a block of steel of volume 12.0 ft^3?

MOTION 3

3-1 *INTRODUCTION*

The part of physics that is concerned with motion is called *mechanics*. The study of motion is very important in almost every area of science and technology.

Automotive technicians are concerned not only with the motion of the entire auto, but also with the motion of the piston, valves, driveshaft, etc. Obviously, the motion of each of the internal parts has a direct and very important effect on the motion of the entire automobile.

The highway engineer must determine the correct banking angle of a curve if he is to design a safe road. This angle is determined from several laws of motion that we will soon study.

In the next few chapters we will develop the skills necessary for you to understand the basic aspects of motion. You will find this knowledge very helpful if you should decide to pursue a career in automotive mechanics, construction technology, electronics technology, mechanical technology, microprecision technology, or other similar fields.

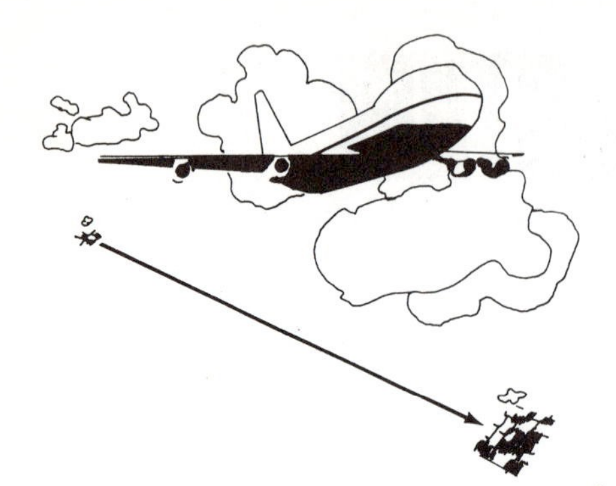

3-2 *DISPLACEMENT*

Motion can be said to be a change of position. An airplane is in motion when it flies through the air because its position is changing as it flies from one city to another. To describe the change of position of some object, such as an airplane, it is necessary to introduce the term *displacement*. Displacement, as a result of motion, is a change of position.

Suppose a friend asked you how to reach your home from school. If you replied that he should walk four blocks, you would not have given him enough information. Obviously, you need to tell him which direction to go. If you had replied, "four blocks north," your friend could then find your home.

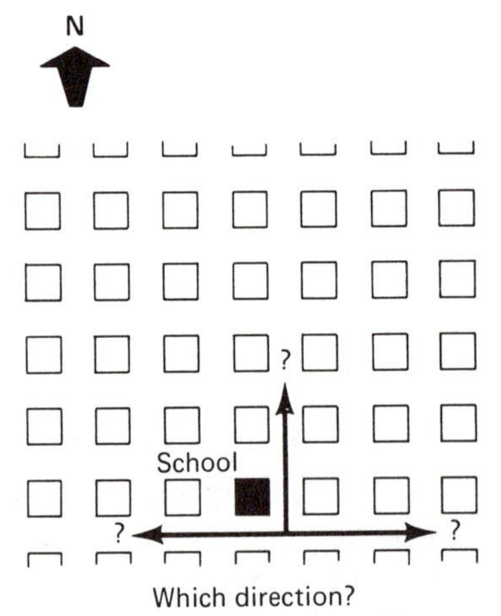

Which direction?

Displacement involves all the necessary information about a change in position, that is, it includes both *distance and direction*. It does not contain any information about the path that has been followed. *The units of displacement are length units* such as feet, meters, or miles. If your friend decides to walk one block west, four blocks north, and then one block east, he will still arrive at your house. This resultant displacement is the same as if he had walked four blocks north.

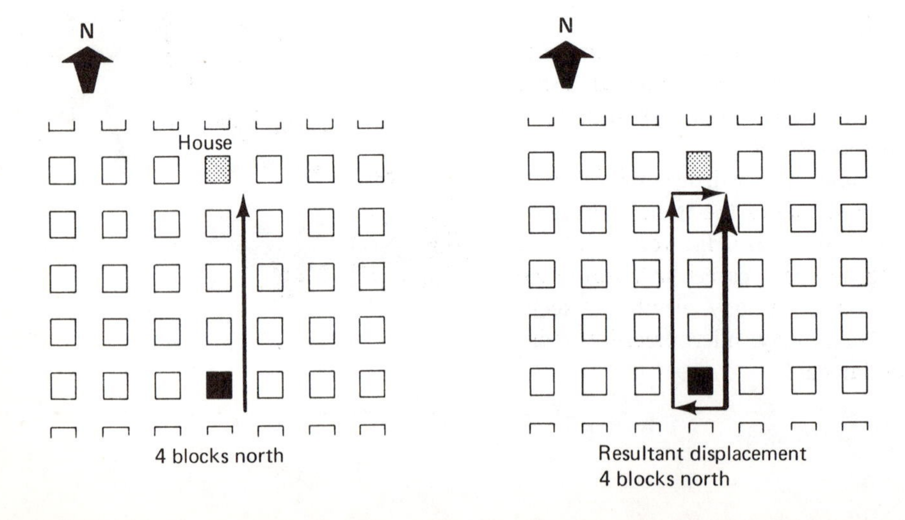

4 blocks north

Resultant displacement
4 blocks north

Displacement problems are examples of a certain type of problem which is easily solved by graphical methods. To solve this type of problem we need to examine the difference between what we call scalar and vector quantities. In Chapter 1 we discussed quantities of length, time, area, and volume. All these quantities can be expressed by a number with the appropriate units. For example, the area of a floor may be expressed as 150 ft². *Quantities such as this which can be completely described by a number and a unit are called scalar quantities or scalars.* They show magnitude only, not direction.

Many quantities such as displacement, force, and velocity must have their direction specified in addition to a number with a unit. To describe such quantities we use vectors. A vector is a quantity which has both size (magnitude) and direction. The magnitude of the displacement vector, "20 miles NE," is 20 miles. *Thus, a vector has both magnitude and direction.*

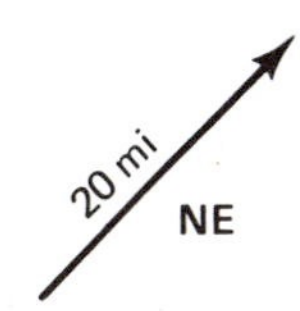

To represent a vector in our diagrams we draw an arrow which points in the correct direction. The magnitude of the vector is indicated by the length of the arrow. We usually choose a scale, such as 1 cm = 25 mi, for this purpose. Thus, a displacement of 100 mi north would be drawn as an arrow (pointing north) 4 cm long since

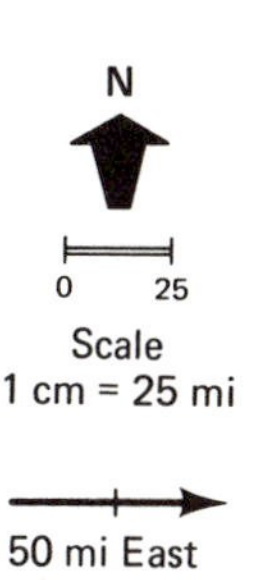

$$100 \text{ mi} \times \frac{1 \text{ cm}}{25 \text{ mi}} = 4 \text{ cm}$$

One end of the vector is called the initial end and the other is called the terminal end as shown.

A useful way of expressing the direction of a displacement vector is to give the bearing. *The bearing of a displacement is the angle measured clockwise from north to the displacement vector.*

See the following diagram for some examples. Always draw a dashed line along the north-south direction for convenience in measuring angles.

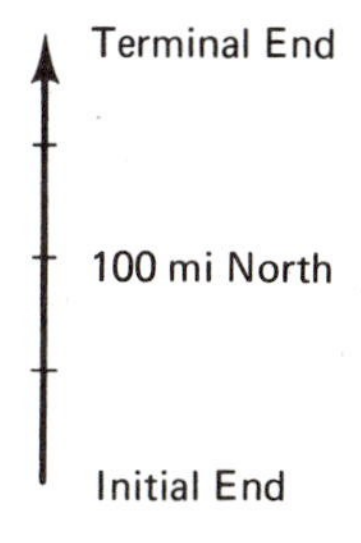

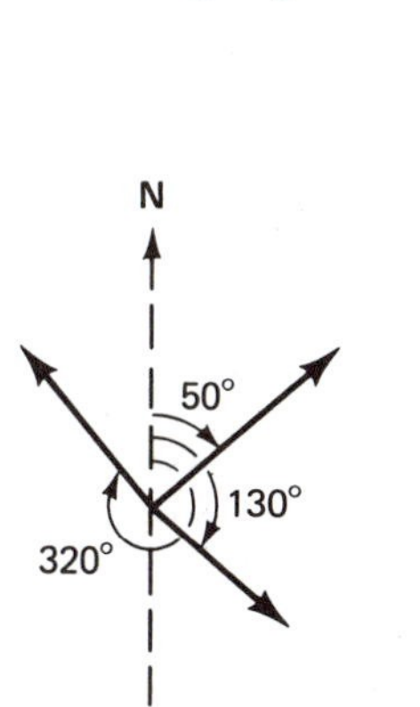

PROBLEMS

1. Using a scale of 1 cm = 25 mi, determine the *length* of the vectors which represents the following displacements.

   (a) displacement 75 mi north length = ______ cm

   (b) displacement 100 mi south length = ______ cm

   (c) displacement 50 mi east length = ______ cm

   (d) displacement 35 mi at 45° length = ______ cm

   (e) displacement 15 mi at 180° length = ______ cm

   (f) displacement 90 mi at 290° length = ______ cm

   (g) displacement 60 mi at 192° length = ______ cm

2. Draw the above vectors using the same scale as indicated above. Be sure to include the direction.

3-4 *GRAPHICAL ADDITION OF VECTORS*

Any given displacement can be the result of many different combinations of displacements.

In the following diagram, the displacement represented by the arrow, labeled *R*, resultant, is the result of either of the two paths shown. This vector is called the *resultant* of the vectors which make up either path 1 or path 2. *The resultant vector is the result (sometimes called the sum) of a set of vectors.* The resultant vector, *R*, in the diagram is the sum of the vectors *A*, *B*, *C* and *D*. It is also the sum of the vectors *E* and *F*.

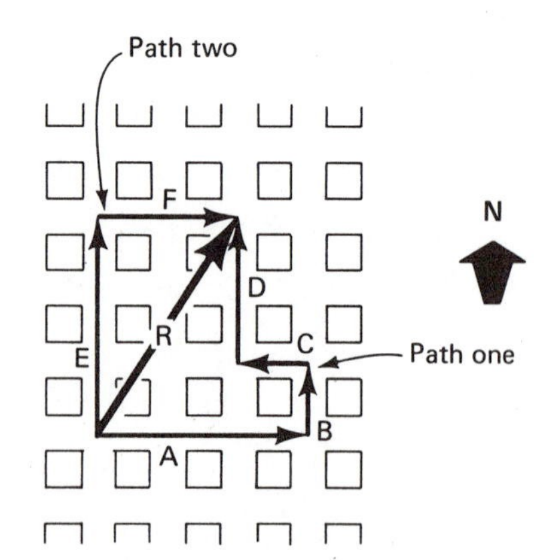

To solve a vector addition problem such as displacement, use the following procedure:

1. Choose a suitable scale and calculate the length of each vector.
2. Draw the north-south reference line. Graph paper should be used.
3. Using a ruler and protractor, draw the first vector and then draw the other vectors so that the initial end of each vector is placed at the terminal end of the previous vector.
4. Draw the sum or resultant vector from the initial end of the first vector to the terminal end of the last vector.

5. Measure the length of the resultant and use the scale to find the magnitude of the vector. Use a protractor to measure the bearing of the resultant.

EXAMPLE 1: Find the resultant displacement of an airplane which flies 20 mi at 90°, then 30 mi at 0°, and 10 mi at 240°.

We choose a scale of 1 cm = 5 mi so that the vectors are large enough to be accurate and small enough to fit on the paper. The length of the first vector is:

$$20 \cancel{\text{mi}} \times \frac{1 \text{ cm}}{5 \cancel{\text{mi}}} = 4 \text{ cm}$$

The length of the second vector is:

$$30 \cancel{\text{mi}} \times \frac{1 \text{ cm}}{5 \cancel{\text{mi}}} = 6 \text{ cm}$$

The length of the third vector is:

$$10 \cancel{\text{mi}} \times \frac{1 \text{ cm}}{5 \cancel{\text{mi}}} = 2 \text{ cm}$$

We draw the north-south reference line and draw the first vector (shown at the right). The second and third vectors are then drawn as shown in the following two diagrams.

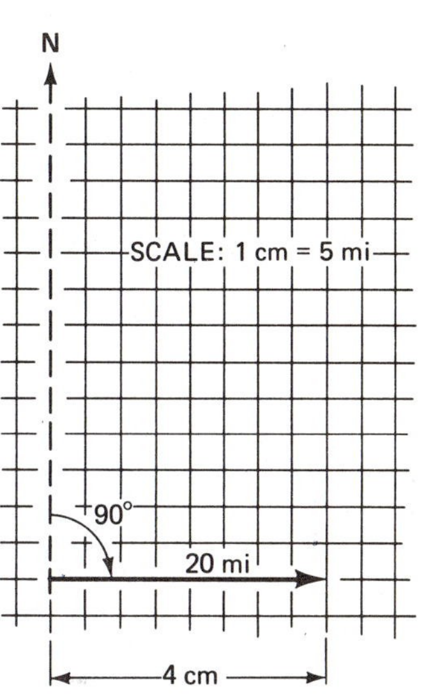

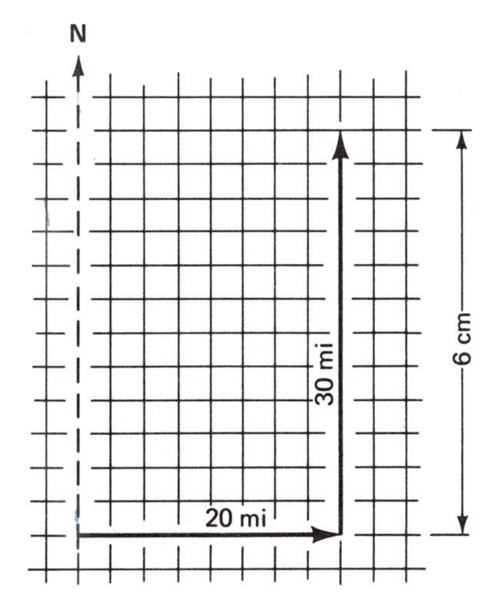

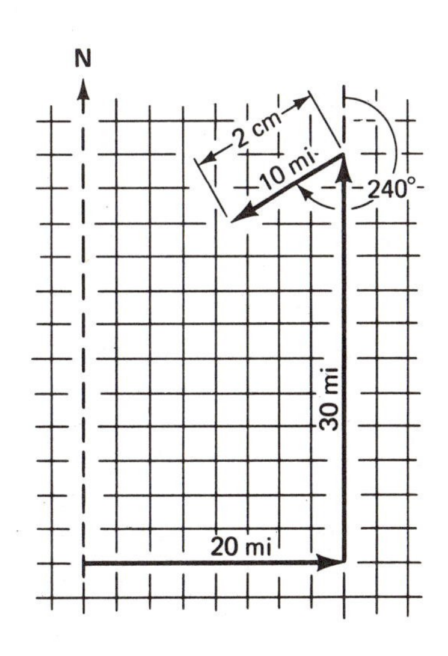

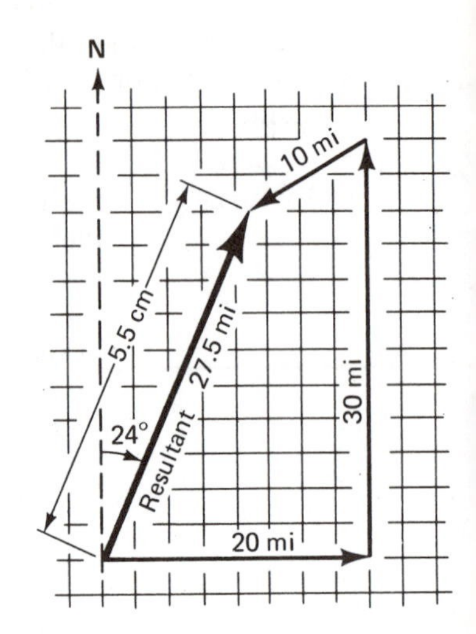

The length of the resultant is measured to be 5.5 cm. Since 1 cm = 5 mi, this represents a displacement with magnitude:

$$5.5\ \cancel{\text{cm}} \times \frac{5\ \text{mi}}{1\ \cancel{\text{cm}}} = 27.5\ \text{mi}$$

The bearing is measured to be 24°, so the resultant is 27.5 mi at 24°.

EXAMPLE 2: Find the resultant of the following displacements: 150 mi at 270°, 200 mi at 90°, and 125 mi at 180°.

Choose a scale of 1 cm = 50 mi.

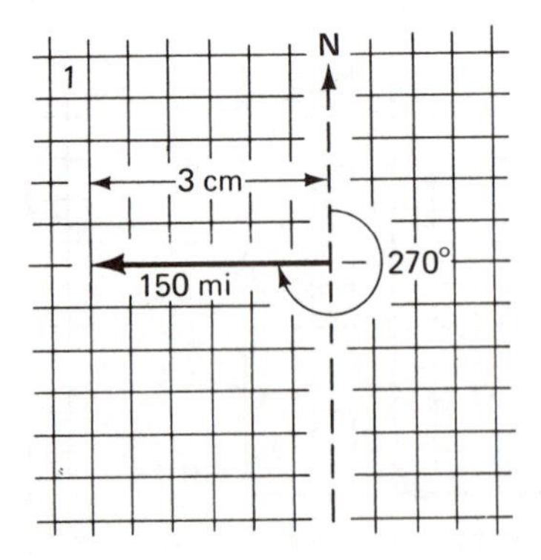

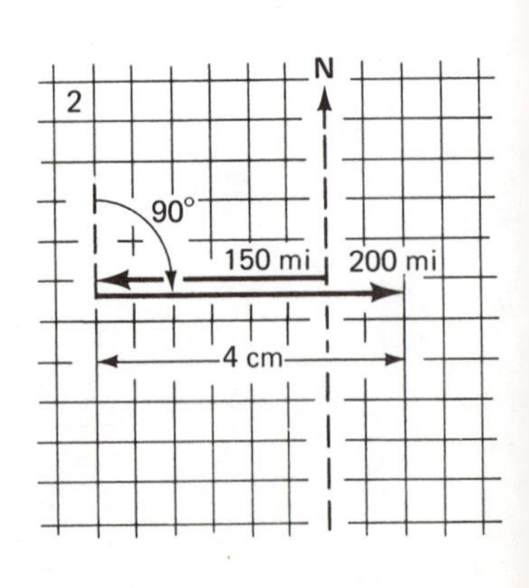

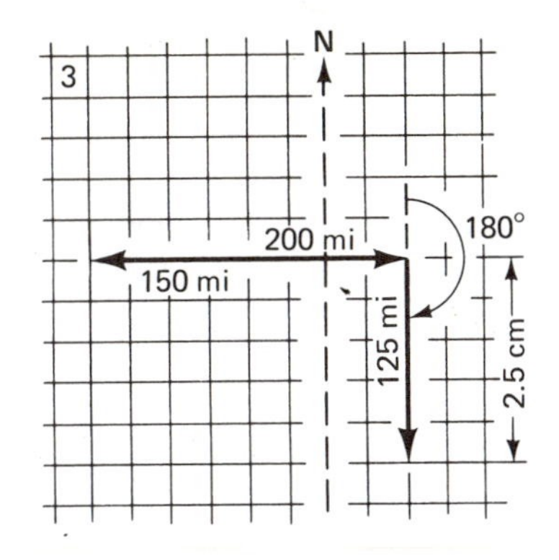

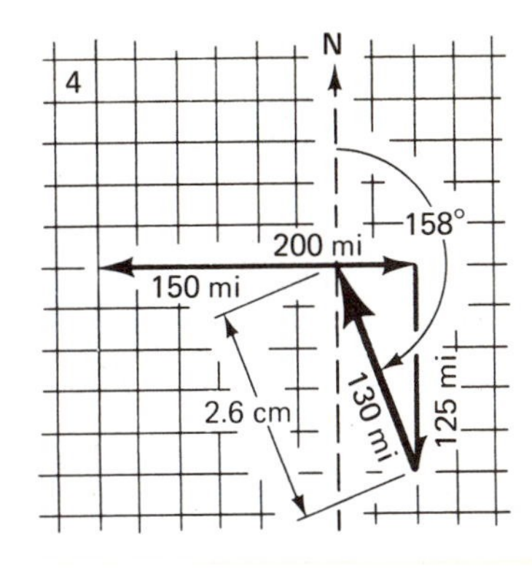

The length of the resultant is 2.6 cm which represents 130 mi at 158°.

PROBLEMS

On graph paper find the resultant of the following displacement pairs:

1. 35 mi at 90° and 50 mi at 0°.
2. 60 mi at 270° and 90 mi at 180°.
3. 500 mi at 75° and 1500 mi at 200°.
4. 20 mi at 87° and 17 mi at 189°.
5. 67 mi at 325° and 46 mi at 115°.
6. 4 mi at 205° and 2 mi at 15°.

Find the resultant of the following combinations of displacements:

7. 60 mi at 180°, 90 mi at 285°, and 75 mi at 45°.
8. 110 mi at 50°, 170 mi at 150°, and 145 mi at 20°.
9. 1700 mi at 0°, 2400 mi at 10°, and 2000 mi at 250°.
10. 90 mi at 350°, 75 mi at 210°, and 55 mi at 160°.
11. 7 mi at 0°, 10 mi at 97°, 15 mi at 190°.
12. 12 mi at 58°, 16 mi at 78°, 10 mi at 45°, and 14 mi at 10°.
13. 1 mi at 195°, 2.7 mi at 35°, 3.1 mi at 5°, and 2.2 mi at 340°.

3-5 *VELOCITY*

When an automobile travels a certain distance, we are interested in how fast the displacement took place. *The distance traveled per unit of time is called the speed.* Speed is a scalar (showing only magnitude, not direction) since it is described by a number and a unit. The unit of time is usually an hour or a second, so that the units of speed are miles per hour (mi/hr), feet per second (ft/s), or meters per second (m/s). We will use the notation mi/hr, ft/s, or m/s when referring to speed.

We are often interested in the direction of travel of an automobile or an airplane. *Velocity is a vector which gives the direction of travel and the distance traveled per unit of time.* The units of velocity are the same as those of speed. (A useful conversion is 60 mi/hr = 88 ft/s.)

EXAMPLE 1: Find the speed of an automobile that travels 125 km in one hour.

The speed is 125 km/hr.

EXAMPLE 2: An airplane flies 3500 miles in 5 hours. What is its speed?

The speed is the number of miles traveled per hour which is

$$\frac{3500 \text{ mi}}{5 \text{ hr}} = 700 \frac{\text{mi}}{\text{hr}}$$

EXAMPLE 3: Find the velocity of an auto which travels 500 miles north in 10 hours.

The magnitude of the velocity is the distance traveled per unit time or

$$\frac{500 \text{ mi}}{10 \text{ hr}} = 50 \frac{\text{mi}}{\text{hr}}$$

The direction of the velocity is north.

PROBLEMS

Find the speed of an auto which travels the following distances in the given times.

1. distance of 80 mi in 1 hour speed = _______ mi/hr
2. distance of 60 ft in 1 second speed = _______ ft/s
3. distance of 90 km in 1 hour speed = _______ km/hr
4. distance of 190 km in $2\frac{1}{2}$ hours speed = _______ km/hr
5. distance of 150 mi in 3 hours speed = _______ mi/hr
6. distance of 30 mi in 0.5 hour speed = _______ mi/hr
7. Find the speed of a racing car which can turn a lap on a one-mile oval track in 30 seconds. Speed = _______ mi/hr.
8. An automobile is traveling 90 mi/hr. Find the speed in ft/s. Find the speed in m/s.

Find the velocity for the following displacements and times.

9. 100 mi north in 3 hours.
10. 160 km east in 2 hours.
11. 31 mi west in 0.5 hour.
12. 1000 mi south in 8 hours.

3-6 *ACCELERATION*

When a racing car travels around a track, its velocity changes. A dragster running the quarter mile has a change in velocity also. Its velocity in the last few feet of the race is much greater than its velocity at the start. The faster the velocity of the dragster changes, the faster his time will be. *Acceleration is the change in velocity per unit time.* The faster the velocity changes, the larger the acceleration will be.

EXAMPLE 1: A dragster starts from rest (velocity = 0 ft/s) and attains a velocity of 150 ft/s in 10 seconds. Find the acceleration.

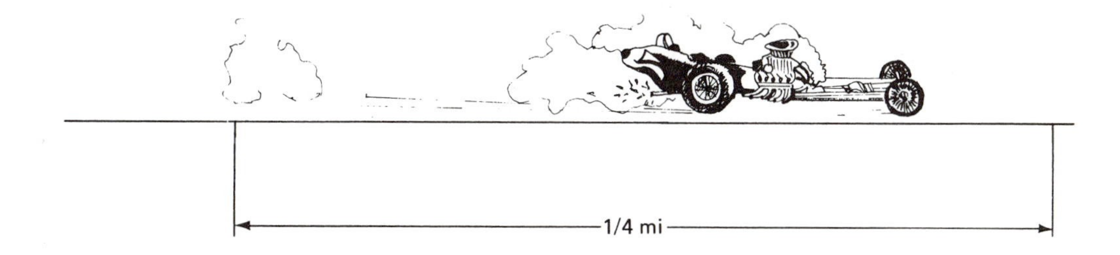

The acceleration is the change in velocity per unit time which is:

$$\frac{\text{Change in velocity}}{\text{Elapsed time}} = \frac{150 \text{ ft/s} - 0 \text{ ft/s}}{10 \text{ s}}$$
$$= \frac{150 \text{ ft/s}}{10\text{s}}$$
$$= \frac{15 \text{ ft/s}}{\text{s}}$$

or 15 feet per second per second.

Recall from arithmetic that to simplify fractions in the form

$$\frac{\frac{a}{b}}{\frac{c}{d}}$$

we divide by the denominator, that is, invert and multiply:

$$\frac{\frac{a}{b}}{\frac{c}{d}} = \frac{a}{b} \div \frac{c}{d} = \frac{a}{b} \cdot \frac{d}{c} = \frac{ad}{bc}$$

Use this idea to simplify the units from above:

$$\frac{\frac{15\text{ ft}}{\text{s}}}{\frac{\text{s}}{1}} = \frac{15\text{ ft}}{\text{s}} \div \frac{\text{s}}{1} = \frac{15\text{ ft}}{\text{s}} \cdot \frac{1}{\text{s}} = \frac{15\text{ ft}}{\text{s}^2}$$

The units of acceleration are usually ft/s^2 or m/s^2.

Certain relations between the quantities that we have discussed in this chapter will be used to solve motion problems. But first we must study equations. This will be done in the next chapter. We will then return to our study of motion in Chapter 5.

PROBLEMS

Find the acceleration of an automobile which changes velocity as indicated below:

1. velocity from 0 to 10 ft/s in 1 second.
2. velocity from 0 to 6 m/s in 1 second.
3. velocity from 60 ft/s to 70 ft/s in 1 second.
4. velocity from 45 m/s to 65 m/s in 2 seconds.
5. velocity from 10 ft/s to 70 ft/s in 3 seconds.
6. velocity from 15 m/s to 60 m/s in 3 seconds.

EQUATIONS 4

4-1 *REVIEW OF SOLVING EQUATIONS*

An equation is a mathematical sentence stating that two quantities are equal.

EXAMPLE 1: $3 + 2 = 5$ is an equation. This particular equation is a *statement* or *closed* equation since there are no variables in either member of the equation.

EXAMPLE 2: $3x + 5 = 7$ is an equation. This equation is an *open* equation since the left member contains a variable, x. Any equation containing a variable is called an open equation, since x may be replaced by any value. Thus, the choices for x are "open."

We are concerned with finding the correct replacement for the variable in an open equation. The value we find that makes the equation a true statement is called the *root* of the equation. When the root of an equation is found, we say we have *solved* the equation.

RULE 1

If $a = b$, then $a + c = b + c$. (If two quantities are equal, then adding the same quantity to each of them maintains the equality.)

To solve an equation using this rule, think first of undoing what has been done to the variable.

EXAMPLE 3: Solve $x + 4 = 29$ for x.

$$x + 4 = 29$$

$$x + 4 + (-4) = 29 + (-4)$$ (Undo the addition by adding -4. Use rule 1.)

$$x + 0 = 25$$ (Add)

$$x = 25$$ (Solution)

EXAMPLE 4: Solve $x - 5 = -9$ for x.

$$x - 5 = -9$$

$$x - 5 + 5 = -9 + 5$$ (Rule 1)

$$x + 0 = -4$$ (Add)

$$x = -4$$ (Solution)

RULE 2

If $a = b$, then $ac = bc$ or $a/c = b/c$ with $c \neq 0$. (If two quantities are equal, then multiplying or dividing both members of the equation by the same number will maintain the equality.)

EXAMPLE 5: Solve $3x = 18$ for x.

$$3x = 18$$

$$\frac{3x}{3} = \frac{18}{3}$$ (Undo the multiplication by dividing by 3. Use rule 2.)

$$1x = 6$$ (Multiply)

$$x = 6$$ (Solution)

EXAMPLE 6: Solve $x/4 = 9$ for x.

$$\frac{x}{4} = 9$$

$$4\left(\frac{x}{4}\right) = 4 \cdot 9$$ (Rule 2)

$$1x = 36$$ (Multiply)

$$x = 36$$ (Solution)

EXAMPLE 7: Solve $3x + 5 = 17$.

In this example more than one operation is indicated on the variable. There is an addition of 5 and a multiplication by 3. *In general, to solve such an equation, undo additions and subtractions first; then undo multiplications and divisions.*

$$3x + 5 = 17$$

$$3x + 5 + (-5) = 17 + (-5) \qquad \text{(Use rule 1)}$$

$$3x = 12 \qquad \text{(Add)}$$

$$\frac{3x}{3} = \frac{12}{3} \qquad \text{(Use rule 2)}$$

$$x = 4 \qquad \text{(Solution)}$$

EXAMPLE 8: Solve $2x - 7 = 10$ for x.

$$2x - 7 + 7 = 10 + 7 \qquad \text{(Use rule 1)}$$

$$2x = 17 \qquad \text{(Add)}$$

$$\frac{2x}{2} = \frac{17}{2} \qquad \text{(Use rule 2)}$$

$$x = \frac{17}{2} = 8.5 \qquad \text{(Solution)}$$

EXAMPLE 9: Solve $x/5 - 10 = 22$ for x.

$$\frac{x}{5} - 10 = 22$$

$$\frac{x}{5} - 10 + 10 = 22 + 10 \qquad \text{(Use rule 1)}$$

$$\frac{x}{5} = 32 \qquad \text{(Add)}$$

$$5\left(\frac{x}{5}\right) = 5 \cdot 32 \qquad \text{(Use rule 2)}$$

$$x = 160 \qquad \text{(Solution)}$$

PROBLEMS

Solve the following equations for x or y:

1. $3x = 4$
2. $y/2 = 10$
3. $x - 5 = 12$
4. $x + 1 = 9$
5. $2x + 10 = 10$
6. $ax = b$
7. $2x - 2 = 33$
8. $4 = x/10$

9. $172 - 43x = 43$

10. $mx + B = c$

11. $Ay - B = 0$

12. $3y + 15 = 75$

13. $15 = 105/y$

14. $6x = x - 15$

15. $2 = \frac{50}{2y}$

16. $9y = 67.5$

17. $ABx - C = D$

18. $10 = \frac{136}{4x}$

19. $2x + 22 = 75$

20. $9x + 10 = x - 26$

4-2 *MORE EQUATIONS*

There are equations written with portions of the equation included in parentheses. To solve these equations, first remove parentheses and then proceed as before. The rules for removing parentheses are the following:

1. If the parentheses are preceded by a plus (+) sign, they may be removed without changing any signs.

   Examples: $2 + (3 - 5) = 2 + 3 - 5$
   $3 + (x + 4) = 3 + x + 4$

2. If the parentheses are preceded by a minus (—) sign, the parentheses may be removed if the signs of the numbers (or letters) within the parentheses are changed.

   Examples: $2 - (3 - 5) = 2 - 3 + 5$
   $5 - (x - 7) = 5 - x + 7$

3. If the parentheses are preceded by a number, the parentheses may be removed if each of the terms inside the parentheses is multiplied by that (signed) number.

   Examples: $2(x + 4) = 2x + 8$
   $-3(x - 5) = -3x + 15$
   $2 - 4(3 - 5) = 2 - 12 + 20$

EXAMPLE 1: Solve $3(x - 4) = 15$ for x.

$3(x - 4) = 15$

$3x - 12 = 15$ (Remove parentheses)

$3x - 12 + 12 = 15 + 12$ (Use rule 1)

$3x = 27$ (Add)

$\frac{3x}{3} = \frac{27}{3}$ (Use rule 2

$1x = 9$ (Multiply)

$x = 9$ (Solution)

EXAMPLE 2: Solve $2x - (3x + 15) = 4x - 1$ for x.

$$2x - (3x + 15) = 4x - 1$$

$$2x - 3x - 15 = 4x - 1 \quad \text{(Remove parentheses)}$$

$$-x - 15 = 4x - 1 \quad \text{(Combine like terms)}$$

$$-x - 15 + x = 4x - 1 + x \quad \text{(Use rule 1)}$$

$$-15 = 5x - 1 \quad \text{(Add)}$$

$$-15 + 1 = 5x - 1 + 1 \quad \text{(Use rule 1)}$$

$$-14 = 5x \quad \text{(Add)}$$

$$\frac{-14}{5} = \frac{5x}{5} \quad \text{(Use rule 2)}$$

$$-2.8 = x \quad \text{(Solution)}$$

PROBLEMS

Solve the following equations.

1. $2y - 10 = 14$
2. $x - 15 = 13$
3. $5x = 275$
4. $13x + 29 = 198$
5. $140 - m = 290$
6. $27k = 0$
7. $4l + 72 = 10$
8. $31 - 3b = 43$
9. $28w - 56 = -8$
10. $5y - 7 = 23$
11. $m/3 - 12 = 21$
12. $4w/2 + 5 = 12$
13. $3k/7 = 18$
14. $11 - (x + 12) = 100$
15. $7h - (13 - 2h) = 5$
16. $x/5 - 2(2x/5 + 1) = 28$
17. $3(x + 117) = 201$
18. $\frac{1}{3}k - 31 = 19$
19. $17y/4 + 7 = 33$
20. $20(7p - 2) = 180$

4-3 *FORMULAS*

Solving equations is closely related to working with formulas. A formula is an equation relating one or more variables. Usually it expresses a relationship between or among physical properties. To solve problems it is necessary to know how to solve a formula for a particular variable in terms of the other known variables.

EXAMPLE 1: $v = s/t$ is a formula expressing v in terms of s and t. Suppose s and v are known, how do we find t?

$$v = \frac{s}{t}$$

$$v \cdot t = \frac{s}{t} \cdot t \qquad \text{(Rule 2)}$$

$$vt = s$$

Now, to find t:

$$\frac{vt}{v} = \frac{s}{v} \qquad \text{(Rule 2)}$$

$$t = \frac{s}{v} \qquad \text{(Solution)}$$

It is often convenient to use the same quantity in more than one way in a formula. For example, we may wish to use a particular measurement of a quantity, such as velocity, at a given time, say at $t = 0$ seconds, and then use the velocity at a later time, say at $t = 6$ seconds. To write out these desired values of the velocity is rather awkward. We simplify this written statement by using *subscripts* (small letters or numbers printed a half space below the printed line but next to the quantity referred to) to shorten what we must write. From the example given, v at time $t = 0$ seconds will be written as v_i (initial velocity); v at time $t = 6$ seconds will be written as v_f (final velocity). Mathematically, v_i and v_f are two different quantities which in most cases are unequal.

v_i and v_f cannot be added as like terms nor multiplied as numbers having the same base. The sum of v_i and v_f is written as $v_i + v_f$. The product of v_i and v_f is written as $v_i v_f$.

The subscript notation is used only to distinguish the general quantity, v, velocity, from the measure of that quantity at certain specified times.

EXAMPLE 2: Solve the formula $x = x_i + v_i t + \frac{1}{2}at^2$ for v_i.

$$x = x_i + v_i t + \tfrac{1}{2}at^2$$

$$-v_i t = x_i + \tfrac{1}{2}at^2 - x \qquad \text{(First express } v_i t \text{ in terms of the other variables)}$$

$$\frac{-v_i t}{t} = \frac{x_i + \frac{1}{2}at^2 - x}{t} \qquad \text{(Use rule 2)}$$

So,

$$-v_i = \frac{x_i + \frac{1}{2}at^2 - x}{t}$$

$$v_i = \frac{-(x_i + \frac{1}{2}at^2 - x)}{t}$$

$$v_i = \frac{-x_i - \frac{1}{2}at^2 + x}{t}$$

EXAMPLE 3: Solve the formula $v_{\text{avg}} = \frac{1}{2}(v_f + v_i)$ for v_f (avg is a subscript meaning average).

$$v_{\text{avg}} = \tfrac{1}{2}(v_f + v_i)$$

$$2v_{\text{avg}} = v_f + v_i$$

$$2v_{\text{avg}} - v_i = v_f$$

So,

$$v_f = 2v_{\text{avg}} - v_i$$

PROBLEMS

Solve the given formula for the quantity given.

1. $v = s/t$ for s
2. $a = v/t$ for v
3. $x = x_i + v_i t + \frac{1}{2}at^2$ for x_i
4. $x = x_i + v_i t + \frac{1}{2}at^2$ for a
5. $2a(s - s_i) = v^2 - v_i^2$ for a
6. $2a(s - s_i) = v^2 - v_i^2$ for s_i
7. $P = w/t$ for t
8. K. E. $= \frac{1}{2}mv^2$ for m
9. K. E. $= \frac{1}{2}mv^2$ for v^2
10. $w = mg$ for m
11. P. E. $= mgh$ for g
12. P. E. $= mgh$ for h
13. $F = ma$ for a
14. $W = Fs$ for s
15. $v_{\text{avg}} = \frac{1}{2}(v_f + v_i)$ for v_i

4-4 *QUADRATIC EQUATIONS OF THE FORM* $ax^2 = b$

EXAMPLE 1: Solve $x^2 = 16$ for x.

To solve a quadratic equation of this type, all we need to do is take the square root of both members of the equation.

$x^2 = 16$ (Take the square root of both sides)

$x = \pm 4$

In general, equations of the form $ax^2 = b$, where $a \neq 0$, can be solved as follows:

$$ax^2 = b$$

$$x^2 = \frac{b}{a} \quad \text{(Take the square root of both sides)}$$

$$x = \pm\sqrt{\frac{b}{a}}$$

EXAMPLE 2: Solve $2x^2 - 18 = 0$ for x.

$$2x^2 - 18 = 0$$

$$2x^2 = 18$$

$$x^2 = 9 \qquad \text{(Take the square root of both sides)}$$

So,

$$x = \pm 3$$

EXAMPLE 3: Solve $5y^2 = 100$ for y.

$$5y^2 = 100$$

$$y^2 = 20$$

$$y = \pm \sqrt{20}$$

$$y = \pm 4.47 \qquad \text{(By slide rule)}$$

EXAMPLE 4: Solve the formula $\text{K. E.} = \frac{1}{2}mv^2$ for v.

If $\text{K. E.} = \frac{1}{2}mv^2$, then

$$2(\text{K. E.}) = mv^2$$

$$\frac{2(\text{K. E.})}{m} = v^2$$

So,

$$v = \pm \sqrt{\frac{2(\text{K. E.})}{m}}$$

EXAMPLE 5: Solve $2a(s - s_i) = v^2 + v_i^2$ for v_i.

$$2a(s - s_i) = v^2 + v_i^2$$

$$2a(s - s_i) - v^2 = v_i^2$$

So,

$$v_i = \pm \sqrt{2a(s - s_i) - v^2}$$

PROBLEMS

Solve as indicated.

1. $x^2 = 36$ for x
2. $y^2 = 100$ for y
3. $2x^2 = 98$ for x
4. $5x^2 = 0.05$ for x
5. $3x^2 - 27 = 0$ for x
6. $2y^2 - 15 = 17$ for y
7. $10x^2 + 4.9 = 11.3$ for x
8. $2a(s - s_i) = v^2 - v_i^2$ for v
9. $2d = at^2$ for t
10. $A = \pi r^2$ for r

PROBLEM SOLVING 5

5-1 *PROBLEM SOLVING*

Problem solving in technical fields is more than plugging into formulas. It is necessary that you develop skill in taking data, analyzing the problems present, and finding the solution in an orderly manner.

Understanding the principle involved in a problem is more important than blindly substituting into a formula. By following an orderly procedure for problem solving, we hope to develop an approach to problem solving you can use in your studies and on the job.

In all problems in the remainder of this text, the method described below will be applied to all problems where appropriate.

Problem Solving Method

1. *Read the Problem Carefully.* This might appear obvious to you, but it is the most important step in solving a problem.
2. *Make a Sketch.* All problems may not lend themselves to a sketch. However, make a sketch whenever it is possible. Many times, seeing the problem before you will show if you have forgotten important parts of the problem and may suggest the solution.

3. *Write Down All Given Information.* This is necessary to get all essential facts in mind before looking for the solution.

4. *Write Down the Unknown or Quantity Called For.* Many students have difficulty solving problems because they don't know what they are looking for and solve for the wrong quantity.

5. *Write Down the Basic Equation or Formula Which Relates the Known and Unknown Quantities.* We find the basic formula or equation to use by studying what we are given and asked to find. Then look for a formula or equation which relates these quantities. Sometimes we may need to use more than one equation or formula in working a problem. Consider the following problem:

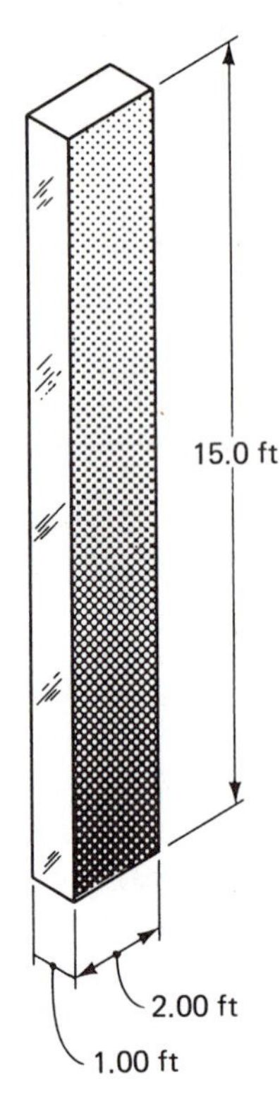

Find the volume of concrete required to fill a bridge abutment whose dimensions are $l = 2.00$ ft, $w = 1.00$ ft, $h = 15.0$ ft.

We know that the volume of a prism is given by $V = Bh$, where $V =$ volume, $B =$ area of the base, and $h =$ height. Our data, however, is in terms of l, w, and h. Therefore, before we can apply $V = Bh$, we must find B. We find B by using a second formula $B = l \times w$. So, substituting our data, $B = 2.00\text{ ft} \times 1.00\text{ ft} = 2.00\text{ ft}^2$. Now we can apply our first formula $V = Bh = 2.00\text{ ft}^2 \times 15.0\text{ ft} = 30.0\text{ ft}^3$.

6. *Find a Working Equation by Solving the Basic Equation or Formula for the Unknown Quantity.* Consider again the example in number 5. Suppose we were given: $V = 30.0\text{ ft}^3$, $B = 2.00\text{ ft}^2$, and were asked to find the height of the prism. Our basic equation is $V = Bh$. Note, however, that we are asked to find h. Therefore, we solve $V = Bh$ for h by dividing *both* sides of the equation by B:

$$\frac{V}{B} = \frac{Bh}{B}; \quad \therefore\ h = \frac{V}{B}.$$

This is the Working Equation. We are now ready to find h by substituting in the given data for B and V.

7. *Substitute the Data in the Working Equation Including the Appropriate Units.* It is important that you *carry the units all the way through the problem* as a check that you have solved the problem correctly. For example, if you are asked to find the weight of an object in newtons and the units of your answer work out to be meters, you need to review your solution for the error.

8. *Perform the Indicated Operations and Work Out the Solution.* Although this will be your final written step in the solution, in every case you should ask yourself: "Is my answer reasonable?". Here and on the job you will be dealing with practical problems. A quick estimate will many times reveal an error in your calculations.

To help you recall the procedure detailed above, with every problem set that follows, you will find the box shown at the right.

SKETCH
DATA
$a = 1,\ b = 2$
$c = ?$
BASIC EQUATION
$a = bc$
WORKING EQUATION
$c = \frac{a}{b}$
SUBSTITUTION
$c = \frac{1}{2}$

It is not meant to be complete, but only an outline to assist you in remembering and following the procedure for solving problems. *You should follow this outline in solving all problems in this course.*

5-2 *MORE ON VELOCITY AND ACCELERATION*

Now that we have reviewed the solution of equations and have developed a procedure for solving problems, we will apply what we've learned to the ideas of displacement, velocity, and acceleration.

Every time a truck speeds up or slows down, its velocity changes. This change of velocity is called acceleration. Acceleration may be an increase or decrease in velocity. A negative (—) acceleration is commonly called deceleration, meaning the object is slowing down.

Note, however, that a driver does not always speed up or slow down at the same rate. If a child jumps out in front of the truck, the driver may have to stop very quickly. His acceleration is not uniform acceleration.

Because we lack the mathematical tools to study all kinds of motion, we must limit our study to one kind—uniformly accelerated motion. The most common example of this kind of motion is that of a freely falling body. Because of the complexity of this kind of problem, we must assume that falling bodies are unaffected by the resistance of the air. Although, in fact, air resistance is an important factor in the design of machines which must move through the atmosphere. In learning to solve motion problems, we will assume air resistance to be negligible. Note also that for freely falling bodies the acceleration (a) due to gravity is $a = 32.2\ \text{ft/s}^2$ (English system) or $a = 9.81\ \text{m/s}^2$ (metric system).

A number of formulas and equations have been discovered which apply to freely falling bodies and uniformly accelerated motion in general.

$$s = v_{avg}t \qquad s = v_i t + \tfrac{1}{2}a_{avg}t^2$$

$$v_{avg} = \frac{v_f + v_i}{2} \qquad v_f = v_i + a_{avg}t$$

$$a_{avg} = \frac{v_f - v_i}{t} \qquad s = \tfrac{1}{2}(v_f + v_i)t$$

$$2a_{avg}s = v_f^2 - v_i^2$$

where: s = displacement, v_f = final velocity, v_i = initial velocity, v_{avg} = average velocity, a_{avg} = average acceleration, t = time

We will now consider some problems using these equations and applying our problem solution procedure.

EXAMPLE 1: The average velocity of a rolling freight car is 7.00 ft/s. How long does it take for the car to roll 54.0 ft?

Solution

SKETCH: None needed.

DATA:

$$s = 54.0 \text{ ft}$$
$$v_{avg} = 7.00 \text{ ft/s}$$
$$t = ?$$

BASIC EQUATION: $s = v_{avg}t$

WORKING EQUATION: $t = \dfrac{s}{v_{avg}}$

SUBSTITUTION:

$$t = \frac{54.0 \text{ ft}}{7.00 \text{ ft/s}}$$
$$= 7.71 \text{ s}$$

EXAMPLE 2: A dragster starting from a dead stop reaches a final velocity of 198 mi/hr. What is its average velocity?

Solution

SKETCH: None needed.

DATA: $v_i = 0, \quad v_f = 198 \dfrac{\text{mi}}{\text{hr}}, \quad v_{avg} = ?$

BASIC EQUATION: $v_{avg} = \dfrac{v_f + v_i}{2}$

WORKING EQUATION: Same

SUBSTITUTION:

$$v_{avg} = \frac{198 \text{ mi/hr} + 0 \text{ mi/hr}}{2}$$
$$= 99.0 \frac{\text{mi}}{\text{hr}}$$

EXAMPLE 3: A rock is thrown straight down from a cliff with an initial velocity of 10.0 ft/s. Its final velocity when it strikes the water below is 310 ft/s. The acceleration due to gravity is 32.2 ft/s^2. How long is the rock in flight?

Solution

SKETCH: None needed.

DATA:

$$v_i = 10.0 \text{ ft/s}, \qquad v_f = 310 \text{ ft/s}$$
$$a = 32.2 \text{ ft/s}^2, \qquad t = ?$$

Note the importance of listing all the data as an aid to finding the basic equation.

BASIC EQUATION: $v_f = v_i + a_{avg}t$ or $a_{avg} = \dfrac{v_f - v_i}{t}$

(two forms of same equation)

WORKING EQUATION: $t = \dfrac{v_f - v_i}{a_{avg}}$

SUBSTITUTION:

$$t = \frac{310 \text{ ft/s} - 10.0 \text{ ft/s}}{32.2 \text{ ft/s}^2}$$
$$= \frac{300 \text{ ft/s}}{32.2 \text{ ft/s}^2}$$
$$= 9.32 \text{ s}$$

EXAMPLE 4: A train slowing to a stop has an average acceleration of -3.00 m/s². (Note that a minus (—) acceleration is commonly called deceleration, meaning that the object is slowing down.) If its initial velocity is 30.0 m/s, how far does it travel in 4.00 seconds?

Solution

SKETCH: None needed.

DATA:

$$a_{avg} = -3.00 \text{ m/s}^2, \qquad t = 4.00 \text{ s}$$
$$v_i = 30.0 \text{ m/s} \qquad s = ?$$

BASIC EQUATION: $s = v_i t + \frac{1}{2}a_{avg}t^2$

WORKING EQUATION: Same

SUBSTITUTION:

$$s = 30.0 \text{ m/s } (4.00 \text{ s}) + \tfrac{1}{2}(-3.00 \text{ m/s}^2)(4.00 \text{ s})^2$$
$$= 120 \text{ m} - 24.0 \text{ m}$$
$$= 96.0 \text{ m}$$

EXAMPLE 5: An automobile is accelerated from 67.0 km/hr to 96.0 km/hr in 7.80 seconds. What is its acceleration?

Solution

SKETCH: None needed.

DATA: $v_f = 96.0$ km/hr, $v_i = 67.0$ km/hr
$t = 7.80$ s, $a = ?$

BASIC EQUATION: $a_{avg} = \dfrac{v_f - v_i}{t}$

WORKING EQUATION: Same

SUBSTITUTION:
$$a_{avg} = \frac{96.0\text{ km/hr} - 67.0\text{ km/hr}}{7.80\text{ s}} = \frac{29.0\text{ km/hr}}{7.80\text{ s}}$$

Recall that the metric unit of acceleration is m/s². Thus far we have $\dfrac{\text{km/hr}}{\text{s}}$ to which we must apply one or more conversion factors as follows:

$$\frac{\text{km/hr}}{\text{s}}\left(\frac{10^3\text{ m}}{1\text{ km}}\right)\left(\frac{1\text{ hr}}{3600\text{ s}}\right)$$

therefore,

$$a = \frac{\dfrac{29.0\text{ km}}{\text{hr}}\left(\dfrac{10^3\text{ m}}{1\text{ km}}\right)\left(\dfrac{1\text{ hr}}{3600\text{ s}}\right)}{7.8\text{ s}} = \frac{(29.0)(10^3\text{ m})}{(7.8\text{ s})(3.6 \times 10^3\text{ s})} = 1.03\text{ m/s}^2$$

PROBLEMS

Substitute in the given equation and find the unknown quantity.

1. Given: $v_{avg} = \dfrac{v_f + v_i}{2}$
$v_f = 6.20$ ft/s
$v_i = 3.90$ ft/s
$v_{avg} = ?$

2. Given: $a_{avg} = \dfrac{v_f - v_i}{t}$
$a_{avg} = 3.07$ m/s²
$v_f = 16.8$ m/s
$t = 4.10$ s
$v_i = ?$

3. Given: $s = v_i t + \frac{1}{2} a_{avg} t^2$

$$t = 3.00 \text{ s}$$
$$a_{avg} = 6.40 \text{ ft/s}^2$$
$$v_i = 33.0 \text{ ft/s}$$
$$s = ?$$

4. Given: $2a_{avg}s = v_f^2 - v_i^2$

$$a_{avg} = 8.41 \text{ m/s}^2$$
$$s = 4.81 \text{ m}$$
$$v_i = 1.24 \text{ m/s}$$
$$v_f = ?$$

5. Given: $v_f = v_i + a_{avg}t$

$$v_f = 10.4 \text{ ft/s}$$
$$v_i = 4.01 \text{ ft/s}$$
$$t = 3.00 \text{ s}$$
$$a_{avg} = ?$$

PROBLEMS

Solve using the outline in the box at the right.

SKETCH
DATA
$a = 1, b = 2,$
$c = ?$
BASIC EQUATION
$a = bc$
WORKING EQUATION
$c = \frac{a}{b}$
SUBSTITUTION
$c = \frac{1}{2}$

1. The average velocity of a mini-bike is 8.00 ft/s. How long does it take for the bike to go 35.0 ft?
2. A sprinter starting from rest reaches a final velocity of 18.0 mi/hr. What is his average velocity?
3. A coin is dropped with no initial velocity. Its final velocity when it strikes the earth below is 50.0 ft/s. The acceleration of gravity is 32.2 ft/s^2. How long does it fall before striking the earth?
4. A rocket lifting off from earth has an average acceleration of 44.0 ft/s^2. Its initial velocity is zero. How far into the atmosphere does it travel during the first 5.00 seconds, assuming it goes straight up?
5. The final velocity of a truck is 74.0 ft/s. If it accelerates at a rate of 2.00 ft/s^2 from an initial velocity of 5.00 ft/s, how long is required for it to attain its final velocity?
6. A truck can be accelerated from 85.0 km/hr to 120 km/hr in 9.20 seconds. What is its acceleration in m/s^2?
7. How long does it take a rock to drop 95.0 m from rest?
8. What final velocity does the rock in problem 7 attain?
9. A ball is thrown downward from the top of a 43.0 ft building with an initial velocity of 62.0 ft/s. What is its final velocity as it strikes the ground?
10. A worm accelerates from rest to a final velocity of 0.070 cm/s. If it takes 0.700 s for him to attain his final velocity, what is his acceleration?

6 FORCES IN ONE DIMENSION

6-1 *INTRODUCTION*

To understand the causes of the various types of motion as studied in technical programs, we need to study forces. Many types of forces are responsible for the motion of an automobile. The force produced by a hot expanding gas on the pistons causes them to move.

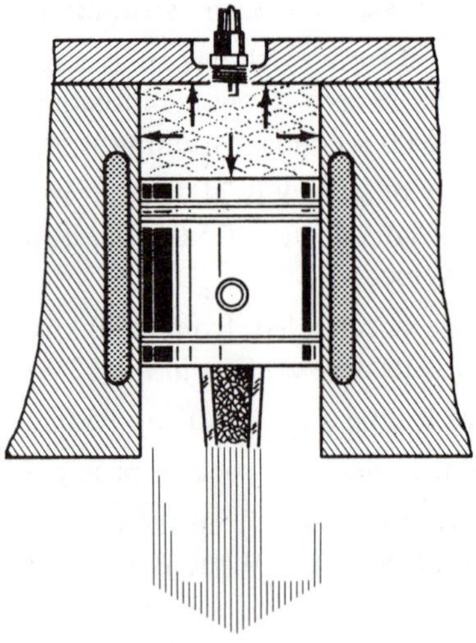

When a structural engineer designs the supports for a bridge, he must allow for the weight of the vehicles on it and also the weight of the bridge itself. These forces do not cause motion but are still very important.

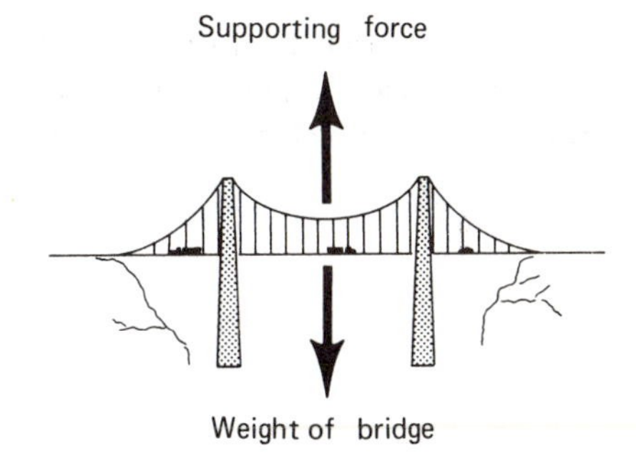

A force is a push or a pull which tends to cause motion. Force is a vector quantity and thus has both magnitude and direction. Some forces, such as the weight of the bridge shown opposite, do not cause motion because they are balanced by other forces. The downward force of the bridge's weight is balanced by the upward force supplied by the supports. If the supports were weakened and could not supply this force, the downward force would no longer be balanced and the bridge would move, that is, it would collapse.

The units for measuring force are the pound (lb) in the English system and the newton (N) in the metric system. The conversion factor is 1 lb = 4.45 N.

6-2 *LAW OF INERTIA*

We now want to examine the relationship between forces and motion. There are three relationships or laws which were discovered by Isaac Newton during the late seventeenth century. The three laws are often called Newton's laws. The first of these is the law of inertia: a body which is in motion continues in motion with the same velocity (at constant speed and in a straight line), and a body at rest continues at rest unless an unbalanced force acts upon it.

Inertia is the property of a body which causes it to remain at rest if it is at rest or to continue moving with a constant velocity unless an unbalanced force acts upon it.

When the accelerating force of an automobile engine is no longer applied to a moving car, it will slow down. This is not a violation of the law of inertia because there are forces being applied to the car through air resistance, friction in the bearings, and the rolling resistance of the tires. If these forces could be removed, the auto would continue moving with a constant velocity.

Anyone who has tried to stop quickly on ice knows the effect of the law of inertia when frictional forces are small.

Some objects tend to resist changes in their motion more than others. It is much easier to push a small automobile than to push a large truck into motion. *Mass is a measure of the resistance a body has to change in its motion.*

The unit of mass is the slug in the English system and the kilogram (kg) in the metric system. We need to look at another law of motion before we can find the conversion factor between slugs and kilograms.

6-3 *THE LAW OF ACCELERATION*

The second law of motion which we will call the law of acceleration relates the applied force and the acceleration of an object. We can state this law as follows: *the total force acting on a body is equal to the mass of the body times its acceleration.* In equation form this is:

$$F = ma$$

where F is the total force, m is the mass, and a is the acceleration.

Let's determine what force is necessary to give an object of mass 1 slug an acceleration of 1 ft/s². The law of acceleration gives us:

$$\begin{aligned} F &= ma \\ &= 1 \text{ slug} \times 1 \text{ ft/s}^2 \\ &= 1 \text{ slug ft/s}^2 \end{aligned}$$

We would expect that the answer should be in the force unit, lb. The units will be correct if we use the fact that the slug is not a basic unit, and it can be expressed in terms of the basic units as:

$$1 \text{ slug} = 1 \frac{\text{lb s}^2}{\text{ft}}$$

Using this as a conversion factor, the above equation becomes:

$$F = 1 \,\cancel{\text{slug}} \frac{\cancel{\text{ft}}}{\cancel{\text{s}^2}} \times \frac{1 \text{ lb } \cancel{\text{s}^2}}{1 \,\cancel{\text{slug}}\,\cancel{\text{ft}}}$$
$$= 1 \text{ lb}$$

So a force of 1 lb gives a mass of 1 slug an acceleration of 1 ft/s^2.

The above conversion factor is often useful when written in the form:

$$1 \text{ lb} = 1 \frac{\text{slug ft}}{\text{s}^2}$$

In the metric system the conversion is:

$$1 \text{ kg} = 1 \frac{\text{N s}^2}{\text{m}}$$

The conversion factor is often useful when written in the form:

$$1 \text{ N} = 1 \frac{\text{kg m}}{\text{s}^2}$$

When the same force is applied to two different masses, these masses will have different accelerations. For example, a much smaller force is required to accelerate a baseball from rest to 60 miles per hour than to accelerate an automobile from rest to 60 miles per hour in the same time period. The reason for this is that the automobile has a much larger mass.

When the same amount of force is applied to two different masses, the smaller in mass will be accelerated more than the larger in mass. Compare samples 1 and 2 and samples 3 and 4 in the computer printout reproduced below which illustrates these principles.

SAMPLE 1:

FORCE = 15.0 LB
MASS = 0.25 SLUG
ACCELERATION = 60.0 FT/S²

TIME	POSITION	VELOCITY
0.0 SEC	0.0 FT	0.0 FT/S
0.1 SEC	0.30 FT	6.00 FT/S
0.2 SEC	1.20 FT	12.00 FT/S
0.3 SEC	2.70 FT	18.00 FT/S
0.4 SEC	4.80 FT	24.00 FT/S
0.5 SEC	7.50 FT	30.00 FT/S
0.6 SEC	10.80 FT	36.00 FT/S
0.7 SEC	14.70 FT	42.00 FT/S
0.8 SEC	19.20 FT	48.00 FT/S
0.9 SEC	24.30 FT	54.00 FT/S
1.0 SEC	30.00 FT	60.00 FT/S
1.1 SEC	36.30 FT	66.00 FT/S
1.2 SEC	43.20 FT	72.00 FT/S
1.3 SEC	50.70 FT	78.00 FT/S
1.4 SEC	58.80 FT	84.00 FT/S

SAMPLE 2:

FORCE = 15.0 LB
MASS = 30.00 SLUG
ACCELERATION = 0.5 FT/S²

TIME	POSITION	VELOCITY
0.0 SEC	0.0 FT	0.0 FT/S
0.1 SEC	0.00 FT	0.05 FT/S
0.2 SEC	0.01 FT	0.10 FT/S
0.3 SEC	0.02 FT	0.15 FT/S
0.4 SEC	0.04 FT	0.20 FT/S
0.5 SEC	0.06 FT	0.25 FT/S
0.6 SEC	0.09 FT	0.30 FT/S
0.7 SEC	0.12 FT	0.35 FT/S
0.8 SEC	0.16 FT	0.40 FT/S
0.9 SEC	0.20 FT	0.45 FT/S
1.0 SEC	0.25 FT	0.50 FT/S
1.1 SEC	0.30 FT	0.55 FT/S
1.2 SEC	0.36 FT	0.60 FT/S
1.3 SEC	0.42 FT	0.65 FT/S
1.4 SEC	0.49 FT	0.70 FT/S

SAMPLE 3:

FORCE = 175.0 LB
MASS = 0.25 SLUG
ACCELERATION = 700.0 FT/S²

TIME	POSITION	VELOCITY
0.0 SEC	0.0 FT	0.0 FT/S
0.1 SEC	3.50 FT	70.00 FT/S
0.2 SEC	14.00 FT	140.00 FT/S
0.3 SEC	31.50 FT	210.00 FT/S
0.4 SEC	56.00 FT	280.00 FT/S
0.5 SEC	87.50 FT	350.00 FT/S
0.6 SEC	126.00 FT	420.00 FT/S
0.7 SEC	171.50 FT	490.00 FT/S
0.8 SEC	224.00 FT	560.00 FT/S
0.9 SEC	283.50 FT	630.00 FT/S
1.0 SEC	350.00 FT	700.00 FT/S
1.1 SEC	423.50 FT	770.00 FT/S
1.2 SEC	504.00 FT	840.00 FT/S
1.3 SEC	591.50 FT	910.00 FT/S
1 4 SEC	686.00 FT	980.00 FT/S

SAMPLE 4:

FORCE = 175.0 LB
MASS = 350.00 SLUG
ACCELERATION = 0.5 FT/S²

TIME	POSITION	VELOCITY
0.0 SEC	0.0 FT	0.0 FT/S
0.1 SEC	0.00 FT	0.05 FT/S
0.2 SEC	0.01 FT	0.10 FT/S
0.3 SEC	0.02 FT	0.15 FT/S
0.4 SEC	0.04 FT	0.20 FT/S
0.5 SEC	0.06 FT	0.25 FT/S
0.6 SEC	0.09 FT	0.30 FT/S
0.7 SEC	0.12 FT	0.35 FT/S
0.8 SEC	0.16 FT	0.40 FT/S
0.9 SEC	0.20 FT	0.45 FT/S
1.0 SEC	0.25 FT	0.50 FT/S
1.1 SEC	0.30 FT	0.55 FT/S
1.2 SEC	0.36 FT	0.60 FT/S
1.3 SEC	0.42 FT	0.65 FT/S
1.4 SEC	0.49 FT	0.70 FT/S

EXAMPLE 1: What total force is necessary to produce an acceleration of 2.00 ft/s² on a mass of 3.00 slugs?

SKETCH:

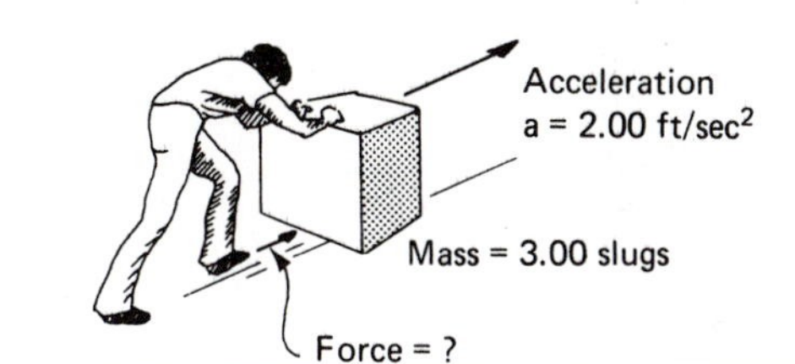

DATA:

$$m = 3.00 \text{ slugs}$$
$$a = 2.00 \frac{\text{ft}}{\text{s}^2}$$
$$F = ?$$

BASIC EQUATION: $F = ma$

WORKING EQUATION: Same

SUBSTITUTION:

$$F = ma = 3.00 \text{ slugs} \times 2.00 \frac{\text{ft}}{\text{s}^2}$$
$$= \frac{6.00 \text{ slug ft}}{\text{s}^2}$$
$$= \frac{6.00 \cancel{\text{slug ft}}}{\cancel{\text{s}^2}} \times \frac{1 \text{ lb} \cancel{\text{s}^2}}{1 \cancel{\text{slug ft}}}$$
$$= 6.00 \text{ lb}$$

(*Note*: We must use a conversion factor to obtain force units.)

EXAMPLE 2: What is the acceleration produced by a total force of 500 N applied to a mass of 20.0 kg?

SKETCH: None

DATA:

$$F = 500 \text{ N}$$
$$m = 20.0 \text{ kg}$$
$$a = ?$$

BASIC EQUATION: $F = ma$

WORKING EQUATION: $a = \frac{F}{m}$

SUBSTITUTION:
$$a = \frac{F}{m}$$
$$= \frac{500\ \text{N}}{20.0\ \text{kg}}$$
$$= 25.0\frac{\text{N}}{\text{kg}}$$
$$= 25.0\frac{\cancel{\text{N}}}{\cancel{\text{kg}}} \times \frac{1\ \cancel{\text{kg}}\ \text{m}}{1\ \cancel{\text{N}}\ \text{s}^2}$$
$$= 25.0\frac{\text{m}}{\text{s}^2}$$

(*Note*: We must use a conversion factor to obtain acceleration units.)

PROBLEMS

Find the total force necessary to give the following masses the given acceleration.

1. $m = 15.0$ kg, $a = 2.00$ m/s² $F =$ ______ N
2. $m = 4.00$ slugs, $a = 0.500$ ft/s² $F =$ ______ lb
3. $m = 111$ slugs, $a = 6.70$ ft/s² $F =$ ______ lb
4. $m = 91.0$ kg, $a = 6.00$ m/s² $F =$ ______ N
5. $m = 28.0$ slugs, $a = 9.00$ ft/s² $F =$ ______ lb
6. $m = 42.0$ kg, $a = 3.00$ m/s² $F =$ ______ N

Find the acceleration of the following masses with the given total force.

7. $m = 7.00$ slugs, $F = 12.0$ lb $a =$ ______ ft/s².
8. $m = 190$ kg, $F = 76.0$ N $a =$ ______ m/s².
9. $m = 3.60$ slugs, $F = 42.0$ lb $a =$ ______ ft/s².
10. $m = 0.790$ slugs, $F = 13.0$ lb $a =$ ______ ft/s².
11. $m = 110$ kg, $F = 57.0$ N $a =$ ______ m/s².
12. $m = 84.0$ kg, $F = 33.0$ N $a =$ ______ m/s².
13. Find the total force (in newtons) necessary to give an automobile of mass 1750 kg an acceleration of 3.00 m/s².
14. Find the acceleration produced by a total force of 93.0 N on a mass of 6.00 kg.
15. Find the total force (in pounds) necessary to give an automobile of mass 120 slugs an acceleration of 11.0 ft/s².
16. Find the total force (in pounds) necessary to give a rocket of mass 25,000 slugs an acceleration of 28.0 ft/s².
17. Find the acceleration produced by a total force of 300 lb on a mass of 0.750 slug.

18. Find the mass of an object that has an acceleration of 15.0 m/s^2 when an unbalanced force of 90.0 N acts on it.
19. An automobile has a mass of 100 slugs. The passengers it carries have a mass of 7.00 slugs each.
    (a) Find the acceleration of the auto and one passenger if the total force acting is 1500 lb.
    (b) Find the acceleration of the auto and six passengers if the total force is again 1500 lb.

6-4 *FRICTION*

When two objects slide across each other, a force which resists the motion is produced. This force is called *friction.* Friction is caused by the irregularities of the two surfaces which tend to catch on each other. Severe engine damage can be caused by friction if proper lubricants are not used. Friction makes it hard to push objects along the floor.

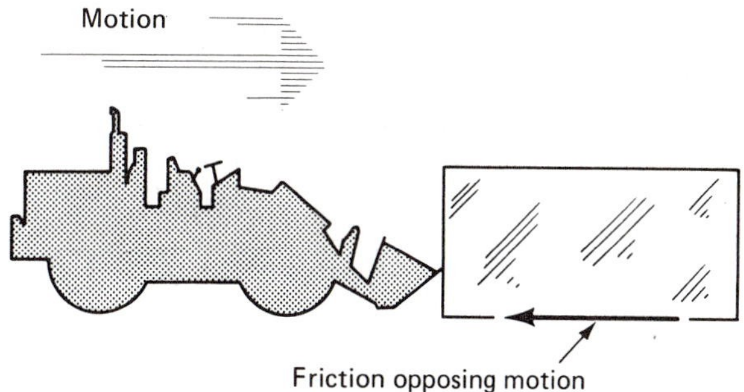

The frictional force depends on the materials and the smoothness of the surfaces involved. In general, the more polished the surfaces, the smaller the frictional force will be.

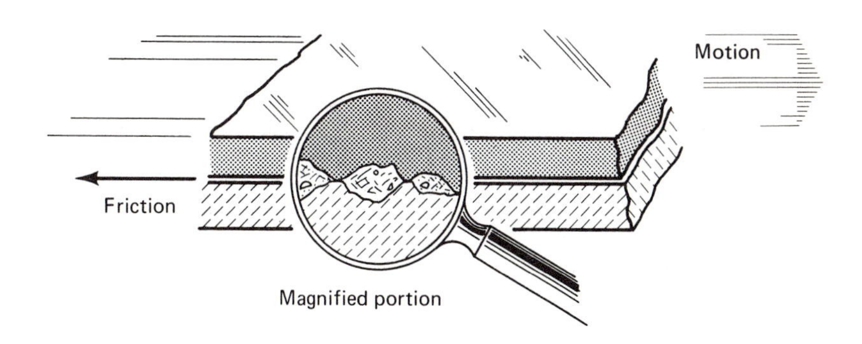

6-5 *TOTAL FORCES IN ONE DIMENSION*

In the examples used to illustrate the law of acceleration, we discussed total forces only. We need to remember that forces are vectors and have magnitude and direction. The total force acting on an object is the resultant of the separate forces. *When forces act in the same or opposite directions (in one dimension), the total force can be found by adding the forces which act in one direction and subtract from that the forces which act in the opposite direction.* It is useful to draw the forces as vectors (arrows) in the sketch before working the problem.

EXAMPLE 1: Two men push in the same direction on a refrigerator. The force exerted by one man is 150 lb. The force by the other is 175 lb. Find the net force.

SKETCH:

Both forces act in the same direction so the total force is the sum of the two.

$$F_{\text{net}} = 150 \text{ lb} + 175 \text{ lb}$$
$$= 325 \text{ lb} \quad \text{to the right}$$

EXAMPLE 2: The same two men push the refrigerator and the motion is opposed by a frictional force of 300 lb. Find the net force.

SKETCH:

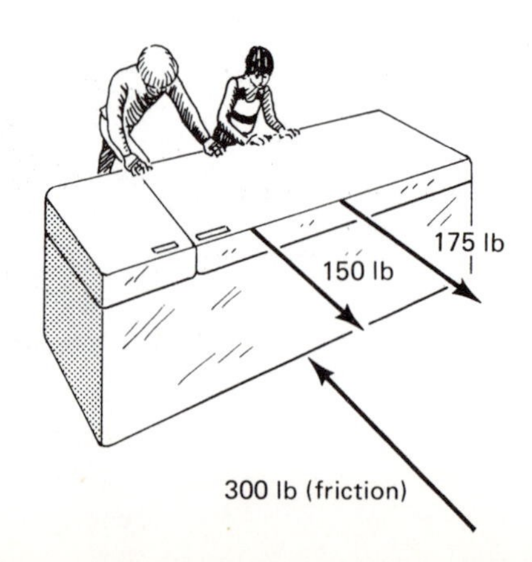

The men push in one direction and friction pushes in the opposite direction, so we add the forces exerted by the men and subtract the frictional force.

$$F_{\text{net}} = 175 \text{ lb} + 150 \text{ lb} - 300 \text{ lb}$$
$$= 25 \text{ lb} \quad \text{to the right}$$

EXAMPLE 3: The refrigerator in Example 2 has a mass of 5.00 slugs. What is its acceleration when the men are pushing against the frictional force.

SKETCH: None

DATA:

$F = 25.0$ lb (from Example 2)
$m = 5.00$ slugs
$a = ?$

BASIC EQUATION: $F = ma$

WORKING EQUATION: $a = \dfrac{F}{m}$

SUBSTITUTION:

$$a = \frac{F}{m}$$
$$= \frac{25.0 \text{ lb}}{5.00 \text{ slugs}}$$
$$= 5.00 \frac{\text{lb}}{\text{slug}}$$
$$= 5.00 \frac{\cancel{\text{lb}}}{\cancel{\text{slug}}} \times \frac{1 \cancel{\text{slug}} \text{ ft}}{1 \cancel{\text{lb}} \text{ s}^2}$$
$$= 5.00 \frac{\text{ft}}{\text{s}^2}$$

(*Note*: We must use a conversion factor to obtain acceleration units.)

EXAMPLE 4: Two men push in the same direction on a large crate. The force exerted by one man is 600 N. The force by the other man is 680 N. The motion is opposed by a frictional force of 1180 N. Find the net force.

$$F_{\text{net}} = 600 \text{ N} + 680 \text{ N} - 1180 \text{ N}$$
$$= 100 \text{ N}$$

PROBLEMS

Find the net force acting when the following forces act in the direction indicated.

1. 100 lb to the left, 75.0 lb to the right, and 10.0 lb to the right.

   Net force = _______ lb

   The direction is _______

2. 265 lb to the left, 40.0 lb to the right.

   Net force = _______ lb

   The direction is _______

3. 17.0 lb to the left, 20.0 lb to the right.

   Net force = _______ lb

   The direction is _______

4. 190 lb to the left, 87.0 lb to the right, and 49.0 lb to the right.

   Net force = _______ lb

   The direction is _______

5. 346 N to the right, 247 N to the left, and 103 N to the left.

   Net force = _______ N

   The direction is _______

6. 37.0 N to the right and 24.0 N to the left.

   Net force = _______ N

   The direction is _______

7. Find the acceleration of an automobile of mass 100 slugs acted upon by a driving force of 500 lb which is opposed by a frictional force of 100 lb.

8. Find the acceleration of an automobile of mass 1500 kg acted upon by a driving force of 2200 N which is opposed by a frictional force of 450 N.
9. A truck of mass 13,100 kg is acted upon by a driving force of 8900 N. The motion is opposed by a frictional force of 2230 N. Find the acceleration.
10. A speed boat of mass 30.0 slugs has a 300 lb force applied by the propellers. The friction of the water on the hull is a force of 100 lb. Find the acceleration.

6-6 *MASS AND WEIGHT*

We have said that the weight of an object is the amount of gravitational pull exerted on an object by the earth. If this force is not balanced by other forces, an acceleration is produced. When you hold a brick in your hand, you exert an upward force on the brick which balances the downward force (weight). If you remove your hand, the brick moves downward due to the unbalanced force. The velocity of the falling brick changes but the acceleration (rate of change of the velocity) is constant.

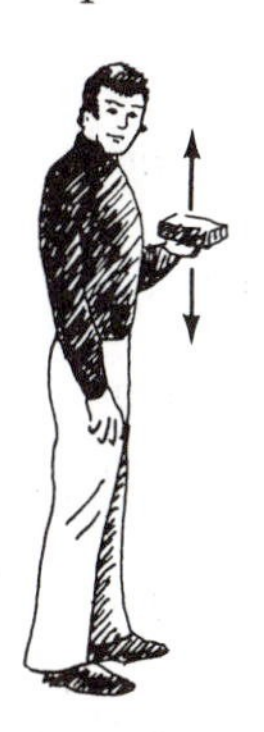

Upward force equals downward force

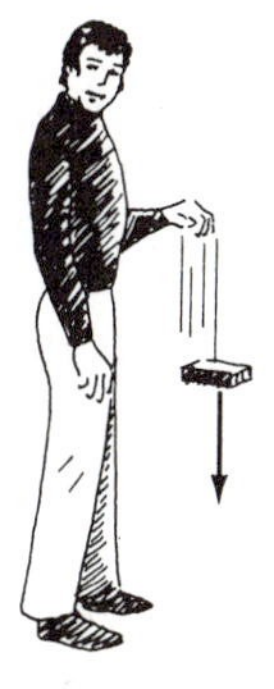

Downward force greater than upward force

The acceleration of all objects which are near the surface of the earth is the same if air resistance is ignored. We call this acceleration due to the gravitational pull of the earth—g. Its value is 32.2 ft/s² in the English system and 9.81 m/s² in the metric system.

The weight of an object is the force which gives the body the acceleration g. This force can be found using $F = ma$, where $a = 9.81\text{ m/s}^2 = 32.2\text{ ft/s}^2 = g$. If we abbreviate weight by F_w, the equation for weight is then:

$$F_w = mg$$

EXAMPLE 1: Find the weight of 1.00 slug.

DATA:

$$m = 1.00\text{ slug}$$
$$a = 32.2\,\frac{\text{ft}}{\text{s}^2}$$
$$F_w = ?$$

BASIC EQUATION: $F_w = mg$

WORKING EQUATION: Same

SUBSTITUTION:

$$F_w = mg$$
$$= 1.00 \text{ slug} \times 32.2 \frac{\text{ft}}{\text{s}^2}$$
$$= \frac{32.2 \text{ slug ft}}{\text{s}^2}$$
$$= \frac{32.2 \cancel{\text{slug ft}}}{\cancel{\text{s}^2}} \times \frac{1 \text{ lb} \cancel{\text{s}^2}}{\cancel{\text{slug ft}}}$$
$$= 32.2 \text{ lb}$$

(*Note*: We must use a conversion factor to obtain force units.)

EXAMPLE 2: Find the weight of 1.00 kg.

DATA:

$$m = 1.00 \text{ kg}$$
$$a = 9.81 \frac{\text{m}}{\text{s}^2}$$
$$F_w = ?$$

BASIC EQUATION: $F_w = mg$

WORKING EQUATION: Same

SUBSTITUTION:

$$F_w = mg$$
$$= 1.00 \text{ kg} \times 9.81 \frac{\text{m}}{\text{s}^2}$$
$$= 9.81 \frac{\text{kg m}}{\text{s}^2}$$
$$= 9.81 \frac{\cancel{\text{kg m}}}{\cancel{\text{s}^2}} \times \frac{1 \text{ N} \cancel{\text{s}^2}}{\cancel{\text{kg m}}}$$
$$= 9.81 \text{ N}$$

(*Note*: We must use a conversion factor to obtain force units.)

PROBLEMS

Find the unknown quantity in the following problems.

1. $m = 10.0$ slugs $F_w =$ _______ lb
2. $m = 9.00$ kg $F_w =$ _______ N
3. $F_w = 17.0$ N $m =$ _______ kg
4. $F_w = 17.0$ N $m =$ _______ slug
5. $F_w = 21.0$ lb $m =$ _______ slug
6. $F_w = 170$ lb $m =$ _______ slug
7. $F_w = 170$ lb $m =$ _______ kg

6-7 *THE LAW OF ACTION AND REACTION*

When an autombile accelerates, we know that a force is being applied to it. What applies this force? You may think that the tires exert this force on the auto. This is not correct since the tires move along with the auto and there must be a force applied to them also. The ground below the tires supplies the force that accelerates the car. This force is called a reaction to the force exerted by the tires on the ground which is called the action force.

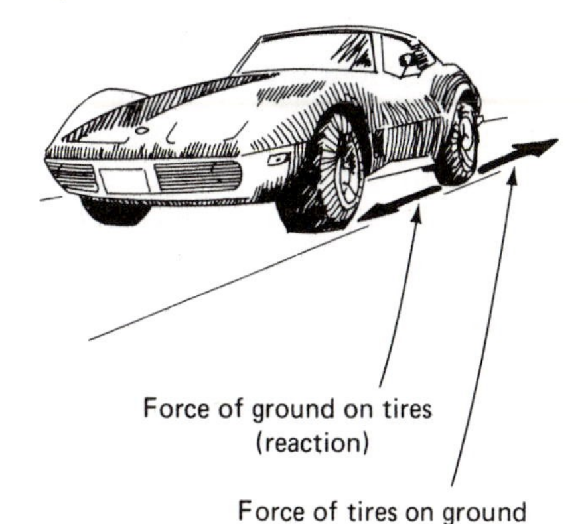

The third law of motion, which is called the law of action and reaction, can be stated as follows: *for every force applied by object* A *to object* B (*action*), *there is a force exerted by object* B *on object* A (*reaction*) *which has the same magnitude but is opposite in direction.*

When a bullet is fired from a handgun (action), the recoil felt is the reaction. These forces are shown in the diagram. Note that the action and reaction forces *never* act on the same object.

6-8 *MOMENTUM*

We all know that if two automobiles are moving with the same velocity and one is heavier than the other, the heavier auto would cause more damage in a head-on collision. The lighter auto can cause as much or more damage if its velocity is greater than that of the heavier auto. *Momentum is a measure of the effect an object would have in a collision brought to rest in a certain amount of time. Momentum is equal to the mass times the velocity of an object.*

$$p = mv$$

where: p = momentum
m = mass
v = velocity.

The units of momentum are $\frac{\text{slug ft}}{\text{s}}$ in the English system and $\frac{\text{kg m}}{\text{s}}$ in the metric system.

EXAMPLE 1: Find the momentum of an auto which has a mass of 110 slugs and a velocity of 60.0 mi/hr.

SKETCH: None

DATA: $m = 110$ slugs
$v = 60.0\ \frac{\text{mi}}{\text{hr}} = 88.0\ \frac{\text{ft}}{\text{s}}$
$p = ?$

BASIC EQUATION: $p = mv$

WORKING EQUATION: Same

SUBSTITUTION:
$$p = mv$$
$$= 110 \text{ slugs} \times 88.0\ \frac{\text{ft}}{\text{s}}$$
$$= 9680\ \frac{\text{slug ft}}{\text{s}}$$

EXAMPLE 2: Find the velocity of a bullet of mass 1.00×10^{-2} kg if it is to have the same momentum as a bullet of mass 1.80×10^{-3} kg and a velocity of 300 m/s.

SKETCH:

$m_1 = 1.00 \times 10^{-2}$ kg $m_2 = 1.80 \times 10^{-3}$ kg

$v_1 = ?$ $v_2 = 300$ m/sec

DATA:

heavier bullet:
$m_1 = 1.00 \times 10^{-2}$ kg
$v_1 = ?$
$p_1 = ?$

lighter bullet:
$m_2 = 1.80 \times 10^{-3}$ kg
$v_2 = 300$ m/s
$p_2 = ?$

BASIC EQUATION: $p_1 = m_1 v_1$
$p_2 = m_2 v_2$
we want:
$p_1 = p_2$
or
$m_1 v_1 = m_2 v_2$

WORKING EQUATION: $v_1 = \frac{m_2}{m_1} v_2$

SUBSTITUTION: $v_1 = \frac{1.80 \times 10^{-3} \text{ kg}}{1.00 \times 10^{-2} \text{ kg}} \times 300 \frac{\text{m}}{\text{s}}$
$= 54.0 \frac{\text{m}}{\text{s}}$

PROBLEMS

Find the momentum for the following objects.

1. $m = 2.00 \text{ kg}, v = 40.0 \frac{m}{s}, p =$ ______ $\frac{\text{kg m}}{\text{s}}$
2. $m = 5.00 \text{ slugs}, v = 90.0 \frac{\text{ft}}{\text{s}}, p =$ ______ $\frac{\text{slug ft}}{\text{s}}$
3. $m = 17.0 \text{ slugs}, v = 45.0 \frac{\text{ft}}{\text{s}}, p =$ ______ $\frac{\text{slug ft}}{\text{s}}$
4. $m = 38.0 \text{ kg}, v = 97.0 \frac{\text{m}}{\text{s}}, p =$ ______ $\frac{\text{kg m}}{\text{s}}$
5. $m = 11.0 \text{ slugs}, v = 82.0 \frac{\text{ft}}{\text{s}}, p =$ ______ $\frac{\text{slug ft}}{\text{s}}$
6. $F_w = 3200 \text{ lb}, v = 60 \frac{\text{mi}}{\text{hr}}$ (change to ft/s), $p =$ ______ $\frac{\text{slug ft}}{\text{s}}$
7. (a) Find the momentum of a heavy American automobile which has a mass of 180 slugs and a velocity of 70.0 ft/s.
   (b) Find the velocity of a light foreign auto of mass 80.0 slugs if it is to have the same momentum as the auto in part (a).
   (c) Find the weight in pounds of the autos in parts (a) and (b).
8. (a) Find the momentum of a bullet of mass 1.00×10^{-3} slugs and velocity 700 ft/s.
   (b) Find the velocity of a bullet of mass 5.00×10^{-4} slugs if it is to have the same momentum as the bullet in part (a).
9. (a) Find the momentum of a heavy American automobile which has a mass of 2630 kg and a velocity of 21.0 m/s.
   (b) Find the velocity of a light foreign auto of mass 1170 kg if it is to have the same momentum as the auto in part (a).

VECTORS AND TRIGONOMETRY 7

7-1 *TRIGONOMETRY*

To this point we have discussed vectors in terms of bearing and magnitude. To study vectors more thoroughly, trigonometry of the right triangle is needed to set up a relationship between bearing and magnitude and the component method.

A right triangle is a triangle with one right angle (90°), two acute angles (less than 90°), two legs, and a hypotenuse (the side opposite the right angle).

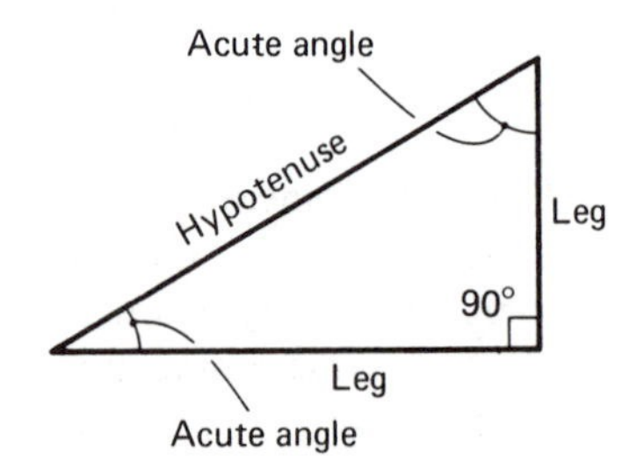

When it is necessary to label a triangle, the vertices are labeled using capital letters and the sides opposite the vertices are labeled using the corresponding small letter.

Note: The side opposite angle A is a.
The side opposite angle B is b.
The side opposite angle C is c.

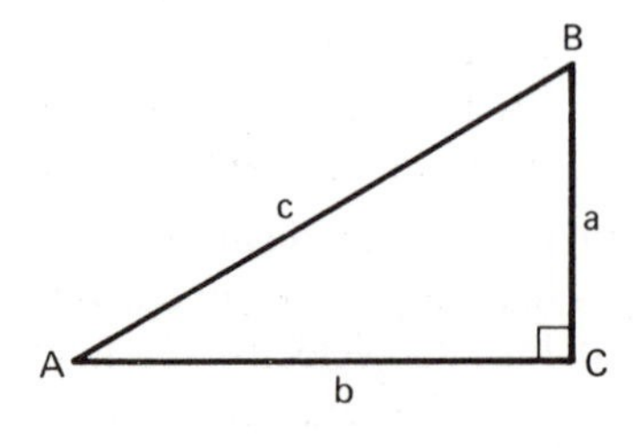

If we consider a certain acute angle of a right triangle, then the two legs can be identified as the side opposite or the side adjacent to an acute angle.

The side opposite angle A is a.
The side adjacent to angle A is b.
The side opposite angle B is b.
The side adjacent to angle B is a.
The hypotenuse is c.

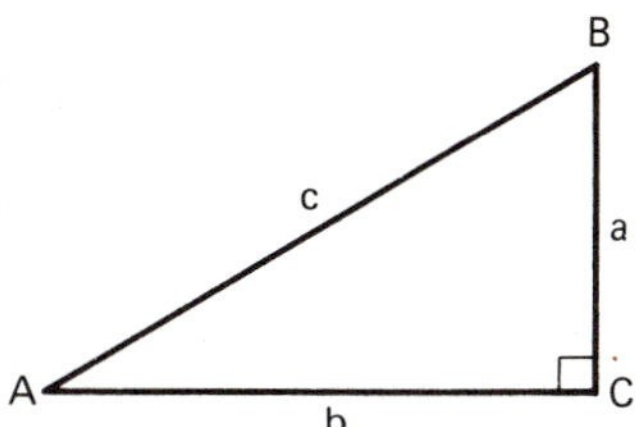

Note: The side opposite angle A is the same as the side adjacent to angle B.
The side adjacent to angle A is the same as the side opposite angle B.
The side opposite angle B is the same as the side adjacent to angle A.
The side adjacent to angle B is the same as the side opposite angle A.

A ratio is a comparison by division of two quantities in the same unit of measure. In a right triangle there are three ratios that are very important. Consider the right triangle below.

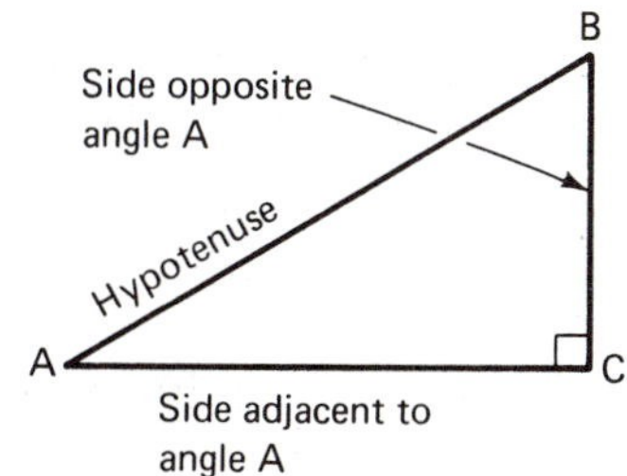

These ratios are:

$\dfrac{\text{side opposite angle } A}{\text{hypotenuse}}$ is called the sine A (abbreviated sin A)

$\dfrac{\text{side adjacent angle } A}{\text{hypotenuse}}$ is called the cosine A (abbreviated cos A)

$\dfrac{\text{side opposite angle } A}{\text{side adjacent to angle } A}$ is called the tangent A (abbreviated tan A)

$$\sin A = \frac{\text{side opposite angle } A}{\text{hypotenuse}}$$

$$\cos A = \frac{\text{side adjacent to angle } A}{\text{hypotenuse}}$$

$$\tan A = \frac{\text{side opposite angle } A}{\text{side adjacent to angle } A}$$

EXAMPLE 1:

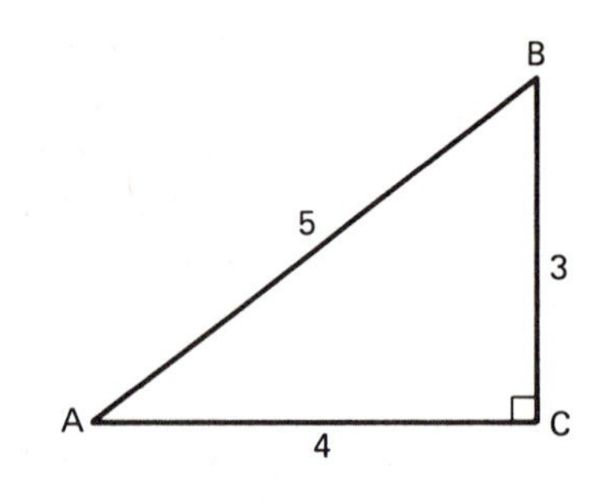

Find the three trigonometric ratios of angle A.

$$\sin A = \frac{\text{side opposite angle } A}{\text{hypotenuse}} = \frac{3}{5} = 0.600$$

$$\cos A = \frac{\text{side adjacent to angle } A}{\text{hypotenuse}} = \frac{4}{5} = 0.800$$

$$\tan A = \frac{\text{side opposite angle } A}{\text{side adjacent to angle } A} = \frac{3}{4} = 0.750$$

EXAMPLE 2:

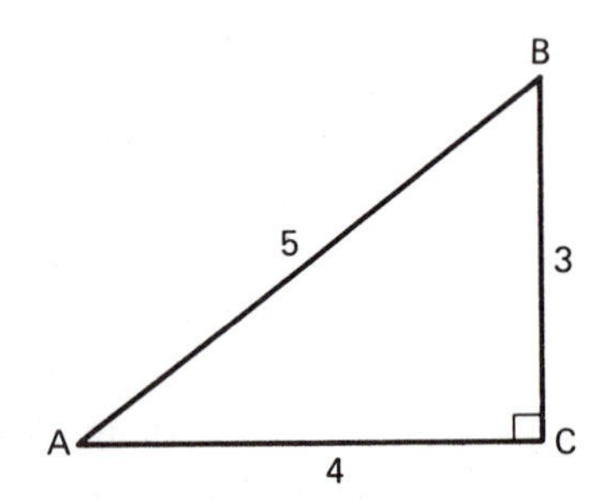

Find the three trigonometric ratios of angle B.

$$\sin B = \frac{\text{side opposite angle } B}{\text{hypotenuse}} = \frac{4}{5} = 0.800$$

$$\cos B = \frac{\text{side adjacent to angle } B}{\text{hypotenuse}} = \frac{3}{5} = 0.600$$

$$\tan B = \frac{\text{side opposite angle } B}{\text{side adjacent to angle } B} = \frac{4}{3} = 1.33$$

Note: Every acute angle has three trigonometric ratios associated with it. In Table T, these angles and their trigonometric ratios are given.

EXAMPLE 3:

Find angle B and side a.

Angle B can be found directly by using the fact that the sum of the angles of a triangle is 180°.

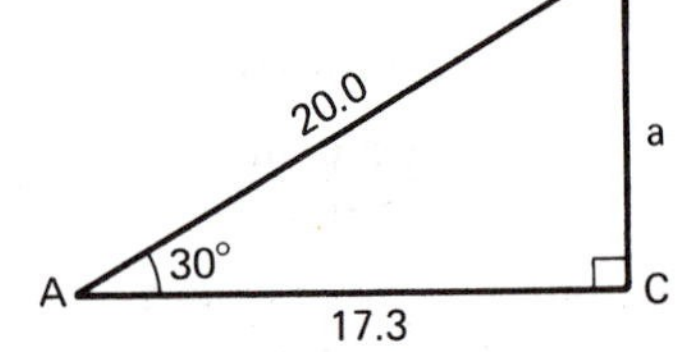

$$30° + 90° + B = 180°$$

$$120° + B = 180°$$

$$120° + B + (-120°) = 180° + (-120°)$$

$$B = 60°$$

To find side a we must use a trigonometric ratio. Note that we are looking for the side opposite angle A and that the hypotenuse is given. The trigonometric ratio having these two quantities is sine.

$$\sin A = \frac{\text{side opposite angle } A}{\text{hypotenuse}} \qquad \sin 30° = \frac{a}{20.0}$$

From Table T, $\sin 30° = 0.500$; therefore,

$$0.500 = \frac{a}{20.0}$$

$$0.500(20.0) = (20.0)\frac{a}{(20.0)}$$

$$10.0 = a$$

EXAMPLE 4:

Find angle A, angle B, and side a.

First, find angle A. The side adjacent to angle A and the hypotenuse are given. Therefore, we use cos A to find angle A because the cos A uses these two quantities:

$$\cos A = \frac{\text{side adjacent to } A}{\text{hypotenuse}}$$

$$\cos A = \frac{13.0}{19.0} = 0.684$$

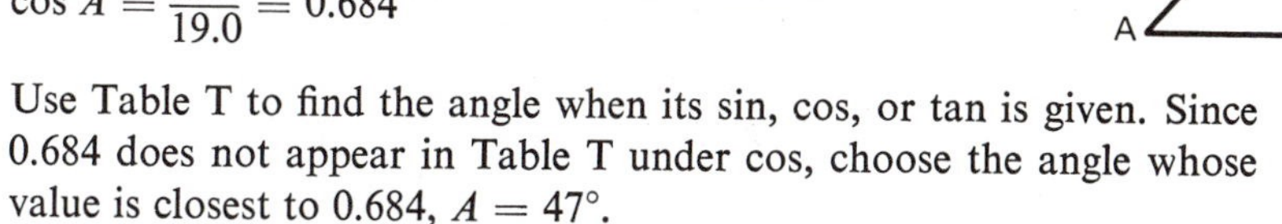

Use Table T to find the angle when its sin, cos, or tan is given. Since 0.684 does not appear in Table T under cos, choose the angle whose value is closest to 0.684, $A = 47°$.

To find angle B, we use the fact that the sum of the angles of a triangle equals 180°.

$$90° + 47° + B = 180°$$

$$137° + B = 180°$$

$$137° + B + (-137°) = 180° + (-137°)$$

$$B = 43°$$

To find side a we use sin A because the hypotenuse is given and side a is opposite angle A.

$$\sin A = \frac{\text{side opposite angle } A}{\text{hypotenuse}}$$

$$\sin 47° = \frac{a}{19.0}$$

From Table T, sin 47° = 0.731
Therefore,

$$0.731 = \frac{a}{19.0}$$

$$(0.731)(19.0) = \frac{a}{(19.0)}(19.0)$$

$$13.9 = a$$

EXAMPLE 5:

Find angle A, angle B, and the hypotenuse in the right triangle below.

To find angle A, use tan A:

$$\tan A = \frac{\text{side opposite angle } A}{\text{side adjacent to angle } A}$$

$$\tan A = \frac{9.00}{19.0} = 0.474$$

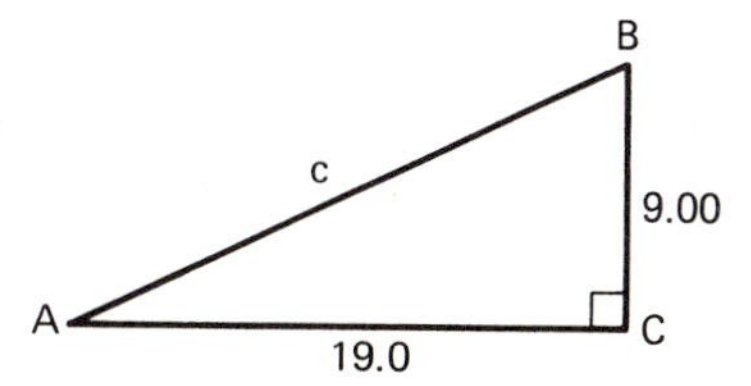

From Table T, $A = 25°$.

To find angle B, use the fact that the sum of the angles of a triangle is 180°.

$$90° + 25° + B = 180°$$
$$115° + B = 180°$$
$$115° + B + (-115°) = 180° + (-115°)$$
$$B = 65°$$

To find the hypotenuse, use sin A.

$$\sin A = \frac{\text{side opposite angle } A}{\text{hypotenuse}}$$
$$\sin 25° = \frac{9.00}{c}$$
$$(\sin 25°)(c) = \frac{9.00}{c}(c)$$
$$c(\sin 25°) = 9.00$$
$$\frac{c(\sin 25°)}{\sin 25°} = \frac{9.00}{\sin 25°}$$
$$c = \frac{9.00}{\sin 25°}$$
$$c = \frac{9.00}{0.423}$$
$$c = 21.3$$

PROBLEMS

Find the following trigonometric ratios using Table T.

1. sin 70°
2. cos 40°
3. tan 21°
4. tan 82°
5. cos 11°
6. sin 79°
7. cos 49°
8. tan 53°
9. tan 17°
10. cos 34°

Find the following angles using Table T.

11. $\sin A = 0.454$
12. $\cos A = 0.574$
13. $\tan A = 0.070$
14. $\tan A = 0.819$
15. $\cos A = 0.697$
16. $\sin A = 0.695$
17. $\cos A = 0.398$
18. $\tan A = 0.292$
19. $\tan A = 1.17$
20. $\cos A = 0.865$

Solve the following triangles (find the missing angles and sides) using trigonometric ratios.

21.

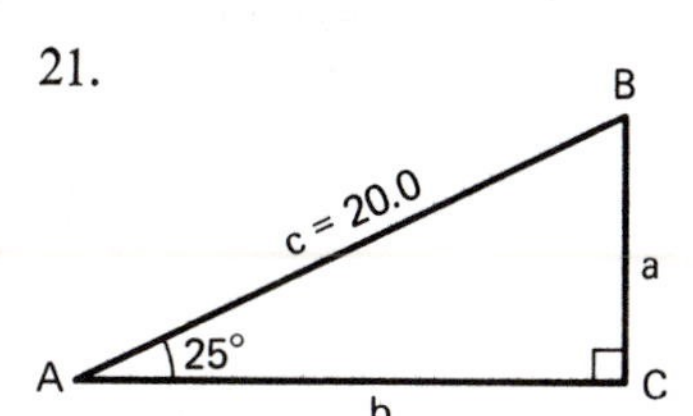

22.

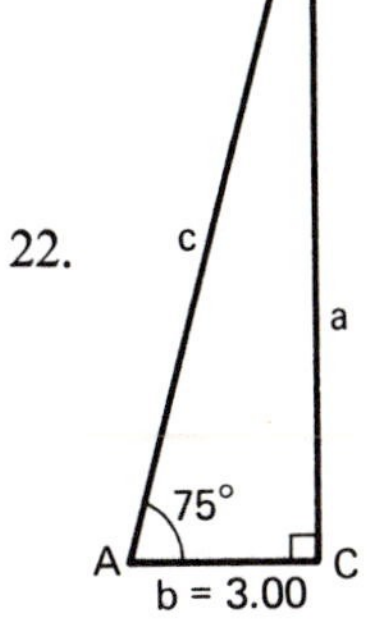

23.

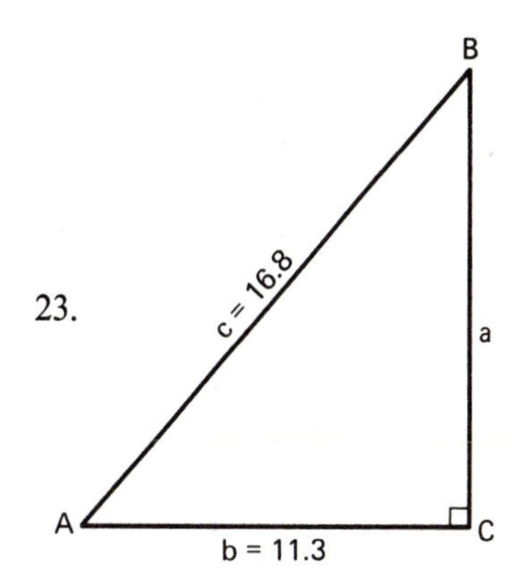

24.

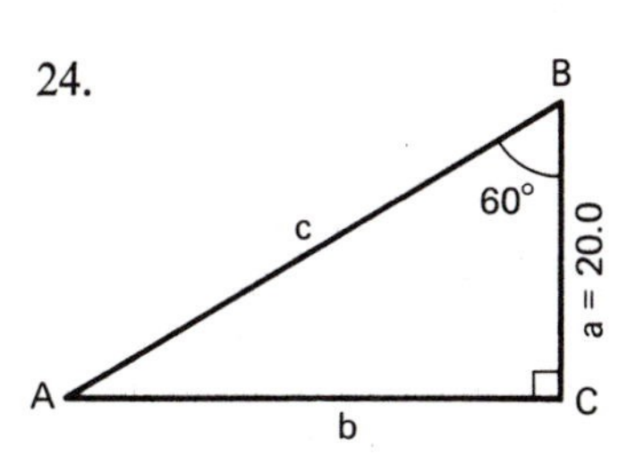

25.

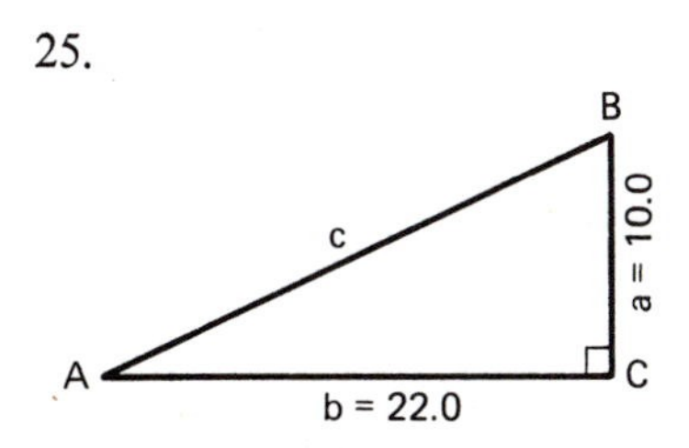

26.

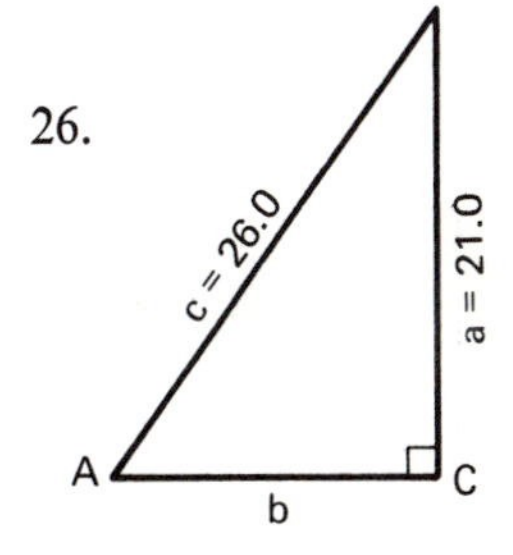

27.

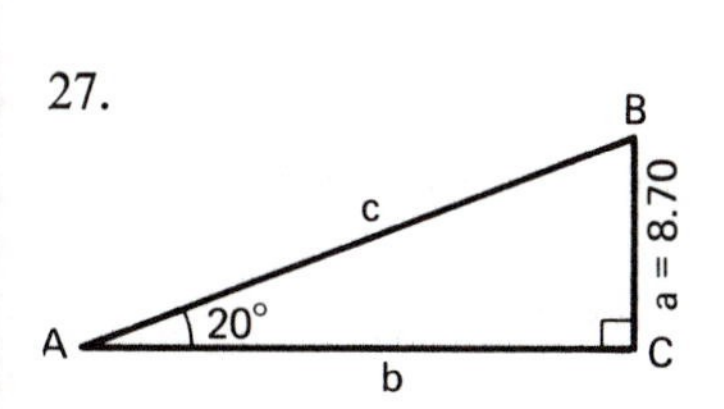

28.

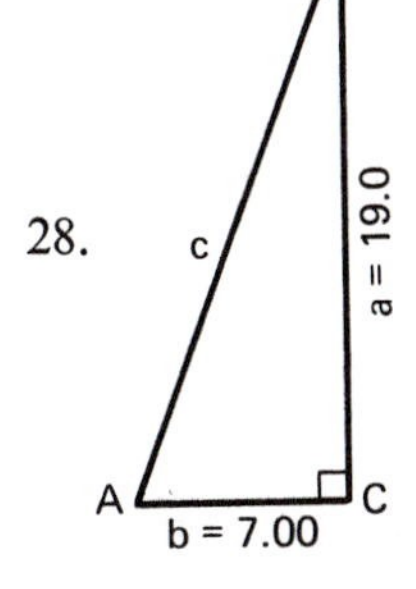

29.

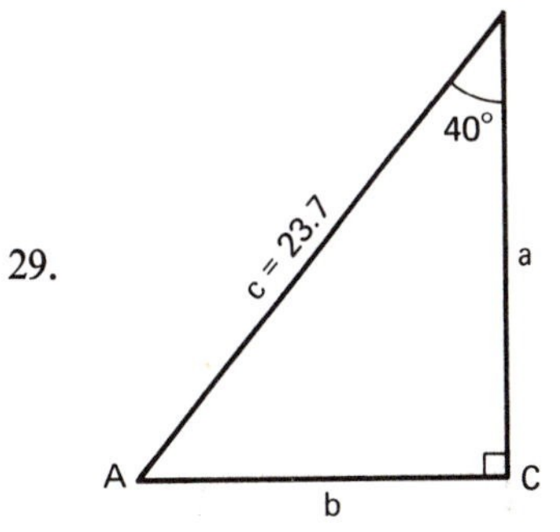

30.

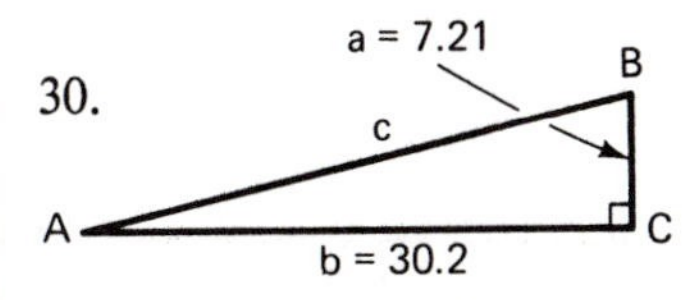

31. Answer the following questions about the round taper shown in the figure:
    (a) What is the value of $\angle BAC$?
    (b) What length is BC?
    (c) What is the diameter of end x?

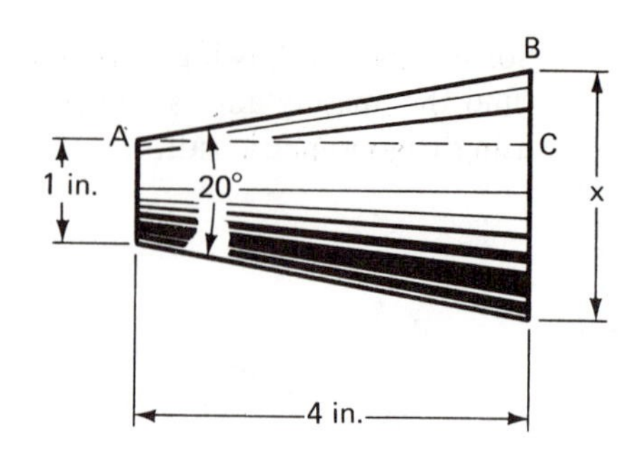

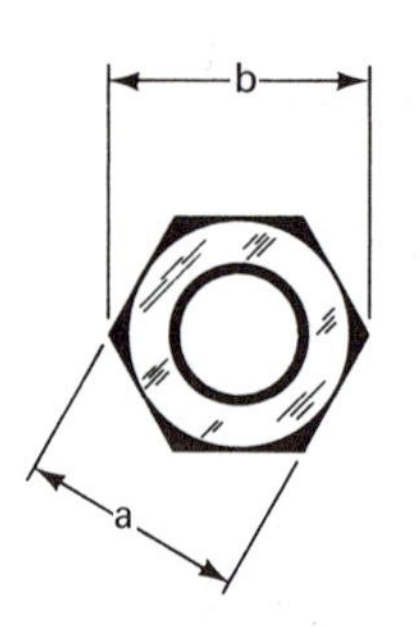

32. Across the flats (a) a hexagonal nut is $\frac{3}{4}$ inch. Calculate the distance across the corners (b).

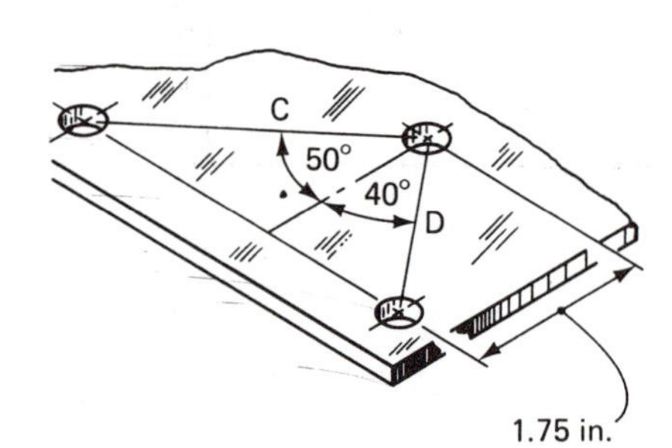

33. What are the distances C and D between the holes of the plate shown in the figure?

7-2 PYTHAGOREAN THEOREM

When given the two legs of a right triangle, the hypotenuse can be found without using trigonometric ratios.

From geometry, the sum of the squares of the legs of a right triangle is equal to the square of the hypotenuse (Pythagorean theorem):

$$a^2 + b^2 = c^2$$

or, by taking the square root of each side of the equation:

$$c = \sqrt{a^2 + b^2}$$

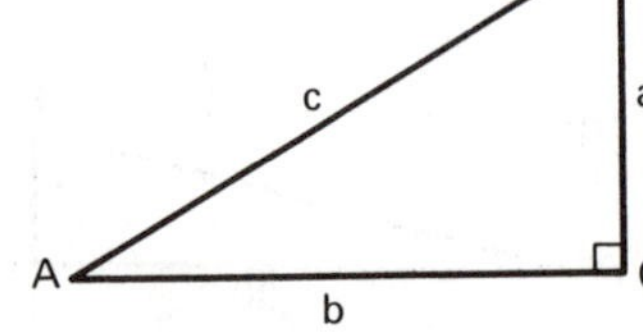

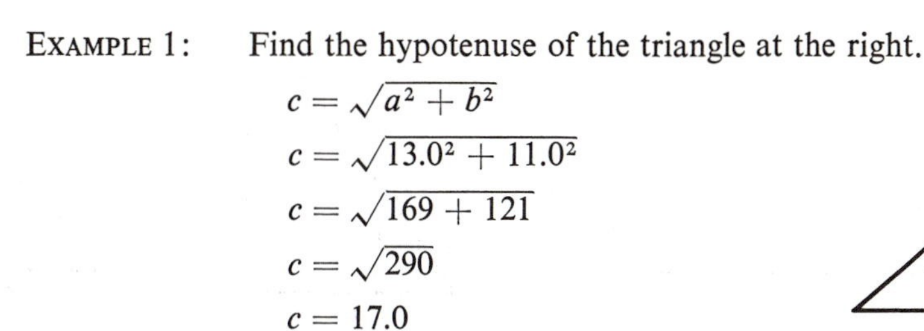

EXAMPLE 1: Find the hypotenuse of the triangle at the right.

$$c = \sqrt{a^2 + b^2}$$
$$c = \sqrt{13.0^2 + 11.0^2}$$
$$c = \sqrt{169 + 121}$$
$$c = \sqrt{290}$$
$$c = 17.0$$

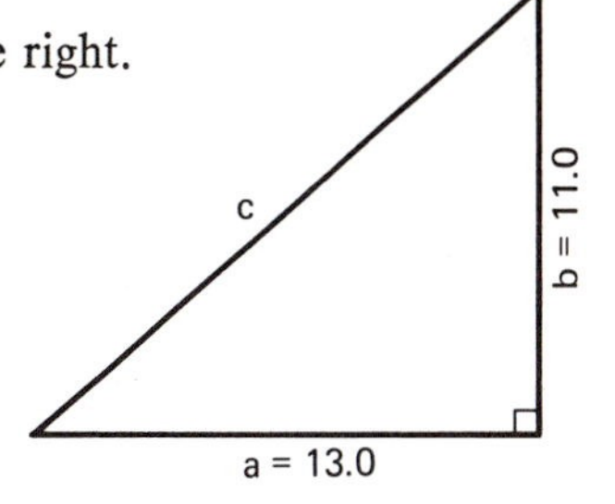

Also, if one leg and the hypotenuse are given, the other leg can be found by:

$$a = \sqrt{c^2 - b^2}$$

or,

$$b = \sqrt{c^2 - a^2}$$

EXAMPLE 2: Find side b in the triangle at the right.

$$b = \sqrt{c^2 - a^2}$$
$$b = \sqrt{12.2^2 - 7.30^2}$$
$$b = \sqrt{149 - 53.3}$$
$$b = \sqrt{95.7}$$
$$b = 9.78$$

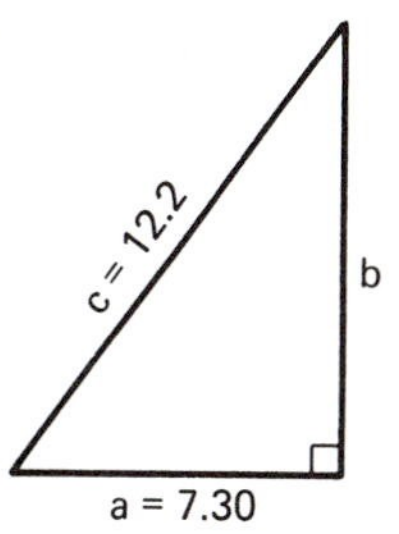

PROBLEMS

Find the missing side in each of the following right triangles using the Pythagorean theorem.

1.

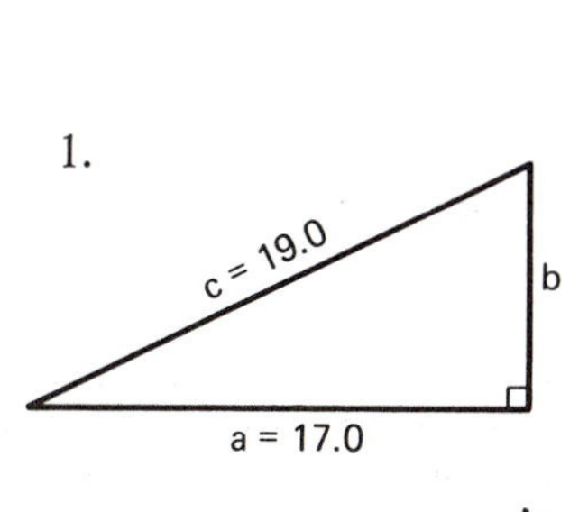

2.

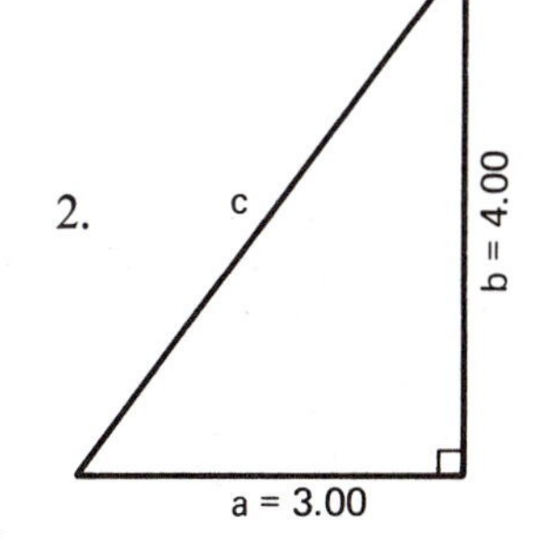

3.

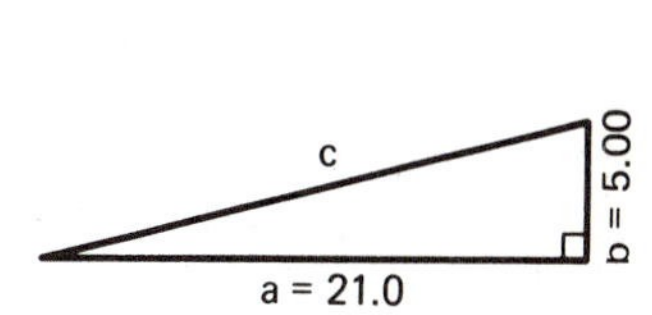

4.

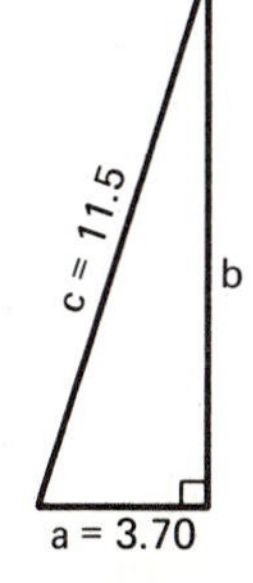

5.

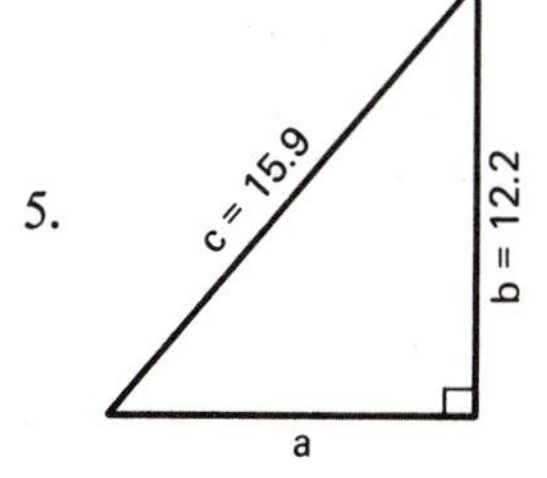

6.

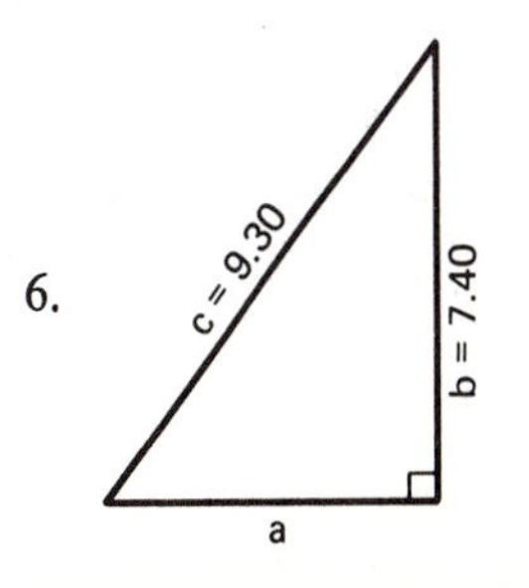

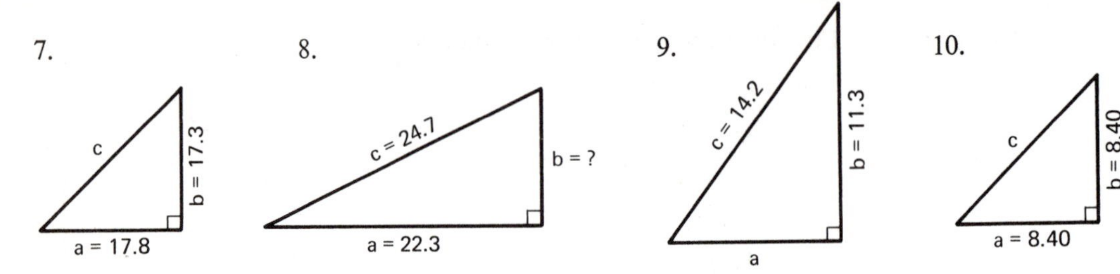

11. What are the distances between holes on the plate shown in the figure?

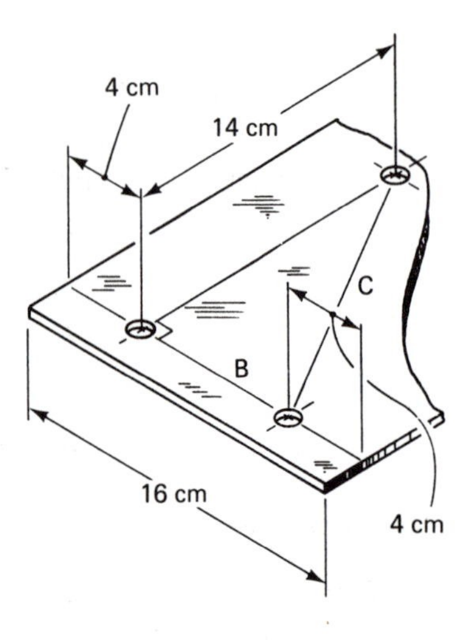

7-3 *THE NUMBER PLANE*

Before further study of vectors, we need to discuss components of vectors. This requires an understanding of the number plane. The number plane is determined by a horizontal line called the x-axis and a vertical line called the y-axis intersecting at right angles as shown below. These two lines divide the number plane into four quadrants which we will label as quadrants A, B, C, and D as illustrated below.

Each axis has a scale, and the intersection of the two axes is called the origin. The x-axis contains positive numbers to the right of the origin and negative numbers to the left of the origin. The y-axis contains positive numbers above the origin and negative numbers below the origin.

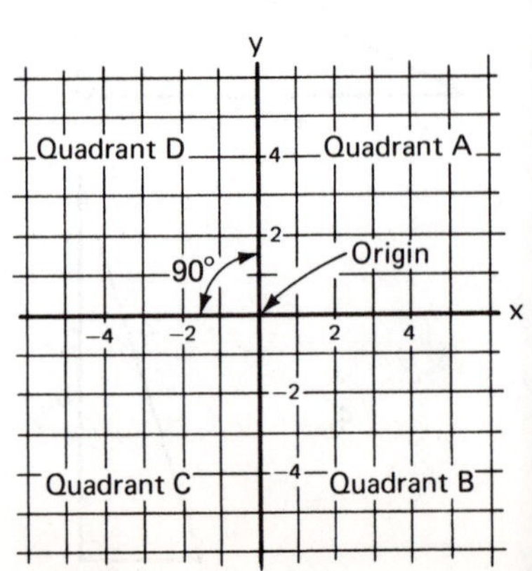

When a vector is expressed graphically as a sum of vectors, the vectors are called *components* of the resultant vector.

The components of vector R are vectors A, B, and C.

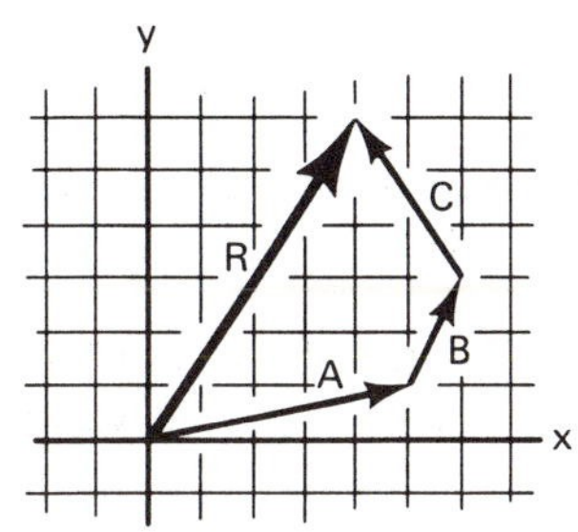

Note: A vector may have more than one set of component vectors.

The components of vector R are vectors E and F.

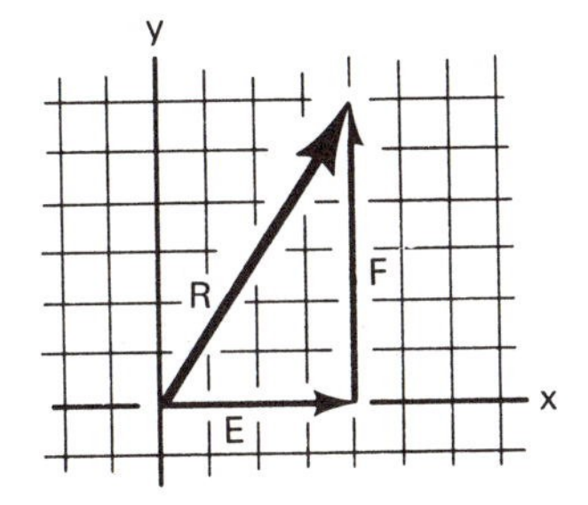

We are interested in the components of a vector which are perpendicular to each other and which are on or parallel to the x and y axes. In particular, we are interested in the type of component vectors we found in the preceding figure (component vectors E and F). The component vector that lies on or is parallel to the x-axis is called the x component. The component vector that lies on or is parallel to the y-axis is called the y component.

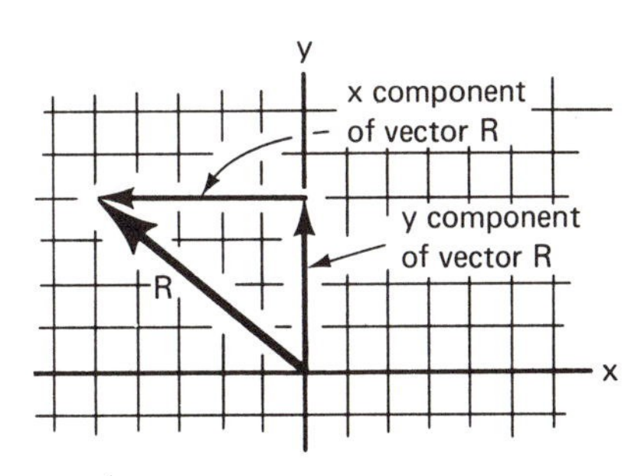

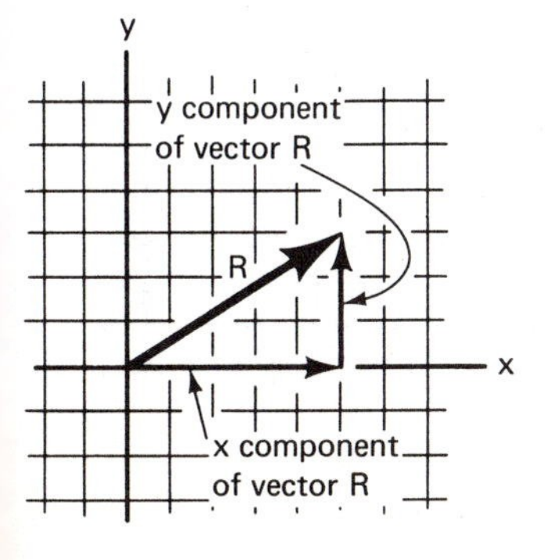

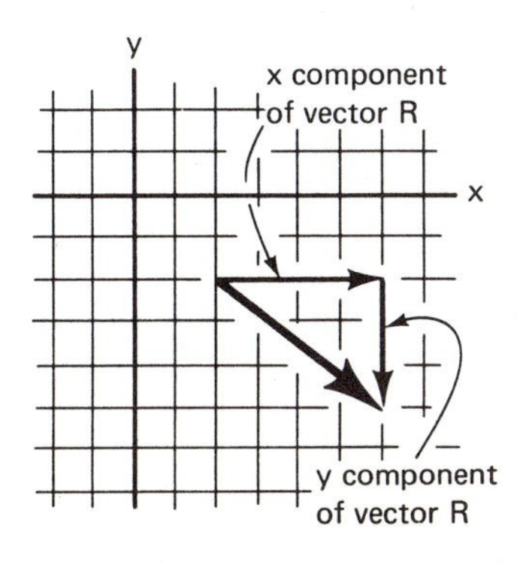

Thus far we have considered the x and y components as vectors. However, they can also be thought of as signed numbers. The sign of the number corresponds to the direction of the components as follows:

x-component	y-component
$+$, if right	$+$, if up
$-$, if left	$-$, if down

The absolute value of the signed number corresponds to the magnitude of the vector.

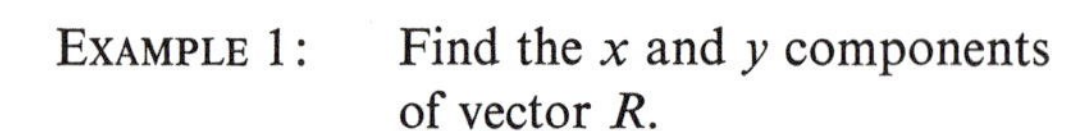

EXAMPLE 1: Find the x and y components of vector R.

x component of $R = +4$
y component of $R = +3$

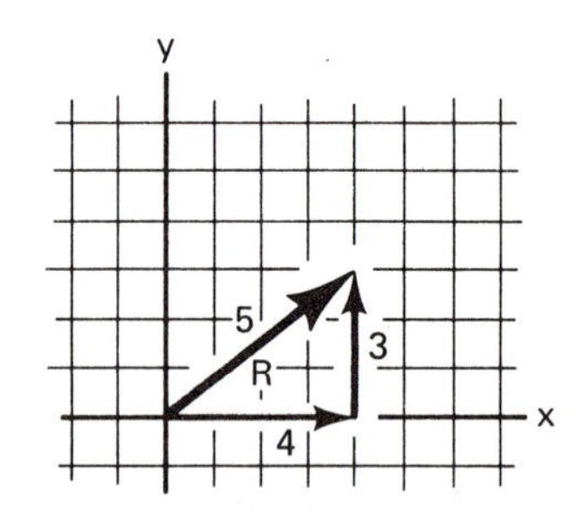

EXAMPLE 2: Find the x and y components of vector R.

x component of $R = +6$
y component of $R = -8$

(y component points in a negative direction)

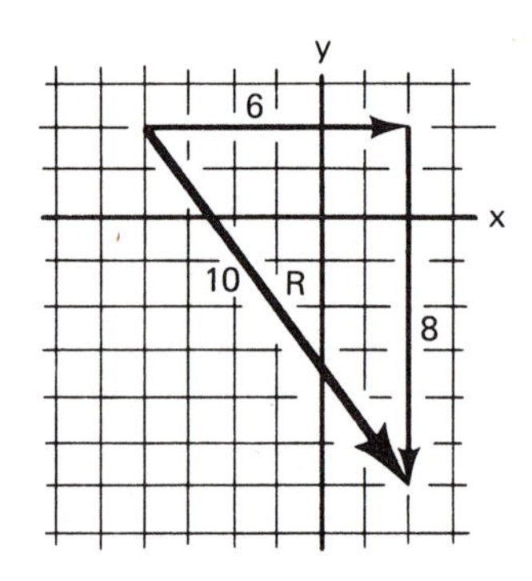

EXAMPLE 3: Find the x and y components of vector R.

x component of $R = -12$
y component of $R = -9$

(both x and y components point in a negative direction)

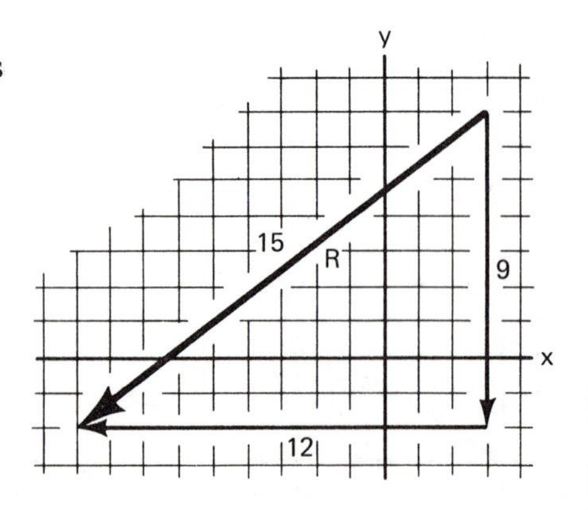

Now that we have considered the x and y components as signed numbers, we can find the resultant vector of several vectors using arithmetic and graphing. To find the resultant vector of several vectors, find the x component of each vector and find the sum of the x components. Then find the y component of each vector and find the sum of the y components. The two sums are the x and y components of the resultant vector. This is shown in the following example.

EXAMPLE 4: Find the x and y components of vector R using the x and y components of vectors A and B.

$A + B = R$

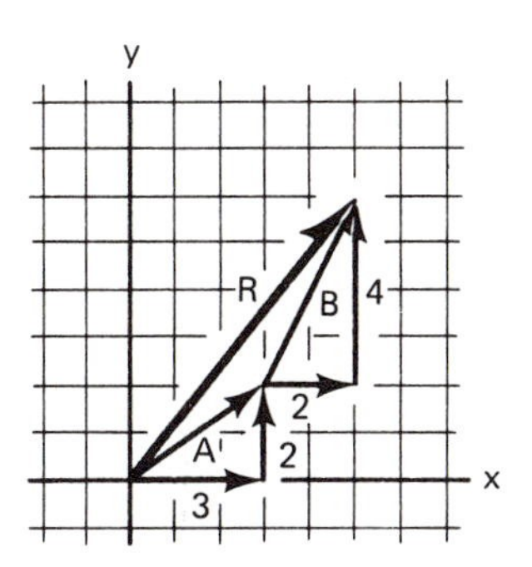

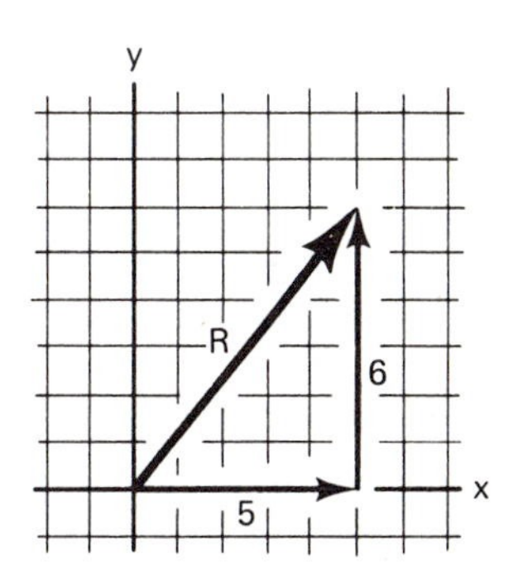

Note: The x component of A is $+3$
The x component of B is $+2$
The x component of R is $(+3) + (+2)$ or $+5$
The y component of A is $+2$
The y component of B is $+4$
The y component of R is $(+2) + (+4)$ or $+6$

EXAMPLE 5: Find the x and y components of vector R using the x and y components of vectors A, B, and C.

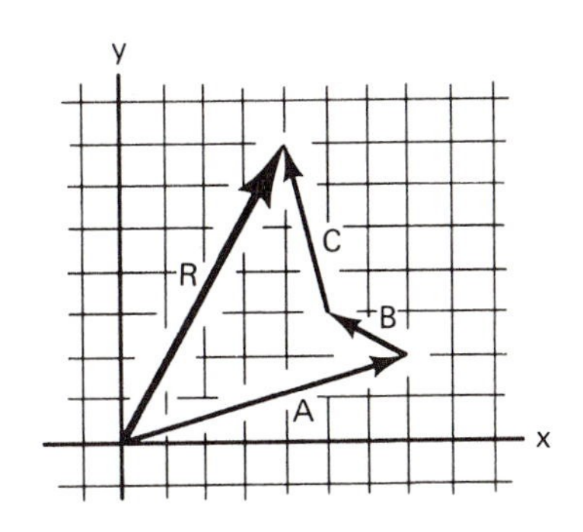

Vector	*x component*	*y component*
A	$+7$	$+2$
B	-2	$+1$
C	-1	$+4$
R	$+4$	$+7$

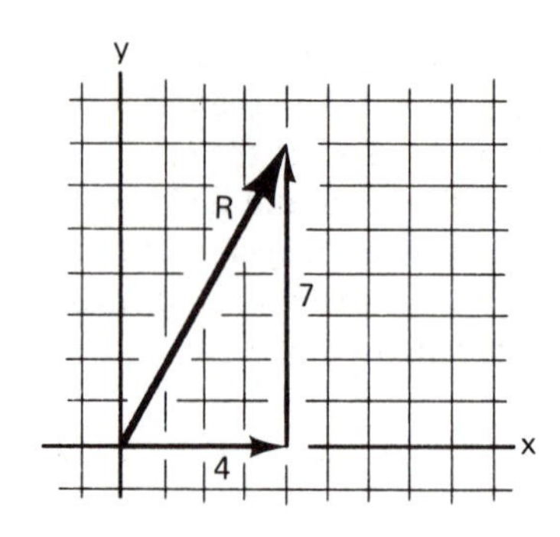

EXAMPLE 6: Find the x and y components of vector R using the x and y components of vectors A, B, C, and D.

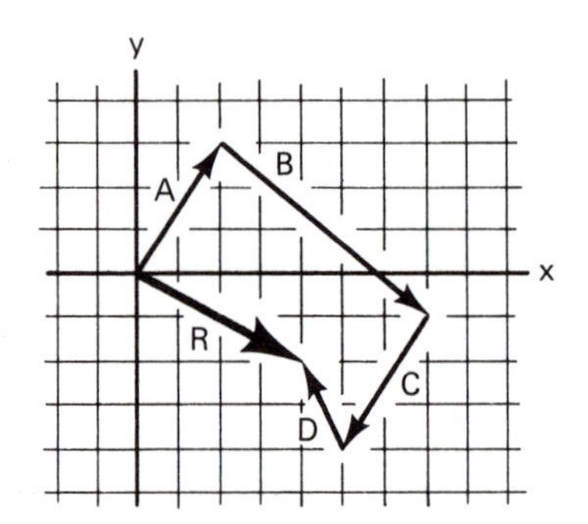

Vector	*x component*	*y component*
A	$+2$	$+3$
B	$+5$	-4
C	-2	-3
D	-1	$+2$
R	$+4$	-2

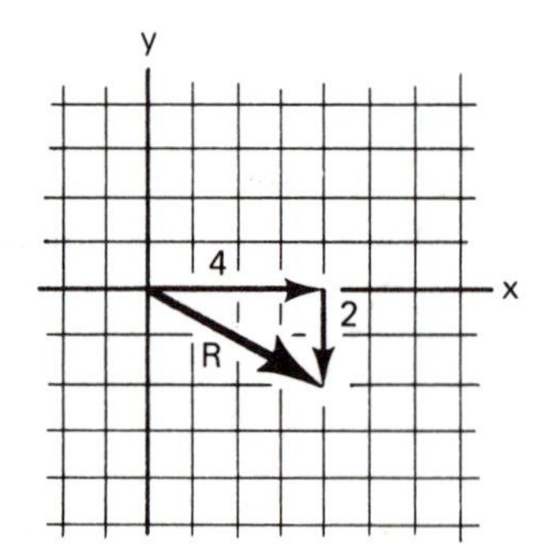

PROBLEMS

Find the x and y components of the following vectors. (Express them as signed numbers and graph them as vectors.)

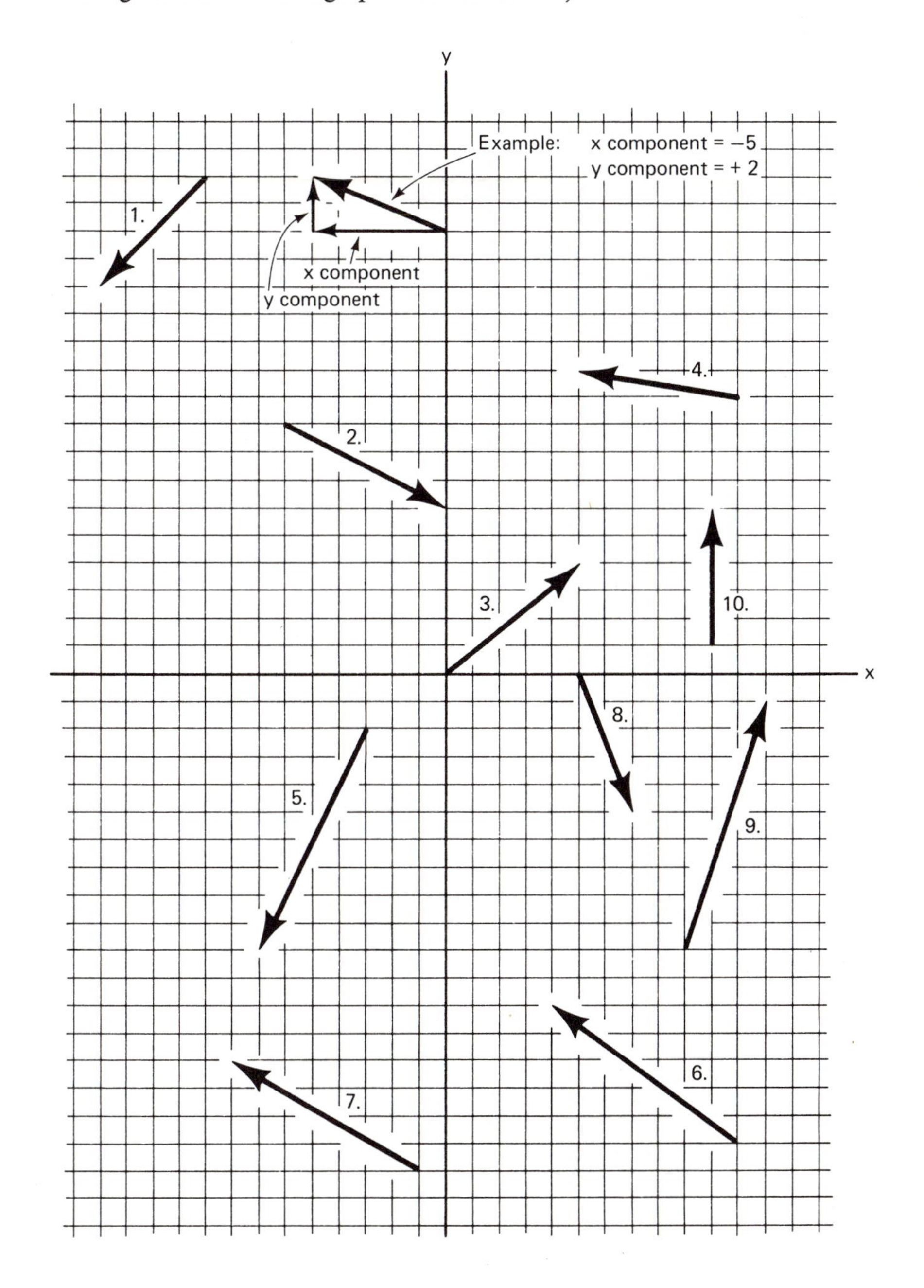

Find the x and y components of the resultant vector R and graph the resultant vector R.

	Vector	*x component*	*y component*
11.	A	+2	+3
	B	+7	+2
	R		
12.	A	+9	−5
	B	−4	−6
	R		
13.	A	−2	+13
	B	−11	+1
	C	+3	−4
	R		
14.	A	+10	−5
	B	−13	−9
	C	+4	+3
	R		
15.	A	+17	+7
	B	−14	+11
	C	+7	+9
	D	−6	−15
	R		
16.	A	+1	+7
	B	+9	−4
	C	−4	+13
	D	−11	−4
	R		
17.	A	+1.5	−1.5
	B	−3	−2
	C	+7.5	−3
	D	+2	+2.5
	R		
18.	A	+1	−1
	B	−4	−2
	C	+2	+4
	D	+5	−3
	E	+3	+5
	R		

	Vector	x component	y component
19.	A	+1.5	+2.5
	B	−2	−3
	C	+3.5	−7.5
	D	−4	+6
	E	−5.5	+2
	R		
20.	A	−7	+15
	B	+13.5	−17.5
	C	−7.5	−20
	D	+6	+13.5
	E	+2.5	+2.5
	F	−11	+11.5
	R		

7-5 *MORE ON VECTORS*

A vector whose initial end is at the origin and points into quadrant B is called a quadrant B vector.

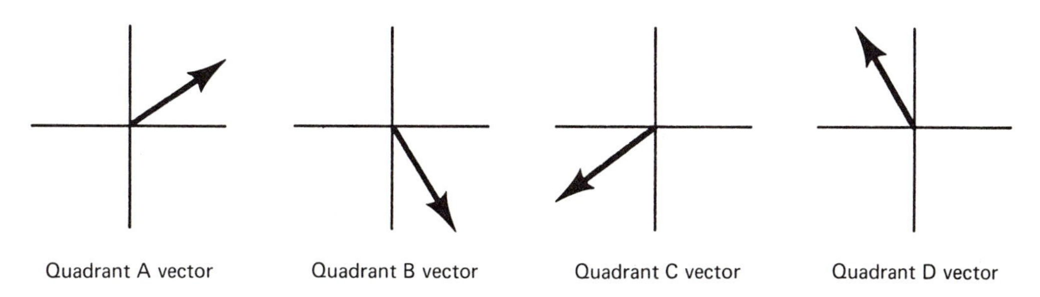

Vectors can be expressed using bearing and magnitude or using x and y components. Now we need to find a way of relating the two.

EXAMPLE 1: Find the x and y components of the vector on the left below.

To find the x and y components, draw a right triangle where the legs represent the x and y components of the vector.

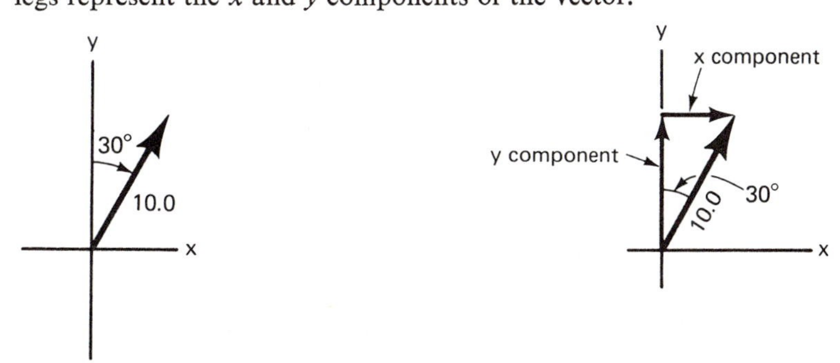

The absolute value of the x component of the vector is the length of the side opposite the 30° angle. Therefore, to find the x component we use:

$$\sin 30^\circ = \frac{\text{side opposite } 30^\circ}{\text{hypotenuse}}$$

$$\sin 30^\circ = \frac{\text{side opposite } 30^\circ}{10.0}$$

From Table T, $\sin 30^\circ = 0.500$

$$0.500 = \frac{\text{side opposite } 30^\circ}{10.0}$$

$$(10.0)(0.500) = (10.0)\frac{\text{side opposite } 30^\circ}{10.0}$$

$$5.00 = \text{side opposite } 30^\circ$$

Since the x component is pointing in the positive x direction, the x component is $+5.00$.

The absolute value of the y component of the vector is the length of the side adjacent to the 30° angle. Therefore, to find the y component we use:

$$\cos 30^\circ = \frac{\text{side adjacent to } 30^\circ}{\text{hypotenuse}}$$

$$\cos 30^\circ = \frac{\text{side adjacent to } 30^\circ}{10.0}$$

From Table T, $\cos 30^\circ = 0.866$

$$0.866 = \frac{\text{side adjacent to } 30^\circ}{10.0}$$

$$(10.0)(0.866) = (10.0)\frac{\text{side adjacent to } 30^\circ}{10.0}$$

$$8.66 = \text{side adjacent to } 30^\circ$$

Since the y component is pointing in the positive y direction, the y component is $+8.66$.

EXAMPLE 2: Find the x and y components of the vector below.

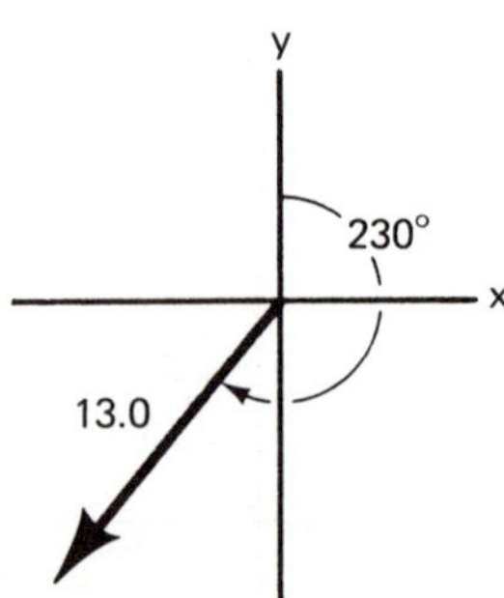

Complete a right triangle with the x and y components being the two legs.

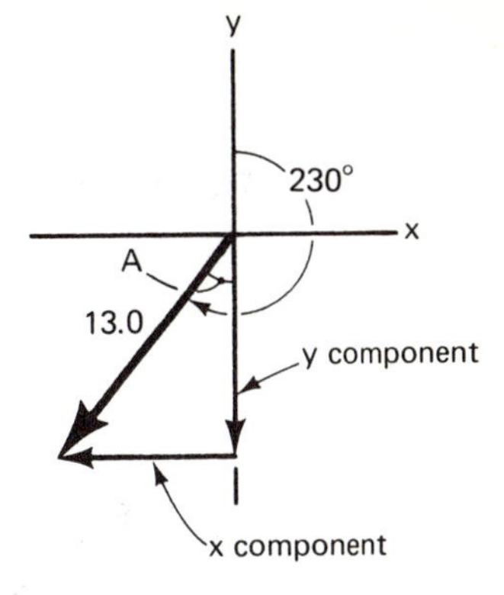

Find angle A as follows:

$$180^\circ + A = 230^\circ \text{ (quadrant } C \text{ vector)}$$

$$180^\circ + A + (-180^\circ) = 230^\circ + (-180^\circ)$$

$$A = 50^\circ$$

To find the y component, find the length of the side adjacent to 50° angle.

$$\cos 50^\circ = \frac{\text{side adjacent to } 50^\circ}{\text{hypotenuse}}$$

$$\cos 50^\circ = \frac{\text{side adjacent to } 50^\circ}{13.0}$$

From Table T, $\cos 50^\circ = 0.643$

$$0.643 = \frac{\text{side adjacent to } 50^\circ}{13.0}$$

$$(13.0)(0.643) = (13.0)\frac{\text{side adjacent to } 50^\circ}{13.0}$$

$$8.36 = \text{side adjacent to } 50^\circ$$

Since the y component is pointing in the negative y direction, the y component is -8.36.

To find the x component, find the length of the side opposite 50° angle.

$$\sin 50^\circ = \frac{\text{side opposite } 50^\circ}{\text{hypotenuse}}$$

$$\sin 50^\circ = \frac{\text{side opposite } 50^\circ}{13.0}$$

From Table T, $\sin 50^\circ = 0.766$

$$0.766 = \frac{\text{side opposite } 50^\circ}{13.0}$$

$$(13.0)(0.766) = (13.0)\frac{\text{side opposite } 50^\circ}{13.0}$$

$$9.96 = \text{side opposite } 50^\circ$$

Since the x component is pointing in the negative x direction, the x component is -9.96.

EXAMPLE 3: Find the x and y components of the vector below.

To find the x and y components, draw a right triangle, with the x and y components being the two legs.

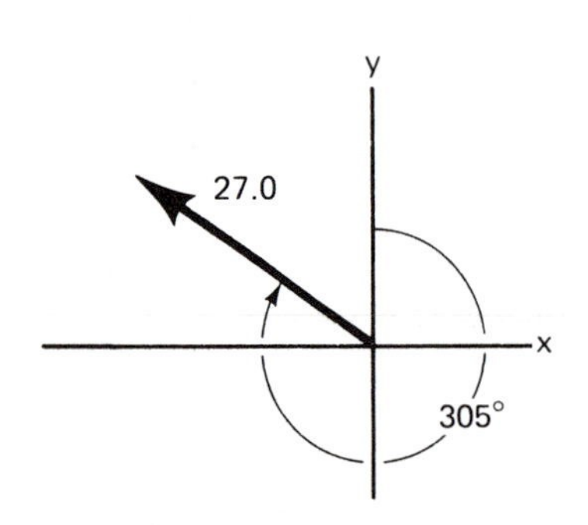

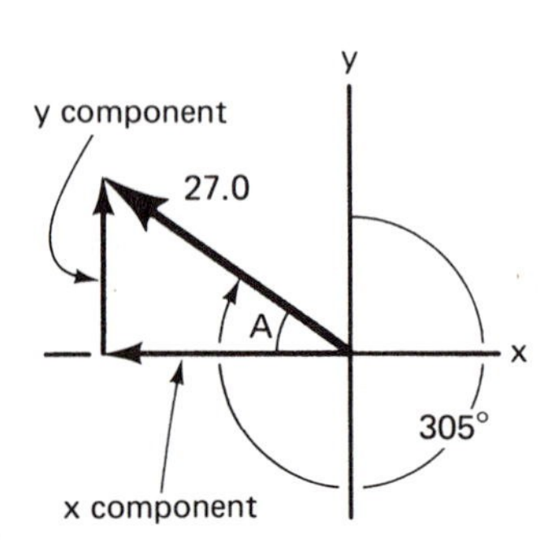

Find angle A as follows:

$$270° + A = 305° \qquad \text{(quadrant } D \text{ vector)}$$

$$270° + A + (-270°) = 305 + (-270°)$$

$$A = 35°$$

To find the x component, find the length of the side adjacent to the 35° angle.

$$\cos 35° = \frac{\text{side adjacent to } 35°}{\text{hypotenuse}}$$

$$\cos 35° = \frac{\text{side adjacent to } 35°}{27.0}$$

From Table T, $\cos 35° = 0.819$

$$0.819 = \frac{\text{side adjacent to } 35°}{27.0}$$

$$(27.0)(0.819) = (27.0)\frac{\text{side adjacent to } 35°}{27.0}$$

$$22.1 = \text{side adjacent to } 35°$$

Since the x component is pointing in the negative x direction, the x component is -22.1.

To find the y component, find the length of the side opposite 35° angle.

$$\sin 35° = \frac{\text{side opposite } 35°}{\text{hypotenuse}}$$

$$\sin 35° = \frac{\text{side opposite } 35°}{27.0}$$

From Table T, $\sin 35° = 0.574$

$$0.574 = \frac{\text{side opposite } 35°}{27.0}$$

$$(27.0)(0.574) = (27.0)\frac{\text{side opposite } 35°}{27.0}$$

$$15.5 = \text{side opposite } 35°$$

Since the y component is pointing in the positive y direction, the y component is $+15.5$

> To find the x and y components of a vector when the bearing and the magnitude are given:
> 1. Complete the right triangle with the legs being the x and y components of the vector.
> 2. Find the lengths of the legs of the right triangle.
> 3. Determine the sign of the x and y components.

EXAMPLE 4: Find the bearing and magnitude of a vector whose x component is $+3$ and y component is $+4$.

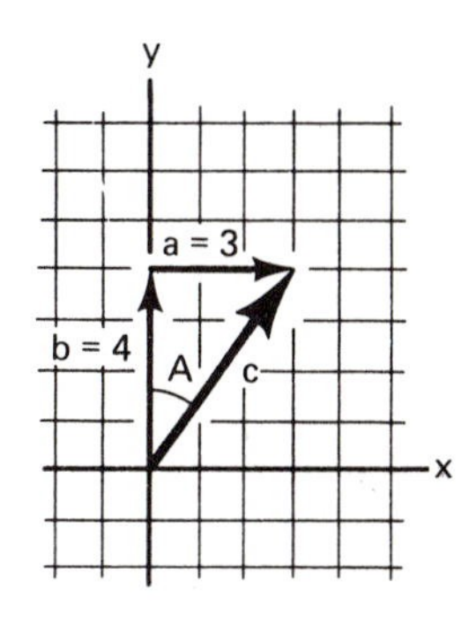

Angle A, the bearing of the vector, can be found by using:

$$\tan A = \frac{\text{side opposite angle } A}{\text{side adjacent to angle } A}$$

$$\tan A = \frac{4}{3} = 0.750$$

Use Table T to find the angle when its tan is given. Since 0.750 does not appear in the Table T under tan, choose the angle whose value is closest to 0.750, $A = 37°$.

The magnitude can be found by using the Pythagorean theorem.

$$c = \sqrt{a^2 + b^2}$$

$$c = \sqrt{3^2 + 4^2}$$

$$c = \sqrt{25}$$

$$c = 5$$

EXAMPLE 5: Find the bearing and magnitude of a vector whose x component is $+7$ and y component is -5.

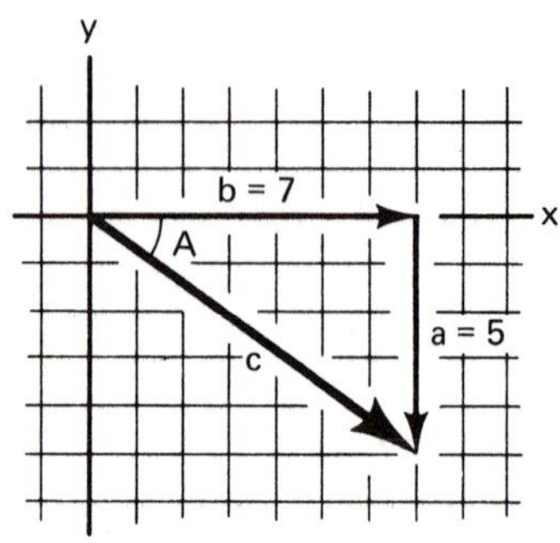

The bearing of the vector is $90° + A$. To find A, use:

$$\tan A = \frac{\text{side opposite angle } A}{\text{side adjacent to angle } A}$$

$$\tan A = \frac{5}{7} = 0.714$$

From Table T, $A = 36°$. Therefore, the bearing is:

$90° + A = 90° + 36° = 126°$ (quadrant B vector)

The magnitude can be found by using the Pythagorean theorem.

$$c = \sqrt{a^2 + b^2}$$

$$c = \sqrt{5^2 + 7^2}$$

$$c = \sqrt{25 + 49}$$

$$c = \sqrt{74}$$

$$c = 8.60$$

EXAMPLE 6: Find the bearing and magnitude of the vector whose x component is -11 and y component is -9.

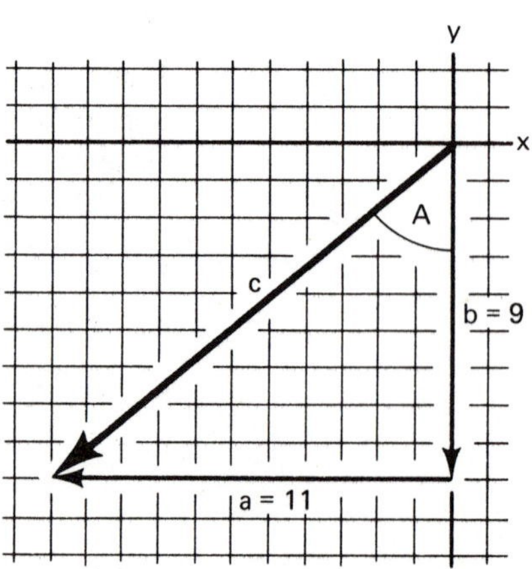

The bearing of the vector is $180° + A$. To find A, use:

$$\tan A = \frac{\text{side opposite angle } A}{\text{side adjacent to angle } A}$$

$$\tan A = \frac{11}{9} = 1.22$$

From Table T, $A = 51°$. Therefore, the bearing is:

$180° + \text{angle } A = 180° + 51° = 231°$ (quadrant C vector)

The magnitude can be found by using the Pythagorean theorem.

$$c = \sqrt{a^2 + b^2}$$
$$c = \sqrt{11^2 + 9^2}$$
$$c = \sqrt{121 + 81}$$
$$c = \sqrt{202}$$
$$c = 14.2$$

To find the bearing and magnitude of a vector when the x and y components are given:

1. Complete the right triangle with the legs being the x and y components of the vector.
2. Find the acute angle A of the right triangle whose vertex is at the origin by using $\tan A$.
3. Find the bearing of the vector as follows:

   Bearing $= 0° + A$ (quadrant A vector)
   Bearing $= 90° + A$ (quadrant B vector)
   Bearing $= 180° + A$ (quadrant C vector)
   Bearing $= 270° + A$ (quadrant D vector)

4. Find the magnitude of the vector by using the Pythagorean theorem

$$c = \sqrt{a^2 + b^2}$$

PROBLEMS

Find the x and y components of the vectors below.

1.

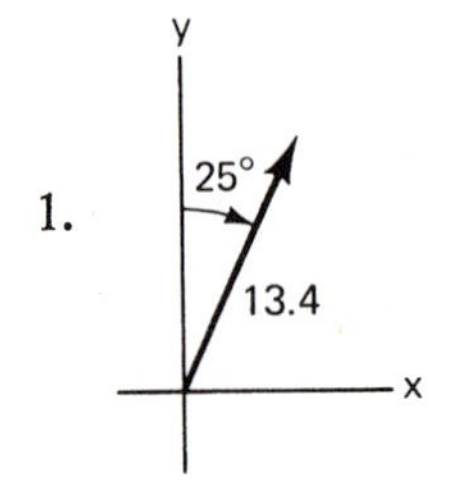

2.

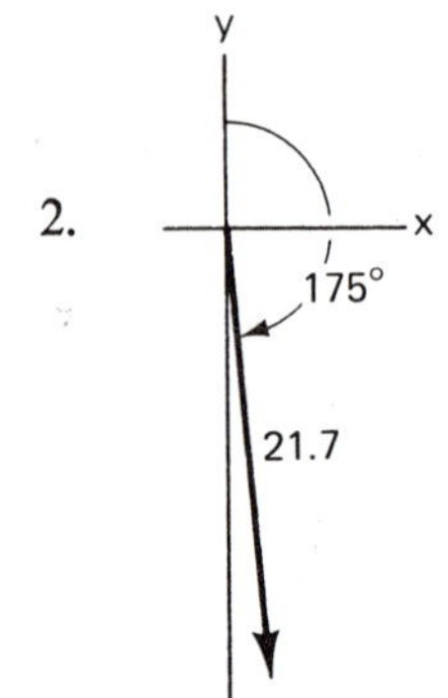

3.

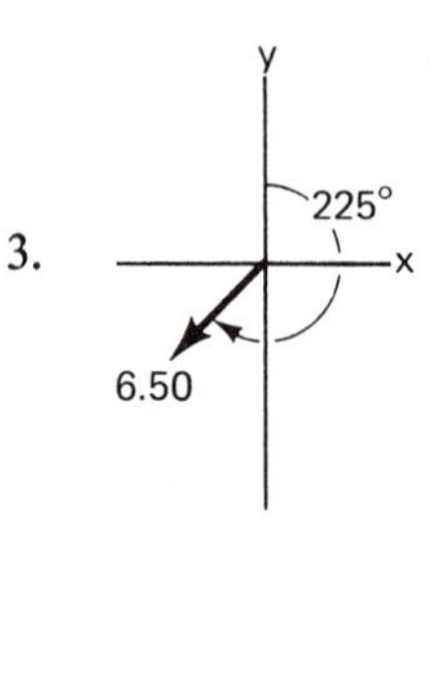

4.

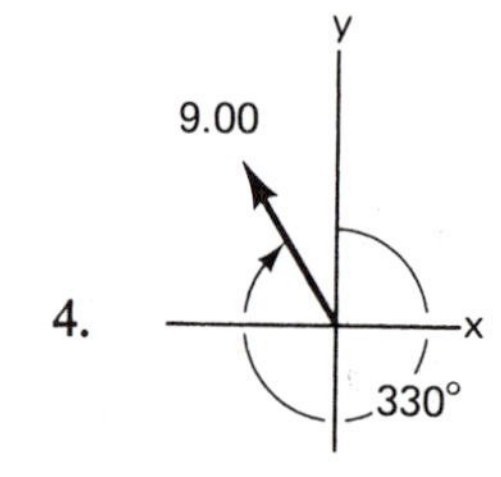

5.

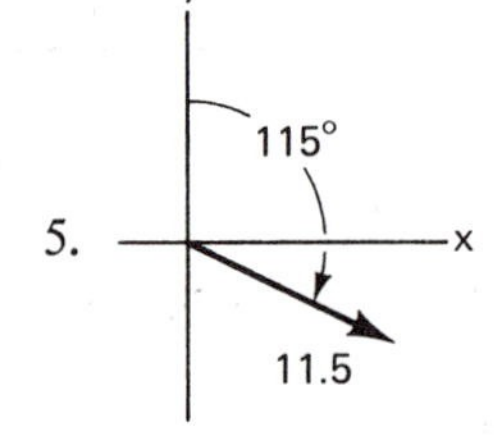

	Bearing	*Magnitude*
6.	68°	34.3
7.	195°	27.2
8.	237°	12.0
9.	290°	5.7
10.	155°	19.4

Find the bearing and magnitude of the vectors below.

11. x component $= +5.00$
    y component $= +8.00$

12. x component $= +13.3$
    y component $= -11.6$

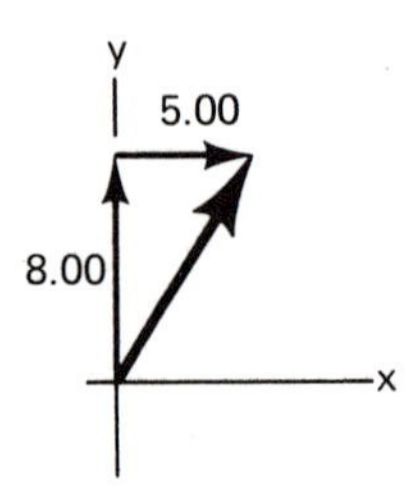

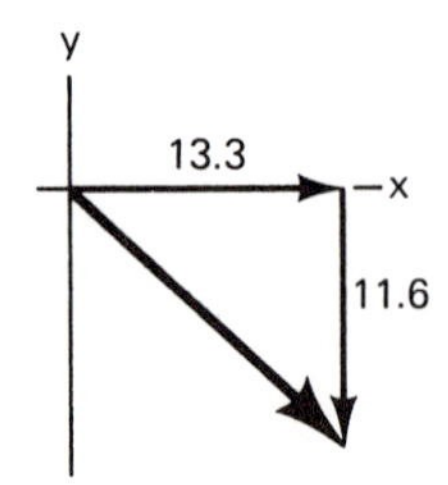

13. x component = −19.2
y component = −7.60

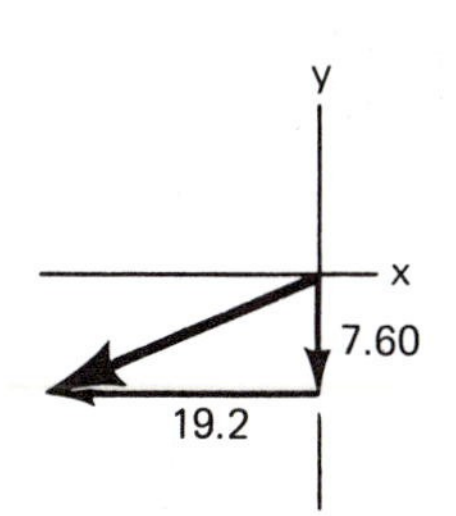

14. x component = −17.6
y component = +33.2

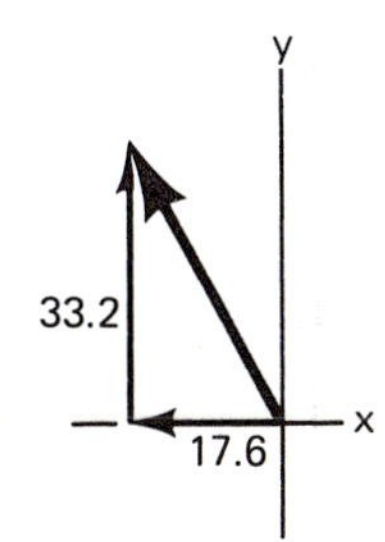

15. x component = +27.3
y component = −29.6

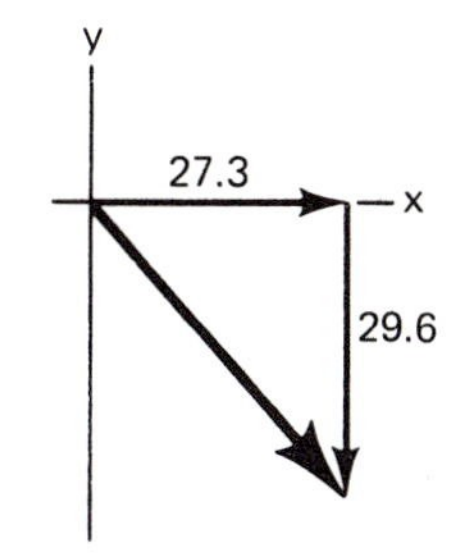

	x component	*y component*
16.	+3.40	+4.90
17.	−37.6	−12.9
18.	−14.4	+6.30
19	+43.7	−22.6
20.	−2.40	+9.90

CONCURRENT FORCES 8

8-1 *INTRODUCTION*

In Chapter 6 we studied forces that actually caused motion. Forces were said to tend to cause motion. We discussed a bridge on which the forces balanced each other and did not actually cause motion. The weight was balanced by the supporting force. Because the forces balanced each other, the net force was zero.

There are many important technical applications in which the net force on an object is zero. *An object is said to be in equilibrium when the net force acting on it is zero.* That is, a body is in equilibrium when it remains at rest or is not accelerating (moving at constant velocity). The study of equilibrium problems is called statics.

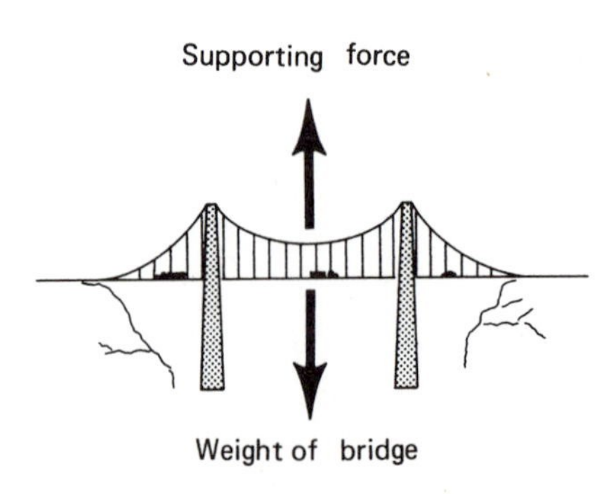

Let the forces applied to an object act in the same direction or in opposite directions (in one dimension). For the net force to be zero, the forces in one direction must equal the forces in the opposite direction.

We can write the equation for equilibrium as:

$$F_+ = F_-$$

where F_+ is the sum of all forces acting in one direction (call it the positive direction) and F_- is the sum of all the forces acting in the opposite (negative) direction.

EXAMPLE 1: A rope supports a large crate of weight 1000 N. What is the upward force on the crate if it is in equilibrium?

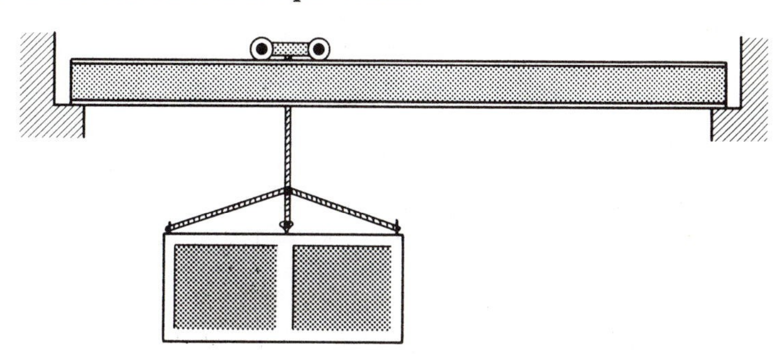

SKETCH: It is helpful to draw a "free-body diagram" of the crate. This is a sketch in which we draw only the object in equilibrium and show the forces which act on it. Note that we call the upward direction positive as indicated by the arrow.

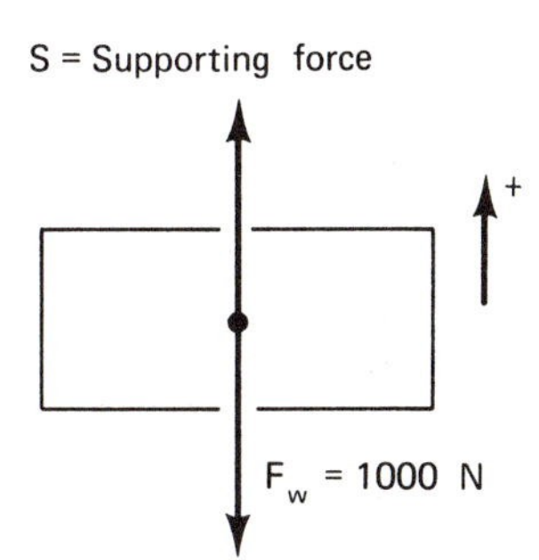

DATA: $F_w = 1000$ N
$S = ?$

BASIC EQUATION: $F_+ = F_-$

WORKING EQUATION: $S = F_w$

SUBSTITUTION: $S = 1000$ N

EXAMPLE 2: A sign of weight 1500 lb is supported by two cables. If one cable has a tension of 600 lb, what is the tension in the other cable?

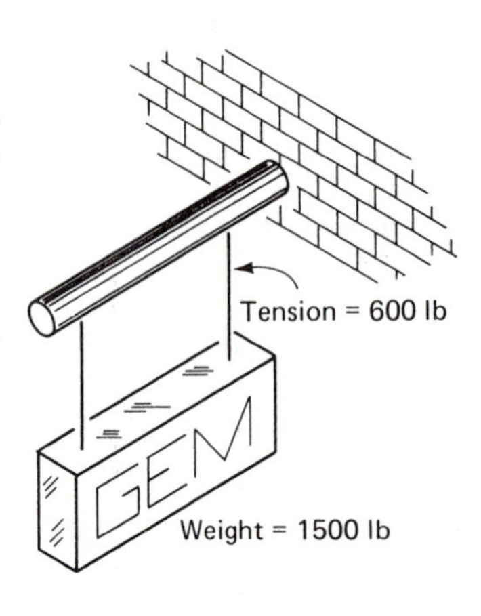

SKETCH: Draw the free-body diagram.

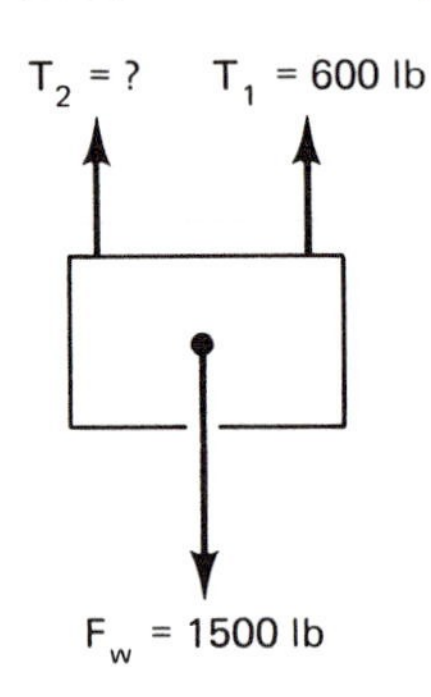

DATA:
$F_w = 1500$ lb
$T_1 = 600$ lb
$T_2 = ?$

BASIC EQUATION: $F_+ = F_-$

WORKING EQUATION:
$T_1 + T_2 = F_w$
$T_2 = F_w - T_1$

SUBSTITUTION:
$T_2 = 1500 \text{ lb} - 600 \text{ lb}$
$T_2 = 900$ lb

PROBLEMS

Find the force F which will produce equilibrium for the free-body diagrams shown below and on the following page. Use the same procedure as in the examples above.

SKETCH
DATA
$a = 1, b = 2,$
$c = ?$
BASIC EQUATION
$a = bc$
WORKING EQUATION
$c = \frac{a}{b}$
SUBSTITUTION
$c = \frac{1}{2}$

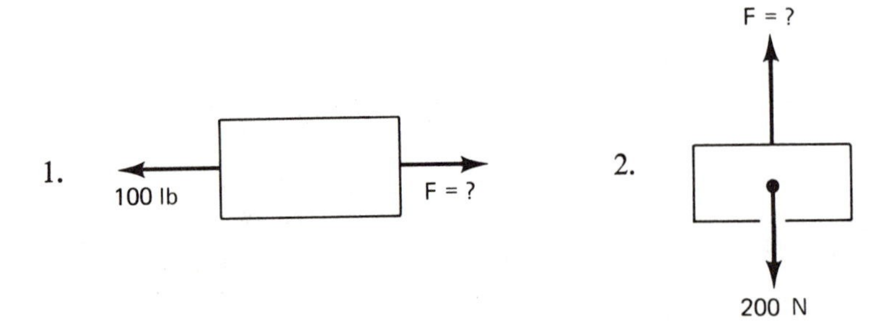

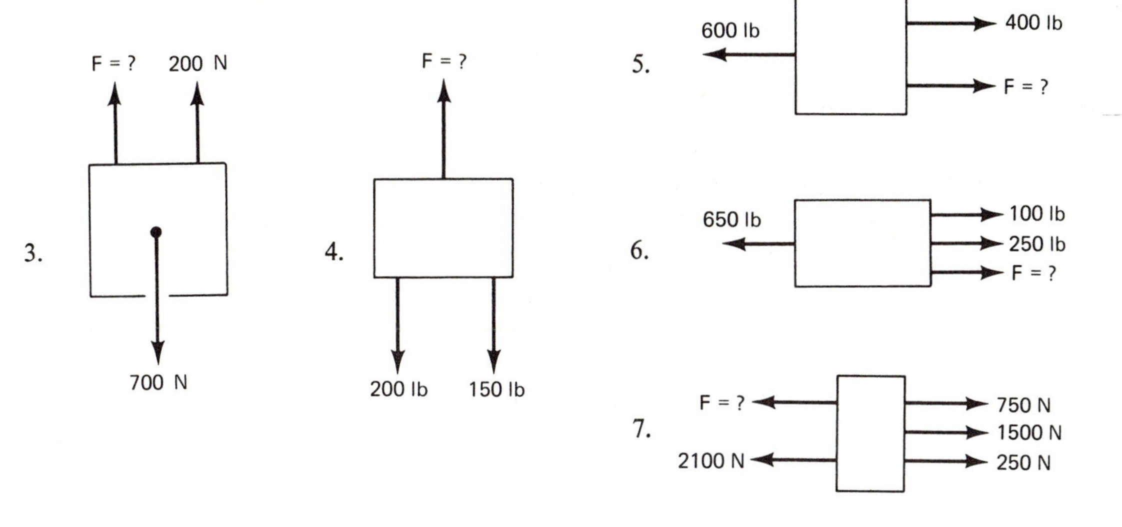

8-3 *CONCURRENT FORCES IN EQUILIBRIUM*

In the construction of buildings and machinery, a technician is often required to solve problems involving the equilibrium of certain parts such as pins and joints. The forces acting on these parts do not always act in one dimension; instead they may act in several directions as shown. These forces are called concurrent because they are all applied at the same point, x.

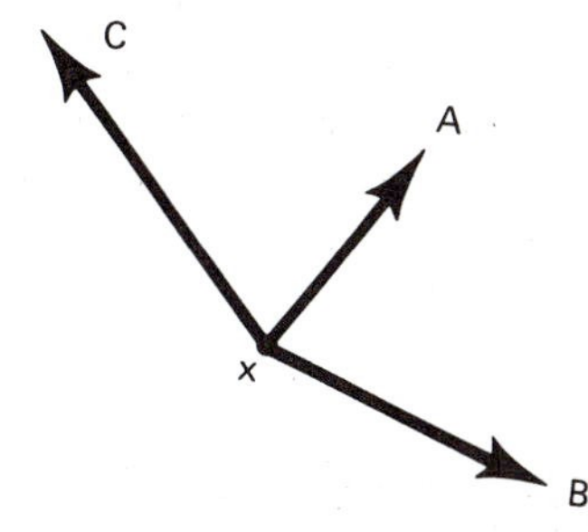

If an object is to be in equilibrium, the net force acting on it must be zero. The separate forces, A, B, and C, are vectors, and the net force is the vector sum of these forces.

We can now apply the component method of adding vectors that we learned in the previous chapter. The x and y components of the net force can be found by adding the x components and the y components of the separate forces. The x component of the net force is:

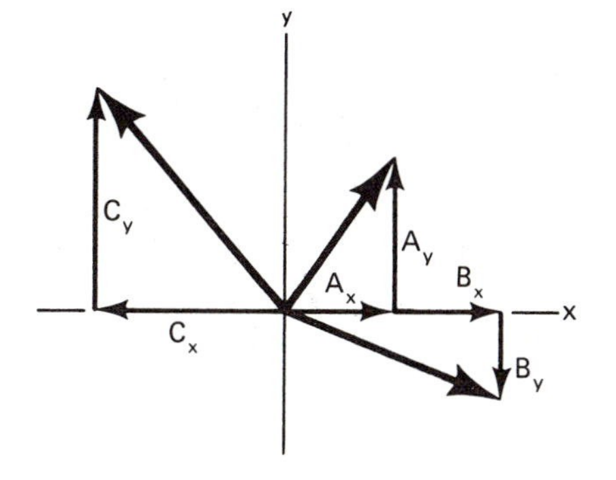

$$R_x = A_x + B_x + C_x = \text{sum of } x \text{ components}$$

The y component of the net force is:

$$R_y = A_y + B_y + C_y = \text{sum of } y \text{ components}$$

If the net force is zero, the sum of the x components and the sum of the y components must also be zero. The conditions for equilibrium then are:

$$\text{Sum of } x \text{ components} = 0$$

$$\text{Sum of } y \text{ components} = 0$$

Use the following procedure to solve equilibrium problems:

(a) Draw the free-body diagram of the point at which the unknown forces act.
(b) Find the x and y component of each force.
(c) Substitute the components in the equations

$$\text{sum of } x \text{ components} = 0$$

$$\text{sum of } y \text{ components} = 0$$

(d) Solve for the unknowns. This may involve two simultaneous equations.
(e) Substitute data into the equations.

In many problems we will be interested in finding the tension or compression in part of a structure, such as in a beam or a cable. *Tension* is a stretching force produced by forces pulling outward on the ends of the object. *Compression* is a compressing force produced by forces pushing inward on the ends of an object. A rubber band being stretched is an example of tension. A valve spring whose ends are pushed together is an example of compression.

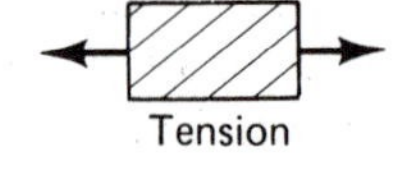
Tension

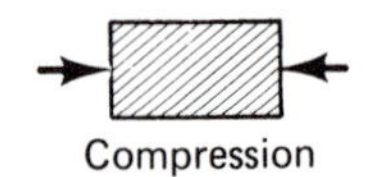
Compression

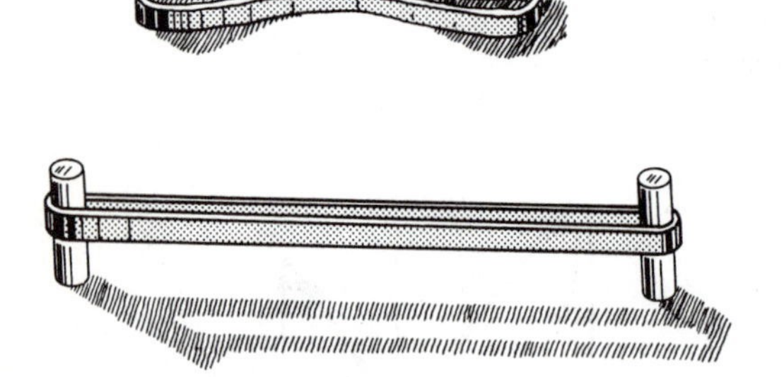

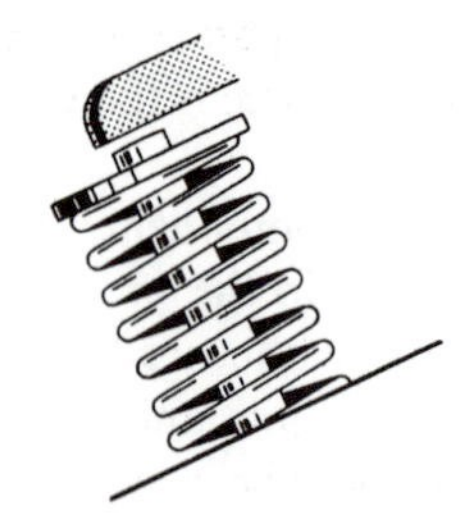

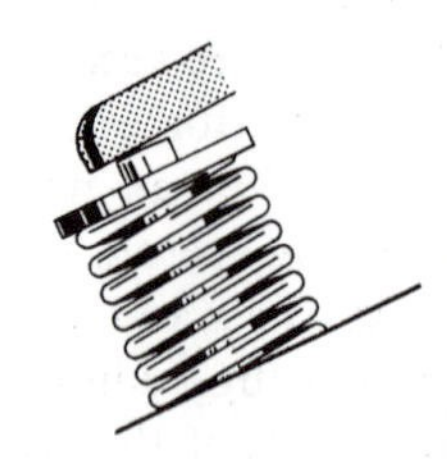

EXAMPLE 1: The crane shown in the figure below is supporting a beam which weighs 6000 N. Find the tension in the horizontal supporting cable and the compression in the boom.

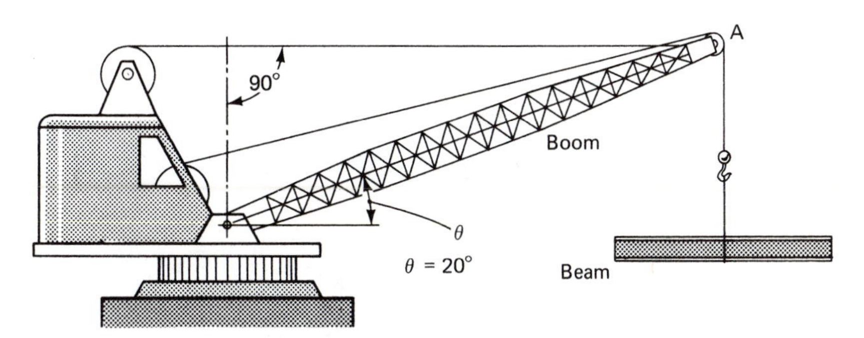

(a) We want to draw the free-body diagram showing the forces *on* the point labeled A. The weight, F_w pulls straight down.
T is the force exerted on A by the horizontal supporting cable.
C is the force exerted by the boom on A.
F_w is the force (weight of the beam) pulling straight down on A.

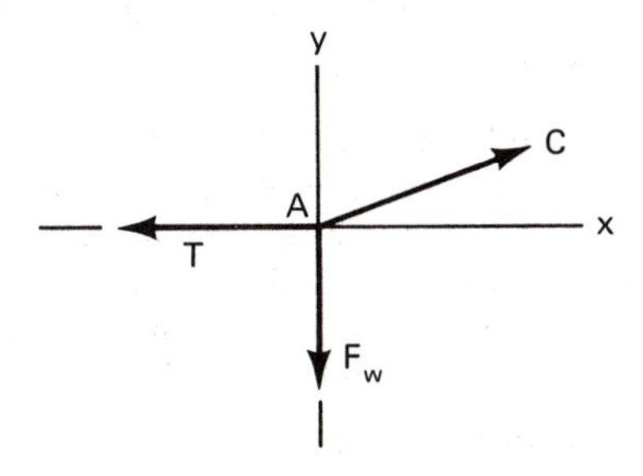

(b) x component of C is $C \cos 20°$ — y component of C is $C \sin 20°$
x component of T is $-T$ — y component of T is 0
x component of F_w is 0 — y component of F_w is $-F_w$

(c) Sum of x components = 0 — Sum of y components = 0
$C \cos 20° + (-T) = 0$ — $C \sin 20° + (-F_w) = 0$

(d) $T = C \cos 20°$

$$C = \frac{F_w}{\sin 20°}$$

$$C = \frac{6000 \text{ N}}{0.342}$$

$$C = 17{,}500 \text{ N}$$

(e) $T = C \cos 20°$
$= (17{,}500 \text{ N})(0.940)$
$= 16{,}500 \text{ N}$

EXAMPLE 2: A man pushes a 40.0-lb lawn mower at a constant velocity. If the frictional force on the mower is 20.0 lb, what force must the man exert on the handle?

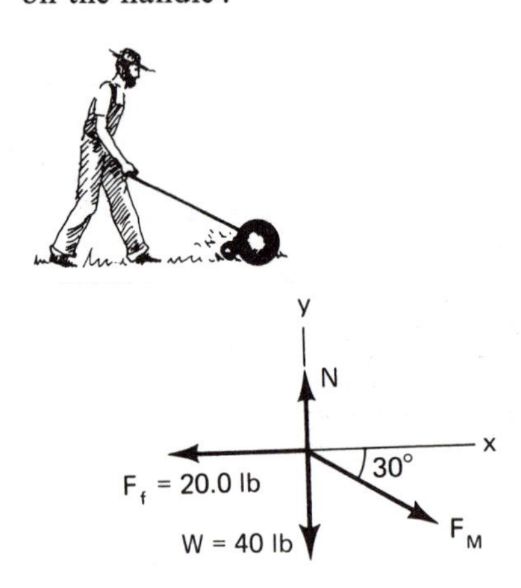

This is an equilibrium problem because the mower is not accelerating and the net force is zero.

(a) Free-body diagram of the mower.
F_m is the force exerted on the mower by the man. This force is directed down along the handle.
N is the force exerted upward on the mower by the ground which keeps the mower from falling through the ground.
F_f is the frictional force which opposes the motion.

(b) x component of N is 0
x component of F_m is $F_m \cos 30°$
x component of F_f is $-F_f$
x component of W is 0

y component of N is N
y component of F_m is $-F_m \sin 30°$
y component of F_f is 0
y component of W is -40

(c) Sum of x components $= 0$
$0 + F_m \cos 30° + (-F_f) = 0$

Sum of y components $= 0$
$N + (-F_m \sin 30°) - 40 = 0$

(d) $F_m \cos 30° = F_f$

$$F_m = \frac{F_f}{\cos 30°}$$

(e)
$$F_m = \frac{20.0 \text{ lb}}{0.866}$$

$$F_m = 23.1 \text{ lb}$$

PROBLEMS

Find the forces F_1 and F_2 necessary to produce equilibrium in the following free-body diagrams.

SKETCH
DATA
$a = 1, b = 2,$
$c = ?$

BASIC
EQUATION
$a = bc$

WORKING
EQUATION
$c = \frac{a}{b}$

SUBSTITUTION
$c = \frac{1}{2}$

1.

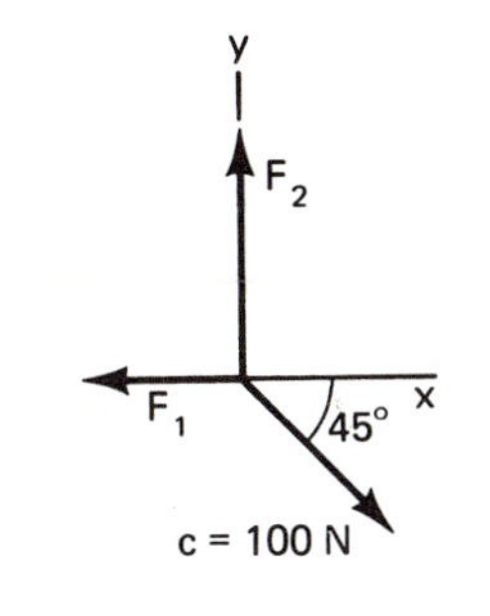

2.

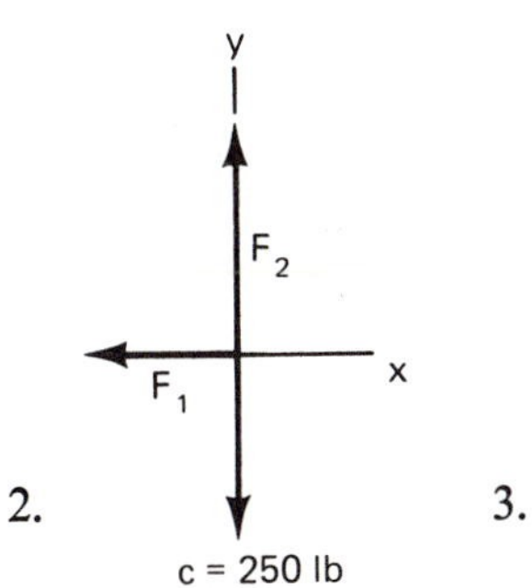

3.

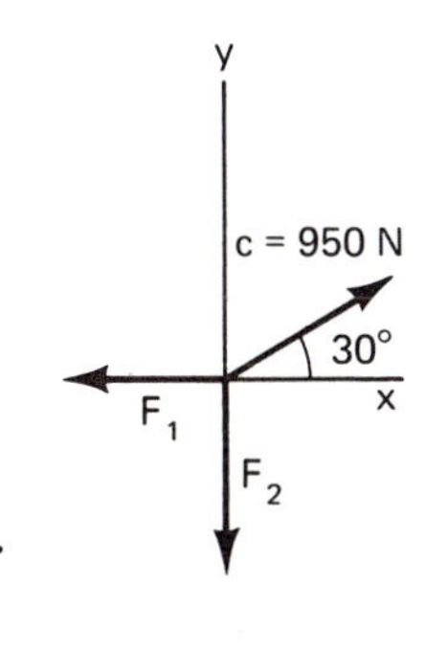

4.

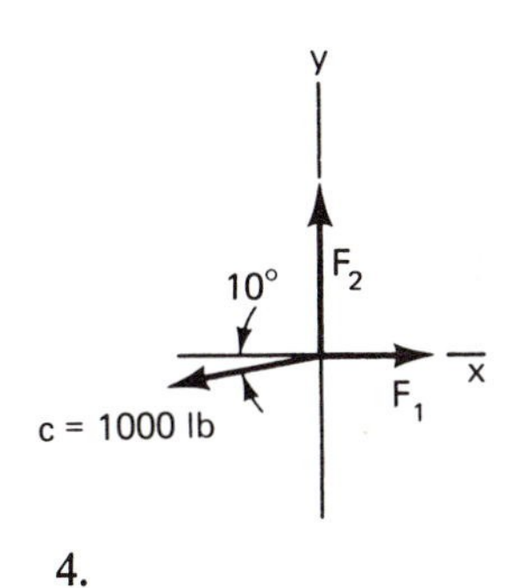

5.

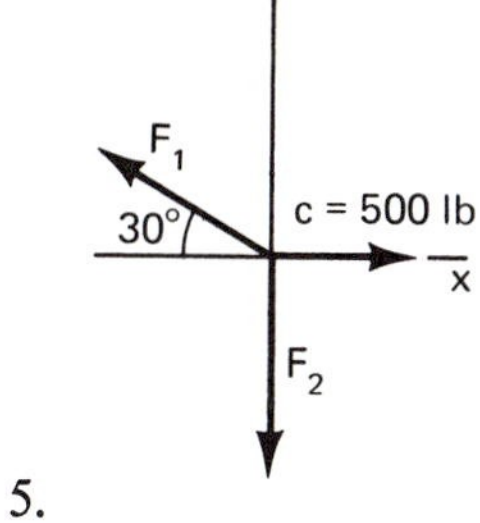

6.

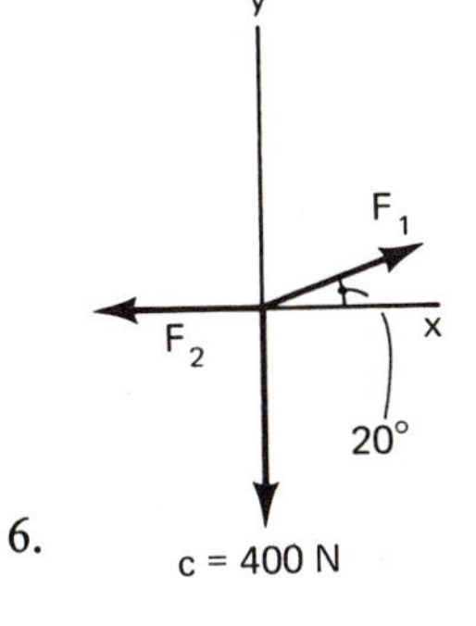

7.

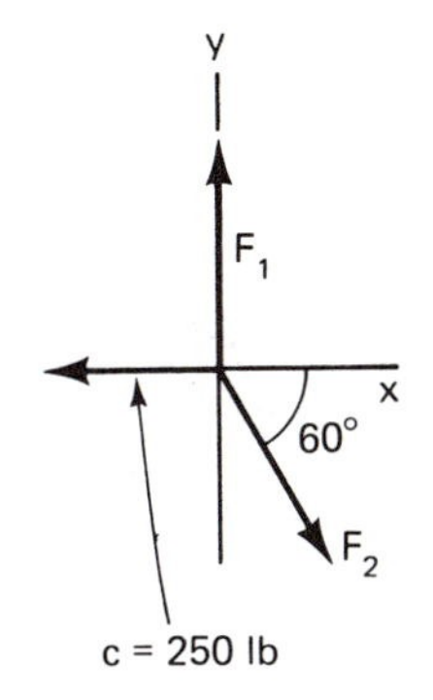

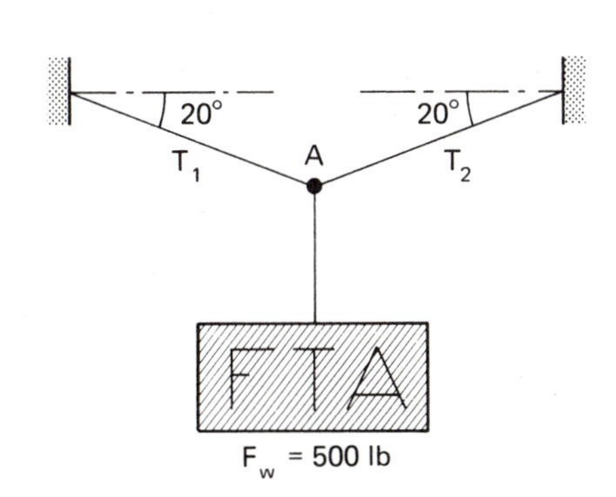

8. The rope shown above is attached to two buildings and supports a 500-lb sign. Find the tension in the two ropes, T_1 and T_2. (*Hint*: Draw the free-body diagram of the point labeled *A*.)

9. If the angle between the horizontal and the ropes in problem 8 is changed to 10°, what are the tensions in the two ropes, T_1 and T_2.

10. Find the tension in the horizontal supporting cable and the compression in the boom of the above crane which supports a 8900-N beam.

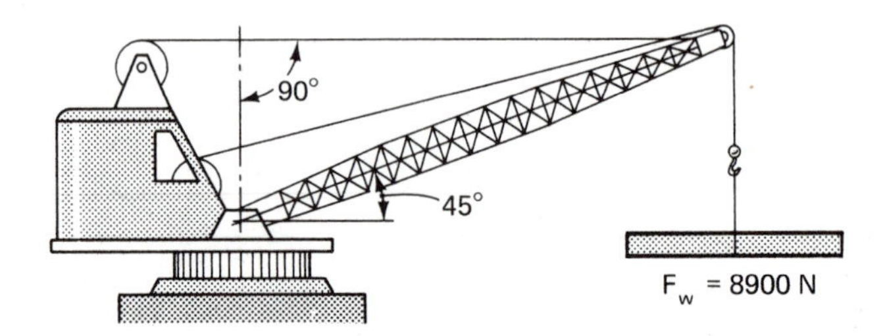

11. Find the tension in the horizontal supporting cable and the compression in the boom of the above crane which supports a 1500-lb beam.

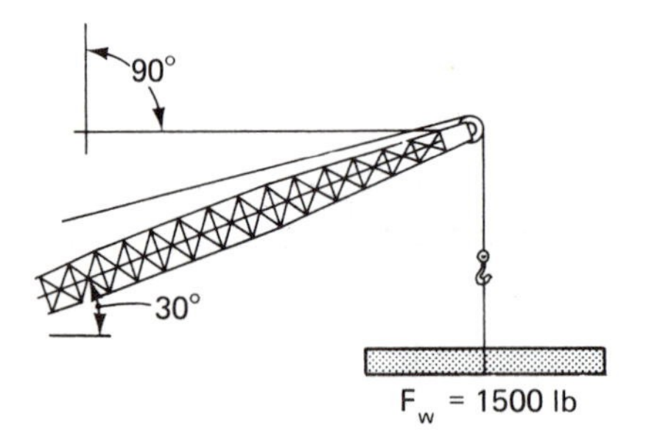

12. The frictional force in the above mower is 20.0 lb. What force must the man exert along the handle to push it at a constant velocity?

13. An automobile which weighs 16,200 N is parked on a 20° hill as shown. What braking force is necessary to keep the auto from rolling downward? (*Hint:* When you draw the free-body diagram of the auto, tilt the *x* and *y* axes as shown. *B* is the braking force directed up the hill and along the *x*-axis.)

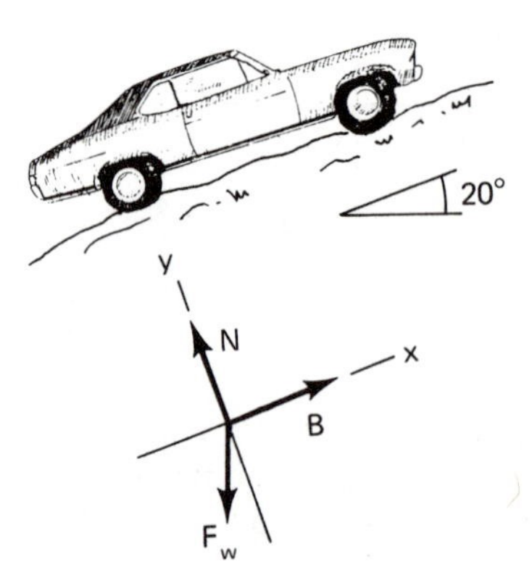

WORK AND ENERGY 9

9-1 *WORK*

What is work? Are we doing work when we try to lift a large crate but it won't budge?

Because we have exerted a force and feel tired, we would probably say we had been working. But is it proper to say that we have done work *on* the crate?

Is work done on the crate if we push the crate across the floor?

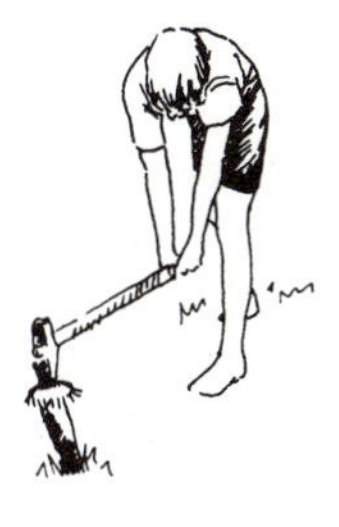

Work does have a technical meaning, however. When a stake is driven into the ground, work is done *by* the moving sledgehammer, and work is done *on* the stake.

When a bulldozer pushes a boulder, work is done *by* the bulldozer, and work is done *on* the boulder.

The examples above show a more limited meaning of work: that work is done when a force acts through a distance. The meaning of work used by scientists and technicians is even narrower: *WORK is the product of the force in the direction of the motion and the displacement.*

$$W = Fs$$

where: W = work
F = force applied
s = displacement *in the direction of the force*

Now, let's apply our technical definition of work to trying unsuccessfully to lift the crate. We applied a force by lifting on the crate. Have we done work? No work has been done. We were unable to move the crate. Therefore, the displacement was zero, and the product of the force and the displacement must also be zero. Therefore, no work was done.

By studying the equation for work, we can ourselves determine the correct units for work. In the metric system, force is expressed in newtons and the displacement in meters.

$$\text{Work} = \text{force} \times \text{displacement} = \text{newton} \times \text{meter} = \text{N m}$$

This unit (N m) has a special name in honor of an English physicist, James P. Joule. It is the joule (J) [pronounced jool].

$$1 \text{ N m} = 1 \text{ joule} = 1 \text{ J}$$

In the English system, force is expressed in pounds (lb) and the displacement in feet (ft).

$$\text{Work} = \text{force} \times \text{displacement} = \text{pounds} \times \text{feet} = \text{ft lb}$$

The English unit of work is called the foot-pound.

Work is not a vector quantity because it has no particular direction. It is a scalar and has only magnitude.

EXAMPLE 1: A man lifting 50.0 lb of bricks to a height of 5.00 ft does 250 ft lb of work.

$W = Fs$

$W = (50.0 \text{ lb})(5.00 \text{ ft})$
$= 250 \text{ ft lb}$

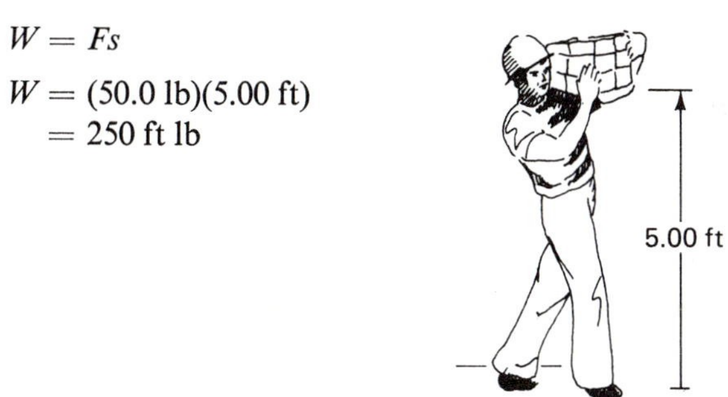

EXAMPLE 2: A man pushing a 60.0-lb pallet (portable platform used in warehouses) a distance of 30.0 ft by exerting a constant force of 10.0 lb does 300 ft lb of work.

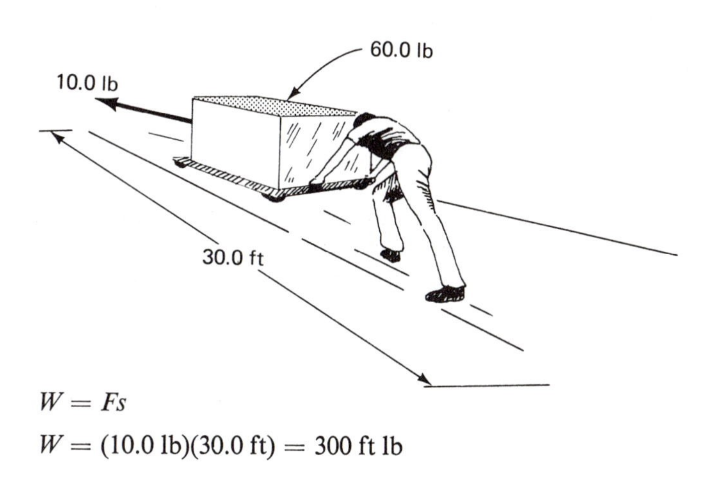

$W = Fs$

$W = (10.0 \text{ lb})(30.0 \text{ ft}) = 300 \text{ ft lb}$

Note in Example 2 that the pallet weighs 60.0 lb. (Recall that the weight of an object is the measure of its gravitational attraction to the earth and is represented by a vertical vector pointing down to the center of the earth.) There is no motion in the direction *this* force is exerted. Therefore, the weight of the box is not the force used to determine the work being done in the problem.

Work is being done by the man pushing the pallet. He is exerting a force of 10.0 lb and there is a resulting displacement in the direction the force is applied. The work done is the product of this force (10.0 lb) and the displacement (30.0 ft) in the direction the force is applied.

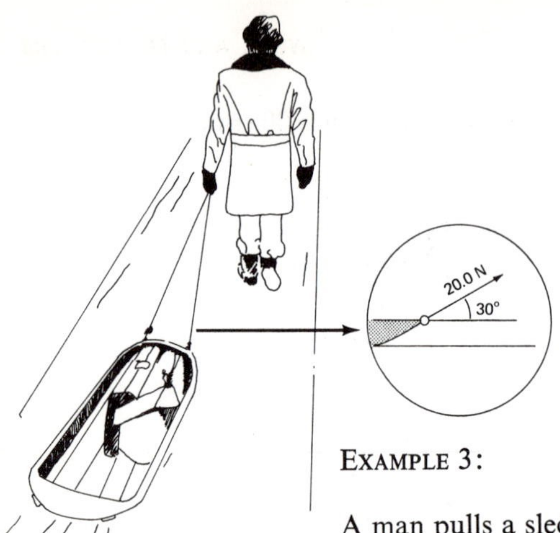

EXAMPLE 3:

A man pulls a sled along level ground for a distance of 15.0 meters by exerting a constant force of 20.0 N at an angle of 30° with the ground. How much work does he do?

Note that in this example the applied force (20.0 N) is *not* in the direction of the displacement. Therefore, we must determine the component of the applied force in the direction of the motion.

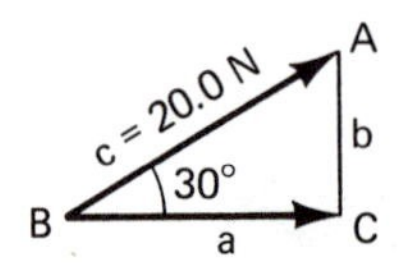

Using the figure above, we can see that we must determine the magnitude of component force vector *a*. We may determine *a* by applying the trigonometric function

$$\cos B = \frac{\text{side adjacent}}{\text{hypotenuse}} = \frac{a}{c}$$

Since $B = 30°$, substituting $\cos 30° = 0.866$ and $c = 20.0$ N, we obtain:

$$\cos B = \frac{a}{c}$$

$$\cos 30° = \frac{a}{20.0\text{ N}}$$

$$0.866 = \frac{a}{20.0\text{ N}}$$

$$\begin{aligned} a &= (0.866)(20.0\text{ N}) \\ &= 17.3\text{ N} \end{aligned}$$

The work done may now be found by multiplying:

$$\begin{aligned} W &= Fs \\ &= (17.3\text{ N})(15.0\text{ m}) \\ &= 260\text{ N m} \\ &= 260\text{ J} \end{aligned}$$

PROBLEMS

Solve using the outline.

SKETCH
DATA
$a = 1, b = 2,$
$c = ?$
BASIC EQUATION
$a = bc$
WORKING EQUATION
$c = \frac{a}{b}$
SUBSTITUTION
$c = \frac{1}{2}$

1. Given: $F = 10.0$ lb
   $s = 3.43$ ft
   $W = ?$
2. Given: $W = 697$ ft lb
   $s = 976$ ft
   $F = ?$
3. Given: $s = 384$ m
   $W = 171$ J
   $F = ?$
4. Given: $F = ma$
   $m = 16.0$ kg
   $a = 9.81$ m/s^2
   $s = 13.0$ m
   $W = ?$
5. How much work is required for a mechanical hoist to lift a 3600-lb automobile to a height of 6.00 ft for repairs?
6. A hay wagon is used to move bales from the field to the barn. The tractor pulling the wagon exerts a constant force of 350 lb. The distance from field to barn is one-half mile. How much work (ft lb) is done in moving one load of hay to the barn?
7. A worker lifts 75 concrete blocks a distance of 1.50 m to the bed of a truck. Each block has a mass of 4.00 kg. How much work must he do to lift all the blocks to the truck bed?
8. The work required to lift eleven 94.0-lb bags of cement from the ground to the back of a truck is 4340 ft lb. What is the distance from the ground to the bed of the truck?
9. A workman carries bricks to a mason 20.0 m away. If the amount of work required to carry one brick is 6.00 J, what force must the workman exert on each brick?
10. A yardman pushes a mower a distance of 900 m in mowing a yard. The handle of the mower makes an angle of 40° with the ground. The man exerts a force of 35.0 N along the handle of the mower. How much work does he do in mowing the lawn?
11. The handle of a vegetable wagon makes an angle of 25° with the horizontal. If the peddler exerts a force of 35.0 lb along the handle, how much work does he do in pulling the cart one mile?

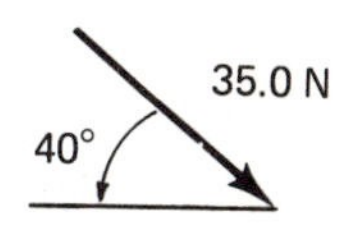

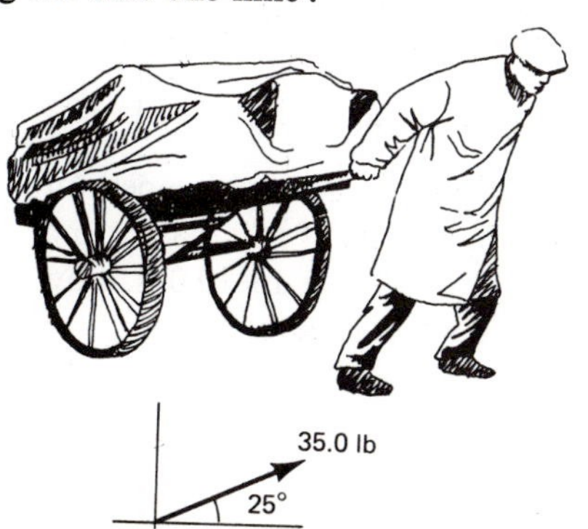

9-2 *POWER*

Many situations arise where we are concerned with not only the amount of work required for a certain task but also the time to do the job. Power also has a special meaning in physics.

> Power is the *rate* of doing work
>
> $$P = \frac{W}{t}$$

where: $P =$ power
$W =$ work
$t =$ time

The units of power are familiar to most of us. In the metric system the unit of power is the watt. Since the unit is inconveniently small, power is usually expressed in kilowatts.

$$P = \frac{W}{t} = \frac{Fs}{t} = \frac{\text{N m}}{\text{s}} = \frac{\text{joule}}{\text{s}} = \text{watt}$$

and

$$1000 \text{ watts (W)} = 1 \text{ kilowatt (kW)}$$

In the English system the unit of power used is either $\frac{\text{ft lb}}{\text{s}}$ or horsepower.

$$P = \frac{W}{t} = \frac{Fs}{t} = \frac{\text{ft lb}}{\text{s}}$$

Horsepower (hp) is a unit derived by James Watt (who also designed the first practical steam engine) to approximate the power delivered by a work horse.

> $$1 \text{ horsepower (hp)} = \frac{550 \text{ ft lb}}{\text{s}} = \frac{33{,}000 \text{ ft lb}}{\text{min}}$$

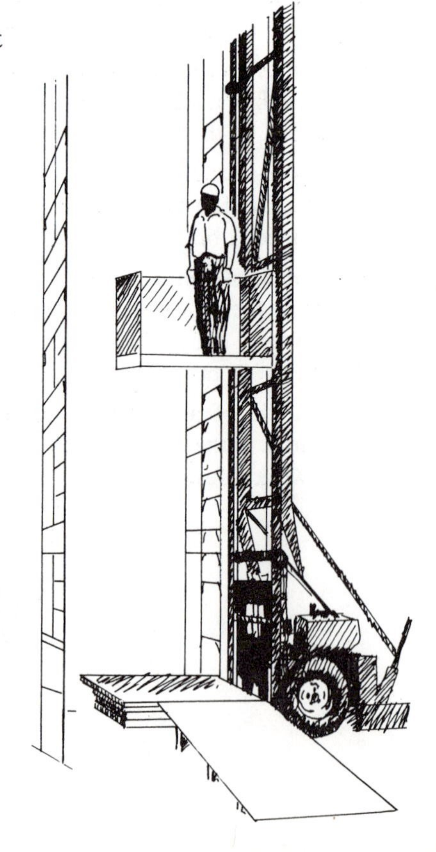

EXAMPLE 1: A freight elevator with operator weighs 500 lb. If it is raised to a height of 50.0 ft in 10.0 s, how much power is developed?

Solution

SKETCH:

DATA: $F = 500 \text{ lb}$
$s = 50.0 \text{ ft}$
$t = 10.0 \text{ s}$
$P = ?$

BASIC EQUATION: $P = \dfrac{W}{t}$

WORKING EQUATION: $P = \dfrac{Fs}{t}$

SUBSTITUTION: $P = \dfrac{(500 \text{ lb})(50.0 \text{ ft})}{10.0 \text{ s}}$

$= 2500 \dfrac{\text{ft lb}}{\text{s}}$

We may find the horsepower developed by the elevator by using the conversion factor

$1 \text{ hp} = \dfrac{550 \text{ ft lb}}{\text{s}}$

THEREFORE: $2500 \dfrac{\cancel{\text{ft lb}}}{\cancel{\text{s}}} \left(\dfrac{1 \text{ hp}}{550 \dfrac{\cancel{\text{ft lb}}}{\cancel{\text{s}}}} \right) = 4.55 \text{ hp}$

EXAMPLE 2: The power expended in lifting a 550-lb girder to the top of a building 100 ft high is 10.0 hp. What time is required to raise the girder?

SKETCH:

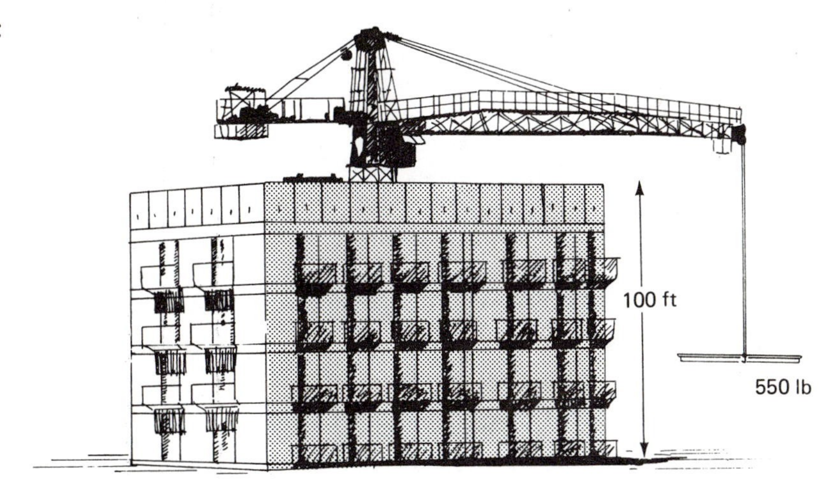

DATA: $F = 550 \text{ lb}$
$s = 100 \text{ ft}$
$P = 10.0 \text{ hp}$
$t = ?$

BASIC EQUATION: $P = \frac{W}{t}$

WORKING EQUATION: $t = \frac{W}{P} = \frac{Fs}{P}$

SUBSTITUTION (1): $t = \frac{(550 \text{ lb})(100 \text{ ft})}{10.0 \text{ hp}}$

Note that in substitution (1) the units do not yield seconds. This indicates to us that we may have made an improper substitution. Studying substitution (1) we see that we have used horsepower directly instead of converting to the fundamental power unit $\frac{\text{ft lb}}{\text{s}}$. Let's substitute again after converting 10.0 hp to $\frac{\text{ft lb}}{\text{s}}$:

$$10 \cancel{\text{hp}} \left(\frac{550 \frac{\text{ft lb}}{\text{s}}}{1 \cancel{\text{hp}}} \right) = 5500 \frac{\text{ft lb}}{\text{s}}$$

SUBSTITUTION (2): $t = \frac{(550 \text{ lb})(100 \text{ ft})}{5500 \frac{\text{ft lb}}{\text{s}}}$

Recall from arithmetic how to simplify fractions of the form $\frac{\frac{a}{b}}{\frac{c}{d}}$.

$$\frac{\frac{a}{b}}{\frac{c}{d}} = \frac{a}{b} \div \frac{c}{d} \quad \text{(divide numerator by denominator)}$$

$$= \frac{a}{b} \cdot \frac{d}{c} \quad \text{(invert and multiply)}$$

$$= \frac{ad}{bc}$$

For example,

$$\frac{\frac{1}{4}}{\frac{3}{5}} = \frac{1}{4} \div \frac{3}{5}$$

$$= \frac{1}{4} \cdot \frac{5}{3}$$

$$= \frac{5}{12}$$

In substitution (2) we must apply this rule to the units to obtain the correct result in seconds:

$$t = \frac{(550 \text{ lb})(100 \text{ ft})}{5500 \frac{\text{ft lb}}{\text{s}}}$$

$$= 10.0 \text{ (lb ft)} \div \left(\frac{\text{ft lb}}{\text{s}} \right)$$

$$= 10.0 \text{ } (\cancel{\text{lb ft}}) \left(\frac{\text{s}}{\cancel{\text{ft lb}}} \right)$$

$$= 10.0 \text{ s}$$

EXAMPLE 3: The mass of a large steel wrecking ball is 2000 kg. What power is used to raise it to a height of 40.0 m if the work is done in 20.0 s?

DATA:

$$m = 2000 \text{ kg}$$
$$s = 40.0 \text{ m}$$
$$t = 20.0 \text{ s}$$
$$P = ?$$

BASIC EQUATION:

$$P = \frac{W}{t} = \frac{Fs}{t}$$

WORKING EQUATION:

$$P = \frac{Fs}{t}$$

Note that we cannot directly substitute into the working equation because our data is in terms of *mass* and we must find *force* to substitute in $P = Fs/t$. Recall the force is the weight of the ball.

$$F = mg = 2000 \text{ kg } (9.81 \text{ m/s}^2) = 19{,}600 \frac{\text{kg m}}{\text{s}^2} = 19{,}600 \text{ N}$$

SUBSTITUTION:

$$P = \frac{(19{,}600 \text{ N})(40 \text{ m})}{20.0 \text{ s}}$$
$$= 39{,}200 \frac{\text{N m}}{\text{s}}$$
$$= 39{,}200 \text{ W}$$

PROBLEMS

1. Given: $W = 132$ ft lb
   $t = 7.00$ s
   $P = ?$
2. Given: $P = 231 \dfrac{\text{ft lb}}{\text{s}}$
   $t = 14.3$ s
   $W = ?$
3. Given: $W = 55.0$ J
   $t = 11.0$ s
   $P = ?$
4. Given: $P = 75.0$ W
   $W = 40.0$ J
   $t = ?$
5. The work required to lift a crate is 310 ft lb. If it can be lifted in 25.0 s, what power is developed?
6. A 3600-lb automobile is pushed by its unhappy driver a quarter of a mile when it runs out of gas. To keep the car rolling he must exert a constant force of 175 lb.

SKETCH
DATA
$a = 1, b = 2,$
$c = ?$

BASIC
EQUATION
$a = bc$

WORKING
EQUATION
$c = \dfrac{a}{b}$

SUBSTITUTION
$c = \dfrac{1}{2}$

(a) How much work does he do?
(b) If it takes him 15.0 minutes, how much power does he exert?
(c) Expressed in horsepower, how much power does he exert?

7. An electric golf cart develops 1250 watts while moving at a constant speed.
(a) Express its power in horsepower.
(b) If the cart travels 200 m in 35.0 s, what force is exerted by the cart?
8. How many seconds would it take a 7.00-hp motor to raise a 475 lb boiler to a platform in a factory 38.0 ft high?
9. How long would it take a 950-W motor to raise a 360-kg mass to a height of 16.0 m?
10. A 1500-lb casting is raised 22 ft in $2\frac{1}{2}$ minutes. Find the required horsepower.
11. What is the rating in watts of a 2-hp motor?
12. A wattmeter shows that a motor is drawing 2200 watts. What horsepower is being delivered?

9-3 *ENERGY*

Physicists and technicians define mechanical energy as the ability to do work. Mechanical energy exists in two forms. One type is energy of motion, the kind of energy an automobile possesses as it moves along the highway. This type is called kinetic energy.

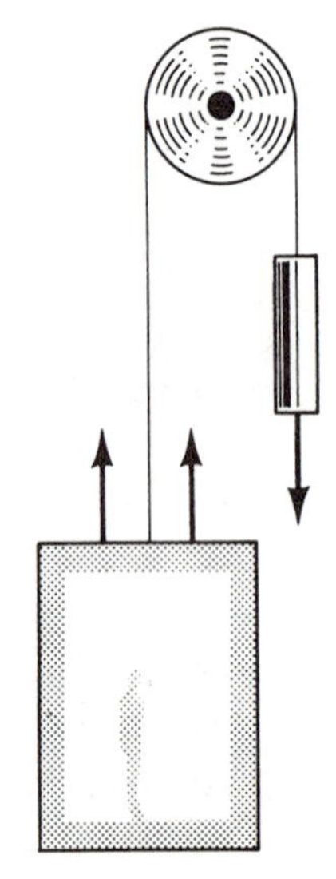

The other type is energy due to an object's condition or position—stored energy. This type is called potential energy. It may be due to internal characteristics of a substance, like gasoline, or its position. The counterweight on an elevator illustrates potential energy of position.

Note that when it is in the raised position (and the elevator is down), the weight has energy available to do work because of the pull of gravity on it. This is called gravitational potential energy.

Since energy is defined as the ability to do work, it is not very surprising that energy is expressed in the same units as work—the joule and the foot pound.

We are now discussing only two kinds of energy—kinetic and potential. Keep in mind that energy exists in many forms—chemical, atomic, electrical, sound, and heat. These forms and the conversion of energy from one form to another will be studied later.

Let's take a closer look now at kinetic energy. The formula for kinetic energy follows:

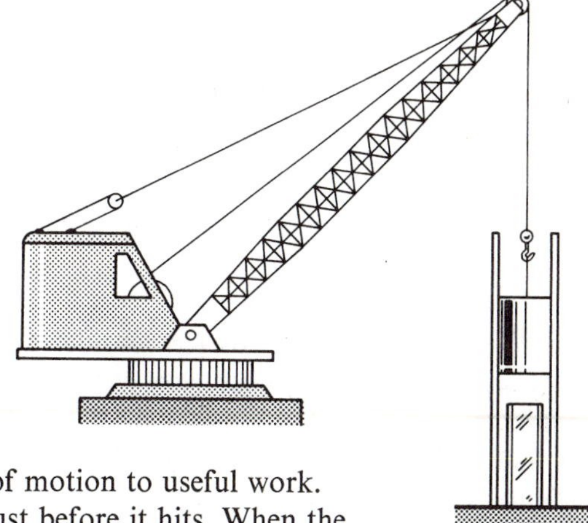

Kinetic Energy (KE) $= \frac{1}{2}mv^2$

where: m = mass of moving object
v = velocity of moving object

A pile driver shows the relation of energy of motion to useful work.

The energy of the driver is kinetic energy just before it hits. When the driver strikes the pile, work is done on the pile and it is forced into the ground. The depth it goes down is determined by the force applied to it. The force applied is determined by the energy of the driver. If all the kinetic energy of the driver is converted to useful work, then $\frac{1}{2}mv^2 = Fs$. Consider the following example.

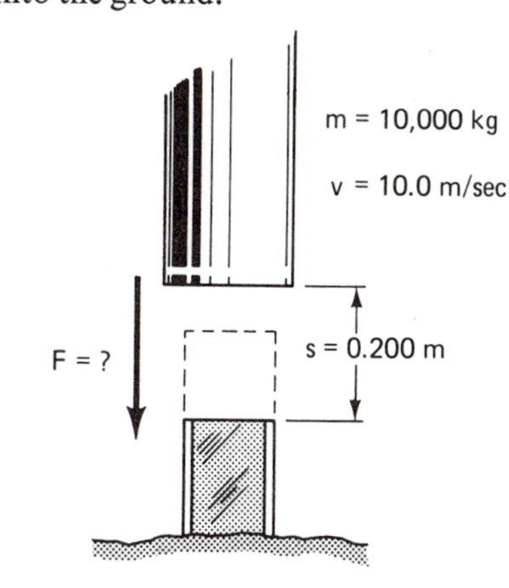

EXAMPLE: The driver has a mass of 10,000 kg and upon striking the pile has a velocity of 10.0 m/s.

(a) What is the kinetic energy of the driver when it strikes the pile?

(b) If the pile is driven 0.200 m into the ground, what force is applied to the pile?

SKETCH: See above.

DATA:
$m = 1.00 \times 10^4$ kg
$v = 10.0$ m/s
$s = 0.200$ m
$F = ?$

(a) BASIC EQUATION: $KE = \frac{1}{2}mv^2$

WORKING EQUATION: Same

SUBSTITUTION:
$$KE = \frac{1}{2}(1.00 \times 10^4 \text{ kg})(10.0 \text{ m/s})^2$$
$$= 5.00 \times 10^5 \frac{\cancel{\text{kg}}\,\cancel{\text{m}^2}}{\cancel{\text{s}^2}} \times \frac{1 \text{ J}\,\cancel{\text{s}^2}}{\cancel{\text{kg}}\,\cancel{\text{m}^2}}$$
$$= 5.00 \times 10^5 \text{ J}$$

(*Note*: Conversion factor is needed for desired units.)

(b) BASIC EQUATIONS: $KE = W = Fs$

WORKING EQUATION: $F = \frac{KE}{s}$ [KE from part (a)]

SUBSTITUTION: $F = \frac{5.00 \times 10^5 \cancel{J}}{0.200 \cancel{m}} \times \frac{N \cancel{m}}{1 \cancel{J}}$ (*Note*: conversion factor)

$= 2.50 \times 10^6$ N

Potential energy will now be considered more fully. Internal potential energy is determined by the nature or condition of the substance, for example, gasoline, a compressed spring, or a stretched rubber band. Gravitational potential energy is determined by the position of the object relative to a particular reference level. The formula for potential energy is:

Potential Energy (PE) $= mgh$

where: m = mass
$g = 9.81$ m/s² or 32.2 ft/s²
h = height above reference level

EXAMPLE 1: In position 1 (see figure below), the crate is at rest on the floor. It has no ability to do work as it is in its lowest position. To raise the crate to position 2, work must be done to lift it. In the raised position, however, it now has stored ability to do work (by falling to the floor!). Its *PE* (potential energy) can be calculated by multiplying the mass of the crate times acceleration of gravity (g) times height above reference level (h). Note that we could calculate the potential energy of the crate with respect to any level we may choose. In this example we have chosen the floor as the zero or lowest reference level.

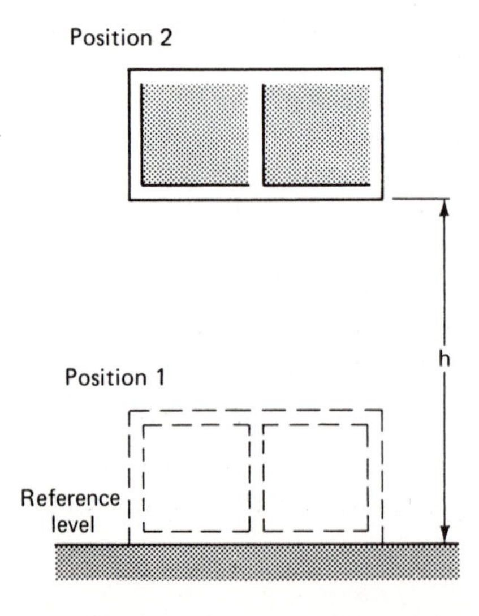

EXAMPLE 2: A wrecking ball of mass 200 kg is poised 4.00 m above a concrete platform whose top is 2.00 m above the ground.

(a) With respect to the platform, what is the potential energy of the ball?

(b) With respect to the ground, what is the potential energy of the ball?

Solution

SKETCH:

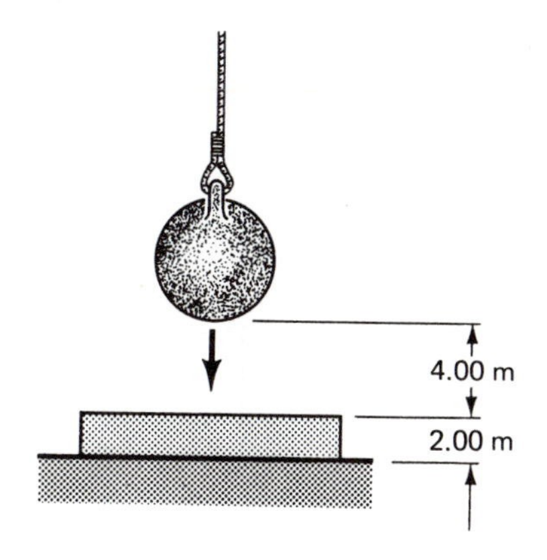

DATA: $m = 200$ kg
$h_1 = 4.00$ m
$h_2 = 2.00$ m

BASIC EQUATION: $PE = mgh$

WORKING EQUATION: Same

(a) SUBSTITUTION: $PE = (200 \text{ kg})\left(9.81 \frac{\text{m}}{\text{s}^2}\right)(4.00 \text{ m})$

$$= 7850 \frac{\text{kg m}^2}{\text{s}^2} \times 1 \frac{\text{J s}^2}{\text{kg m}^2}$$

$$= 7850 \text{ J}$$

(b) SUBSTITUTION: $PE = (200 \text{ kg})\left(9.81 \frac{\text{m}}{\text{s}^2}\right)(6.00 \text{ m})$

$$= 11{,}800 \frac{\text{kg m}^2}{\text{s}^2} \times \frac{1 \text{ J s}^2}{\text{kg m}^2}$$

$$= 11{,}800 \text{ J}$$

Now let's consider kinetic and potential energy together. Can you see how the two might be related? They are, in fact, related by an important principle called the *law of conservation of mechanical energy: the sum of the kinetic energy and potential energy in a system is constant.*

A pile driver shows this energy conservation. When the driver is at its highest position, the potential energy is maximum and the kinetic energy is

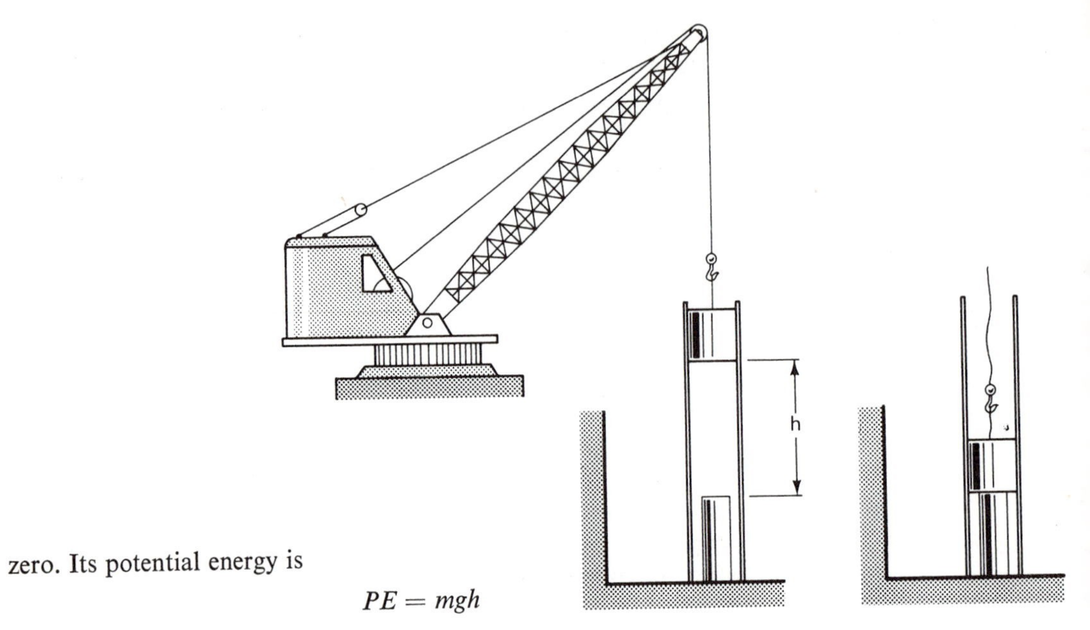

zero. Its potential energy is

$$PE = mgh$$

and its kinetic energy is

$$KE = \tfrac{1}{2}mv^2 = \tfrac{1}{2}m(0)^2 = 0$$

When the driver hits the top of the pile, it has its maximum kinetic energy and the potential energy is

$$PE = mgh = mg(0) = 0$$

Since the total energy in the system must remain constant, the maximum potential energy must equal the maximum kinetic energy.

$$PE_{\text{max}} = KE_{\text{max}}$$

$$mgh = \tfrac{1}{2}mv^2$$

Solving for the velocity of the driver just before it hits the pile, we get

$$v = \sqrt{2gh}$$

PROBLEMS

1. Given: $m = 11.4$ slugs
   $g = 32.2$ ft/s^2
   $h = 22.0$ ft
   $PE = ?$

2. Given: $m = 3.50$ kg
   $g = 9.81$ m/s^2
   $h = 15.0$ m
   $PE = ?$

3. Given: $m = 4.70$ kg
   $v = 9.60$ m/s
   $KE = ?$

4. Given: $PE = 93.6$ J
   $g = 9.81$ m/s^2
   $m = 2.30$ kg
   $h = ?$

SKETCH
DATA
$a = 1, b = 2,$
$c = ?$

BASIC
EQUATION
$a = bc$

WORKING
EQUATION
$c = \dfrac{a}{b}$

SUBSTITUTION
$c = \dfrac{1}{2}$

5. A truck is going along a highway with a velocity of 55.0 miles per hour.
   (a) What is its velocity in ft/s?
   The mass of the truck is 950 slugs.
   (b) What is the kinetic energy of the truck?
6. A bullet travels at 415 m/s. If it has a mass of 12.0 g, what is its kinetic energy? (*Hint*: Convert 12.0 g to kg.)
7. A bicycle and rider together have a mass of 74.0 slugs. If the kinetic energy is 742 ft lb, what is the velocity?
8. A crate of mass 475 kg is raised to a height of 17.0 m from the floor. What potential energy has it acquired with respect to the floor?
9. A shop manual weighing 3.00 lb is on a shelf 4.40 ft above a table top which is itself 2.70 ft above the floor.
   (a) What is the potential energy of the book with respect to the table top?
   (b) What is the potential energy of the book with respect to the floor?
10. The potential energy possessed by a girder after being lifted to the top of a new building is 5.17×10^5 ft lb. If the mass of the girder is 173 slugs, how high is the girder?
11. A pile driver falls through a distance of 12.0 ft before hitting a pile. What is the velocity of the driver just before it hits the pile?
12. A sky diver jumps out of a plane at a height of 5000 ft. If his parachute does not open until he reaches 1000 ft, what is his velocity if air resistance can be neglected?
13. A piece of shattered glass falls from the 82nd floor of a skyscraper, 270 m above the ground. What is the velocity of the glass when it hits the ground, if air resistance can be neglected?

SIMPLE MACHINES 10

10-1 *USES OF MACHINES*

Machines are used to transfer energy from one place to another. By using a pulley system one person can easily lift an engine from an automobile.

Pliers allow a person to cut a wire with the strength of his hand. Machines are sometimes used to multiply force.

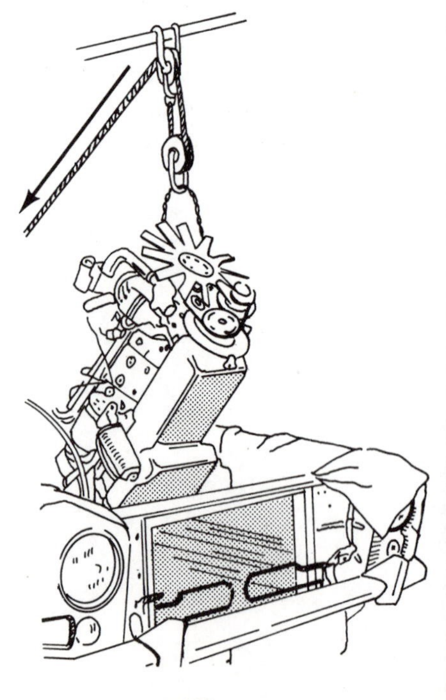

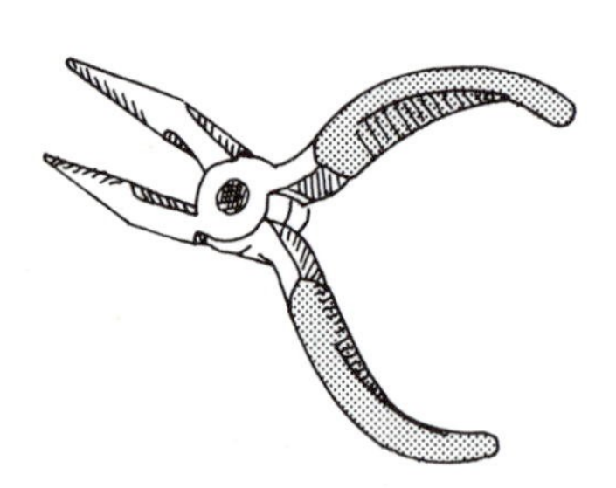

By pushing with a small force, we can use a machine to jack up an automobile.

Machines are sometimes used to multiply speed. The gears on a bicycle are used to multiply speed.

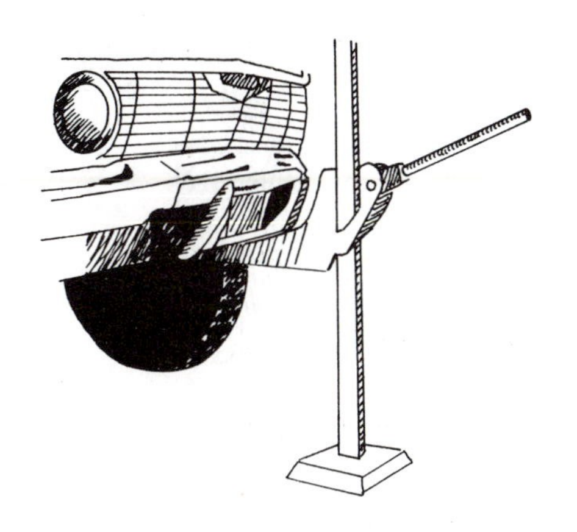

10-2 *THE SIMPLE MACHINES*

There are six basic or simple machines:

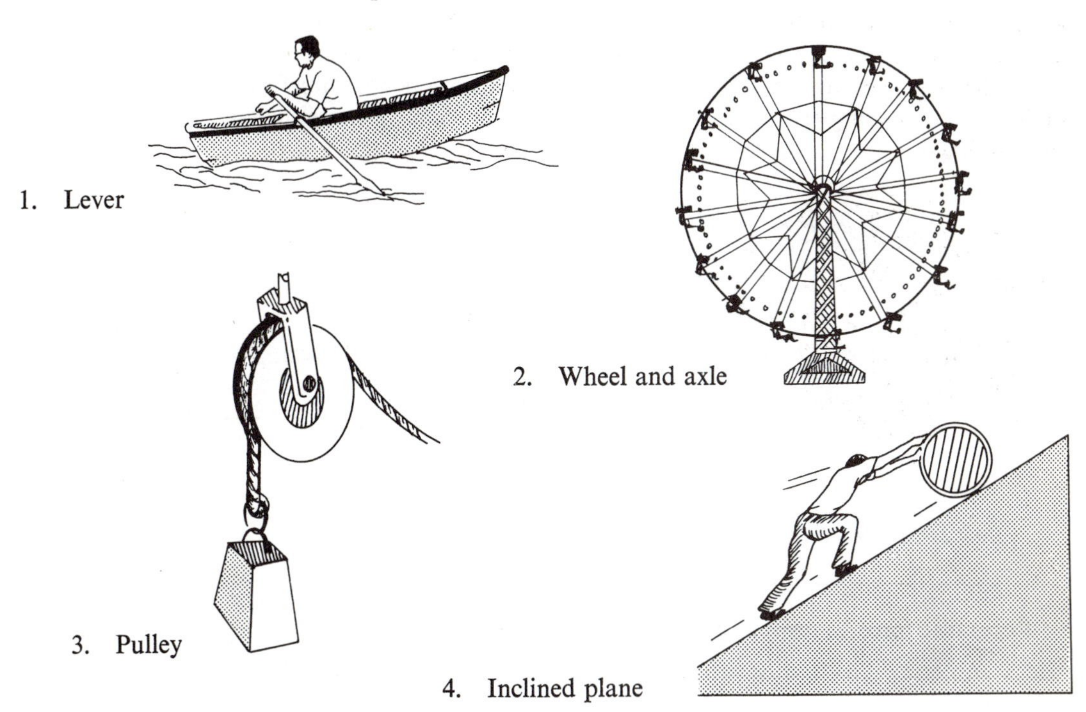

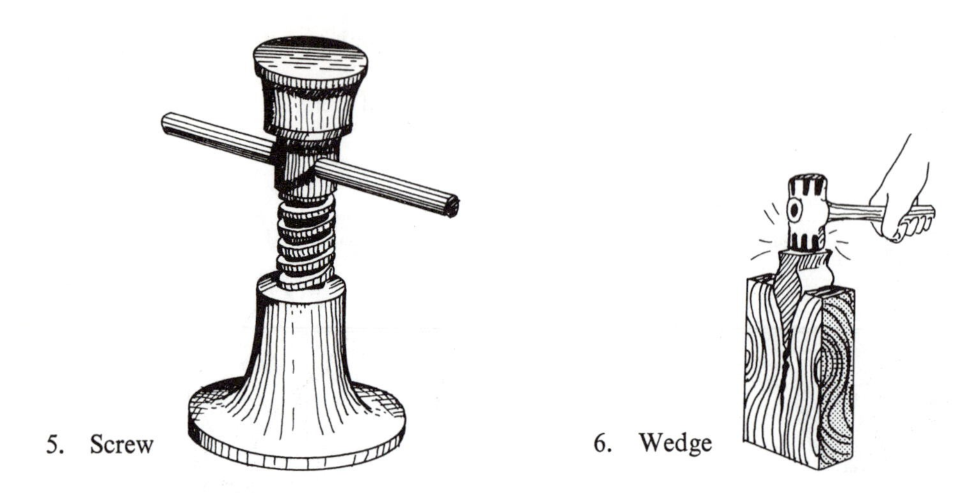

5. Screw

6. Wedge

All other machines—no matter how complex—are combinations of two or more of these simple machines.

10-3 *EFFORT AND RESISTANCE*

In every machine we are concerned with two forces—effort and resistance. The *effort* is the force applied *to* the machine. The *resistance* is the force overcome *by* the machine.

The man below applies 30 lb on the jack handle to produce a lifting force of 900 lb on the camper's bumper. The effort force is 30 lb. The resistance force is 900 lb.

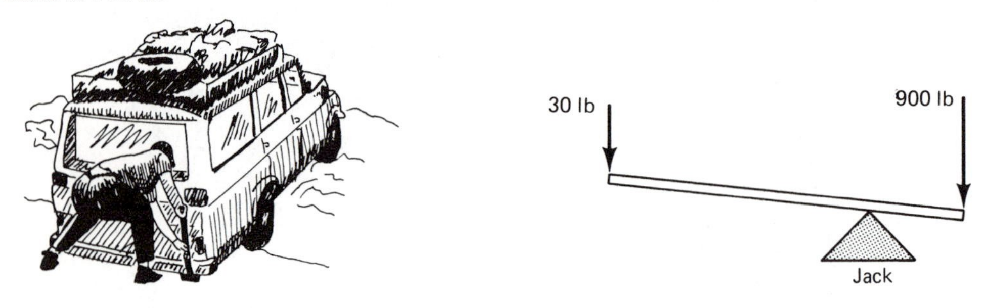

Law of simple machines:

Resistance force $\times$ resistance distance $=$ effort force $\times$ effort distance

10-4 *MECHANICAL ADVANTAGE*

The mechanical advantage is the ratio of the resistance force to the effort force. By formula:

$$MA = \frac{\text{resistance force}}{\text{effort force}}$$

The *MA* of the jack in the previous example is found as follows:

$$MA = \frac{\text{resistance force}}{\text{effort force}} = \frac{900 \not{\text{lb}}}{30 \not{\text{lb}}} = \frac{30}{1}$$

This *MA* means that, for each pound applied by the mechanic, he lifts 30 pounds.

10-5 *EFFICIENCY*

Each time a machine is used, part of the energy or effort applied to the machine is lost due to friction.

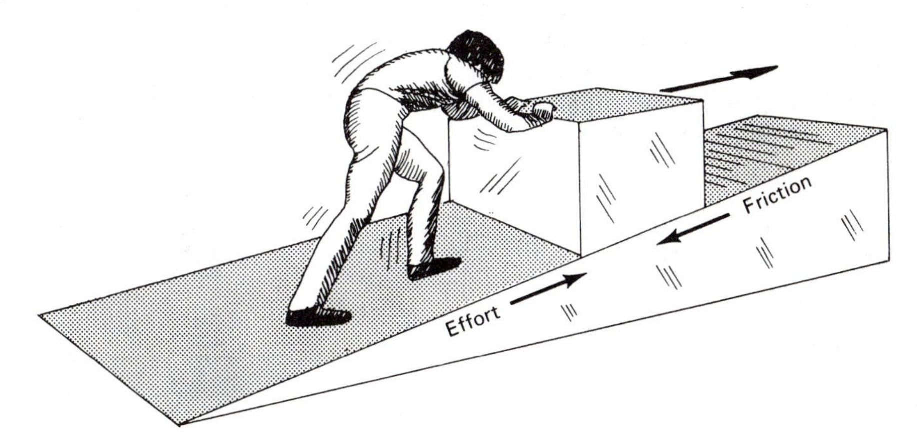

The efficiency of a machine is the ratio of the work output to the work input. By formula:

$$\text{Efficiency} = \frac{\text{work output}}{\text{work input}} \times 100\% = \frac{F_{\text{output}} \times s_{\text{output}}}{F_{\text{input}} \times s_{\text{input}}} \times 100\%$$

A *lever* consists of a rigid bar free to turn on a pivot called the fulcrum.

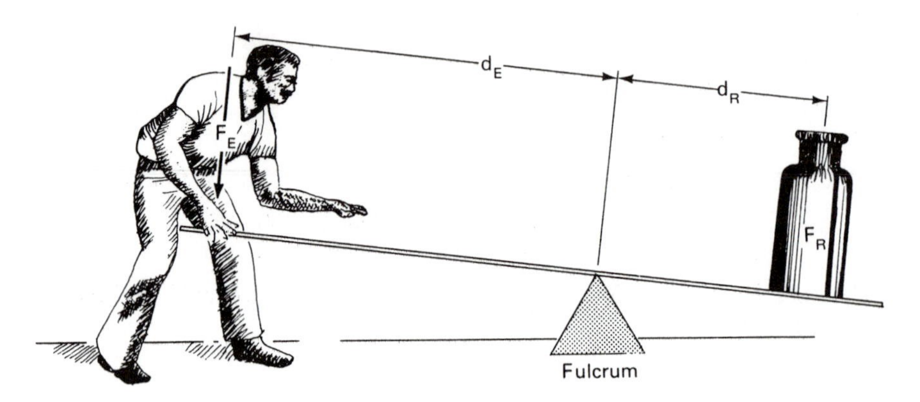

The mechanical advantage (MA) is the ratio of the effort arm (d_E) to the resistance arm (d_R).

$$MA_{\text{lever}} = \frac{\text{effort arm}}{\text{resistance arm}} = \frac{d_E}{d_R}$$

The *effort arm* is the distance from the effort to the fulcrum. The *resistance arm* is the distance from the fulcrum to the resistance.

There are three types or classes of levers to study:

First class: The fulcrum is between the resistance force (F_R) and the effort force (F_E).

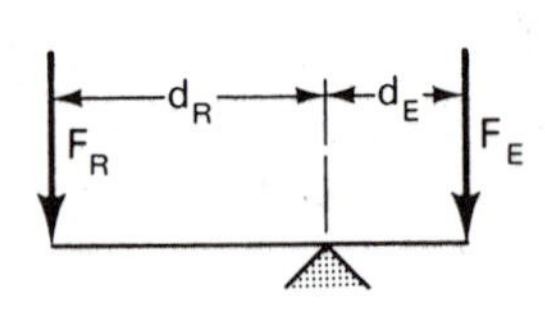

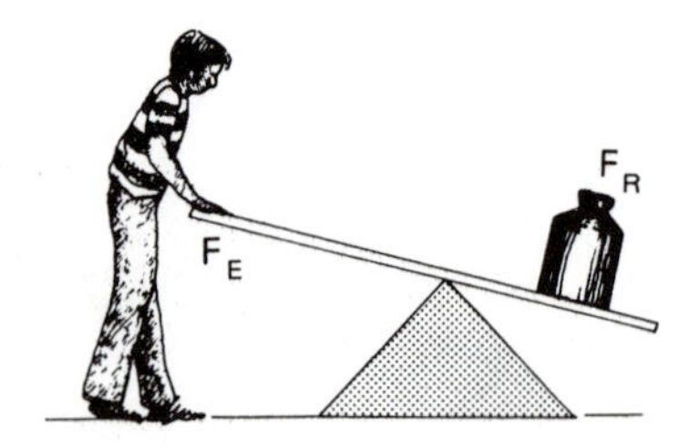

Second class: The resistance force (F_R) is between the fulcrum and the effort force (F_E).

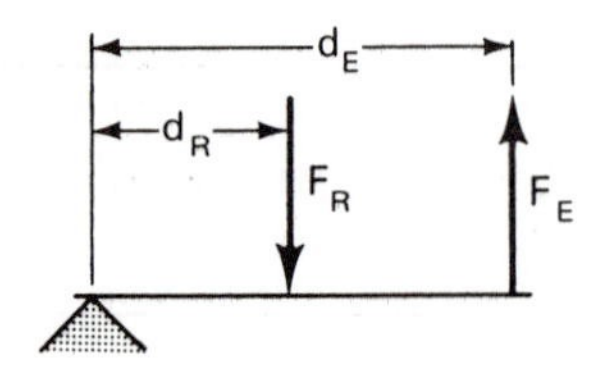

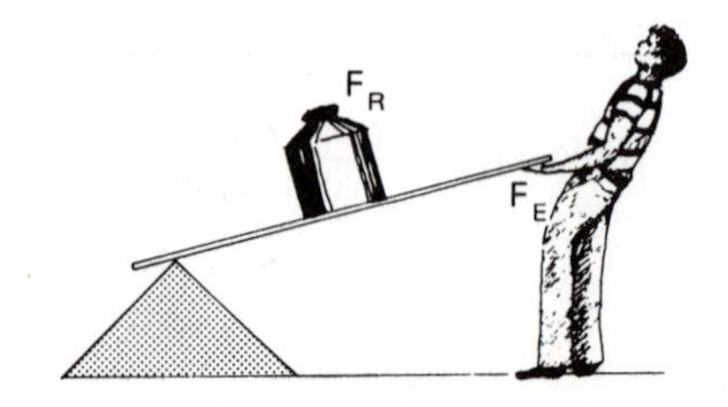

Third class: The effort force (F_E) is between the fulcrum and the resistance force (F_R).

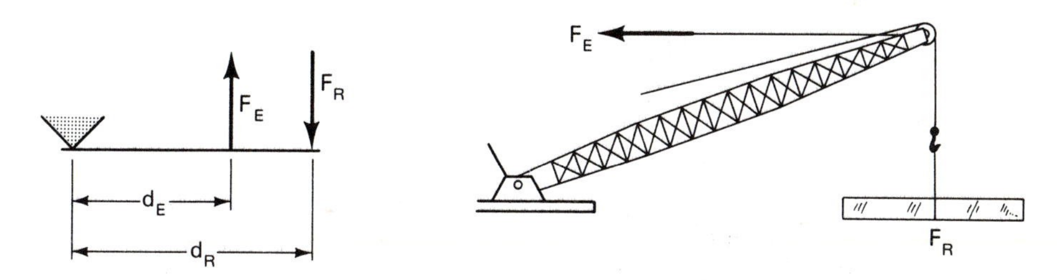

Law of simple machines as applied to levers (basic equation):

$$F_R \cdot d_R = F_E \cdot d_E$$

EXAMPLE 1: A crowbar is used to raise an 1800-lb stone. The pivot is placed 9.00 inches from the stone. The man pushes 108 inches from the pivot. What is the mechanical advantage? What force does he exert?

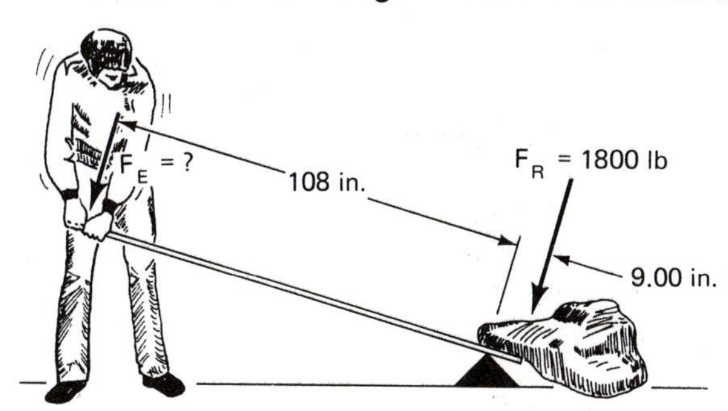

$$MA = \frac{d_E}{d_R} = \frac{108 \text{ in.}}{9.00 \text{ in.}} = \frac{12}{1}$$

To find the force:

DATA: $d_E = 108$ in.
$d_R = 9.00$ in.
$F_R = 1800$ lb
$F_E = ?$

BASIC EQUATION: $F_R \cdot d_R = F_E \cdot d_E$

WORKING EQUATION: $F_E = \frac{F_R \cdot d_R}{d_E}$

SUBSTITUTION: $F_E = \frac{(1800 \text{ lb})(9.00 \text{ in.})}{108 \text{ in.}} = 150 \text{ lb}$

EXAMPLE 2: A wheelbarrow 2.00 m long has a 900 N load 0.500 m from the axle. What is the *MA*? What force is needed to lift the wheelbarrow?

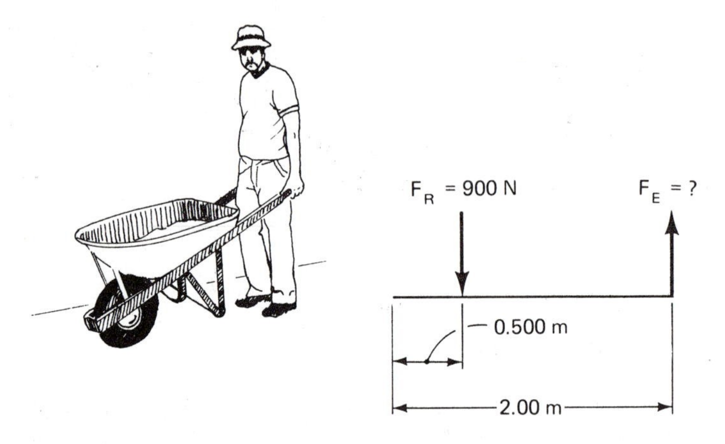

$$MA = \frac{d_E}{d_R} = \frac{2.00\,\text{m}}{0.500\,\text{m}} = 4$$

To find the force:

DATA:

$d_E = 2.00$ m
$d_R = 0.500$ m
$F_R = 900$ N
$F_E = ?$

BASIC EQUATION: $F_R \cdot d_R = F_E \cdot d_E$

WORKING EQUATION: $F_E = \dfrac{F_R \cdot d_R}{d_E}$

SUBSTITUTION: $F_E = \dfrac{(900\text{ N})(0.500\text{ m})}{2.00\text{ m}} = 250\text{ N}$

EXAMPLE 3: The *MA* of a pair of pliers is 6/1. A force of 8.00 lb is exerted on the handle. What force is exerted on a wire in the pliers?

$MA = 6/1$ means that, for each pound of force applied on the handle, 6 pounds are exerted on the wire. Therefore, if a force of 8.00 pounds is applied on the handle, a force of (6)(8.00 lb) or 48.0 lb is exerted on the wire.

PROBLEMS

Given $F_R \cdot d_R = F_E \cdot d_E$, find the missing quantities.

	F_R	F_E	d_R	d_E
1.	20.0	5.00	3.70	____
2.	____	17.6	49.2	76.3
3.	37.0	12.0	____	112
4.	23.4	9.80	____	53.9
5.	119	____	29.7	67.4

Given $MA = F_R/F_E$, find the missing quantities.

	MA	F_R	F_E
6.	____	20.0	5.00
7.	____	23.4	9.80
8.	7.00	119	____
9.	4.00	____	12.2
10.	____	37.0	12.0

Given $MA = d_E/d_R$, find the missing quantities.

	MA	d_R	d_E
11.	____	49.2	76.3
12.	7.00	29.7	____
13.	____	29.7	67.4
14.	4.00	____	67.4
15.	3.00	13.7	____

Solve using the given outline.

16. A pole is used to lift a car which fell off a jack. The pivot is 2.00 ft from the car. Two men exert 275 lb of force 8.00 ft from the pivot. What force is applied to the car?

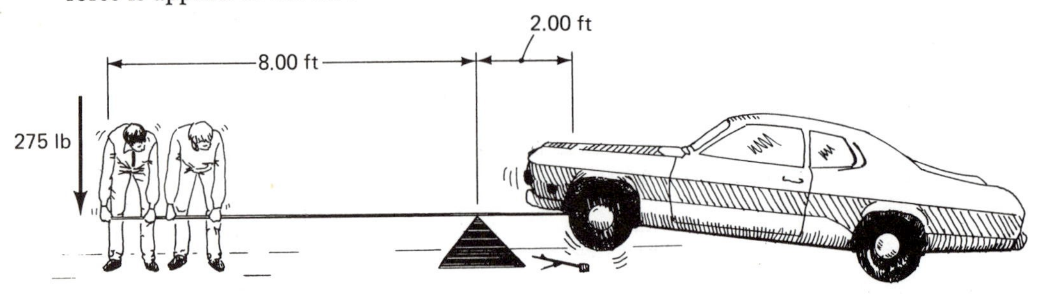

17. Calculate the *MA* of problem 16.
18. A crowbar is used to lift a 100-kg block of concrete. The pivot is 1.00 m from the block. If the man pushes down on the other end of the bar a distance of 2.50 m from the pivot, what force must he apply?
19. Calculate the *MA* of problem 18.
20. A wheelbarrow 6.00 ft long is to be used to haul a 180-lb load. How far from the wheel is the load placed so that a man can lift the load with a force of 45.0 lb?
21. Calculate the *MA* of problem 20.
22. Find the force, F_E, pulling up on the beam holding the sign below.

SKETCH
DATA
$a = 1, b = 2,$
$c = ?$

BASIC EQUATION
$a = bc$

WORKING EQUATION
$c = \frac{a}{b}$

SUBSTITUTION
$c = \frac{1}{2}$

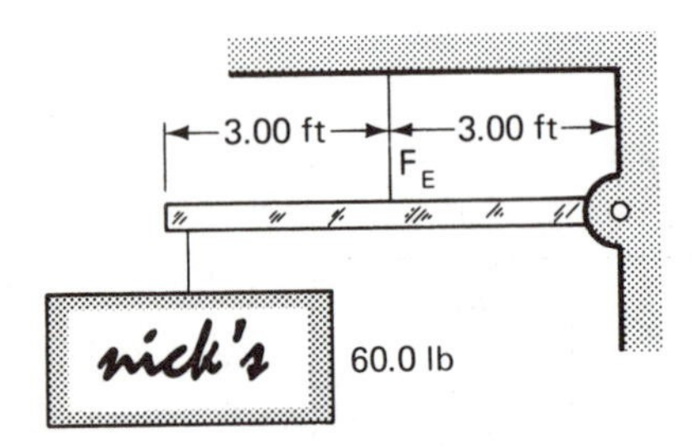

23. Calculate the *MA* of problem 22.

10-7 WHEEL AND AXLE

This simple machine consists of a large wheel attached to an axle so that both turn together.

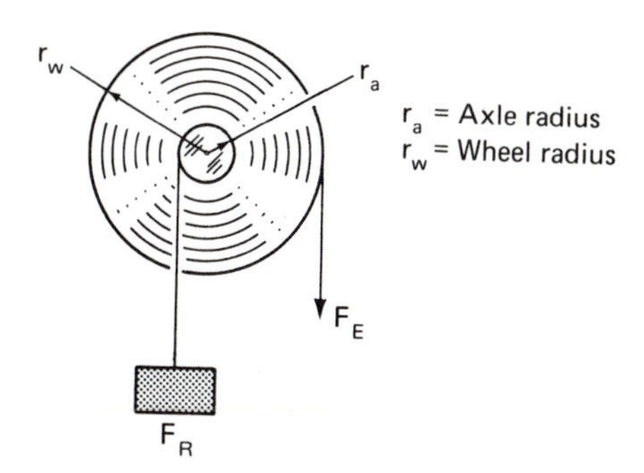

Steering Wheel

$$MA = \frac{\text{radius of wheel}}{\text{radius of axle}} = \frac{r_w}{r_a}$$

Bicycle sprocket
and pedal assembly

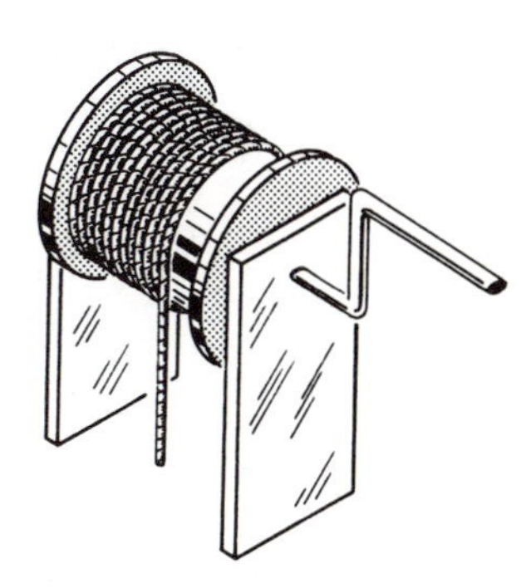

Windlass

Law of simple machines as applied to the wheel and axle (basic equation):

$$F_R \cdot r_a = F_E \cdot r_w$$

EXAMPLE 1: The winch shown at the right has a handle which turns in a radius of 9.00 in. The radius of the drum or axle is 3.00 in. Find the force required to lift a bucket weighing 90.0 lb.

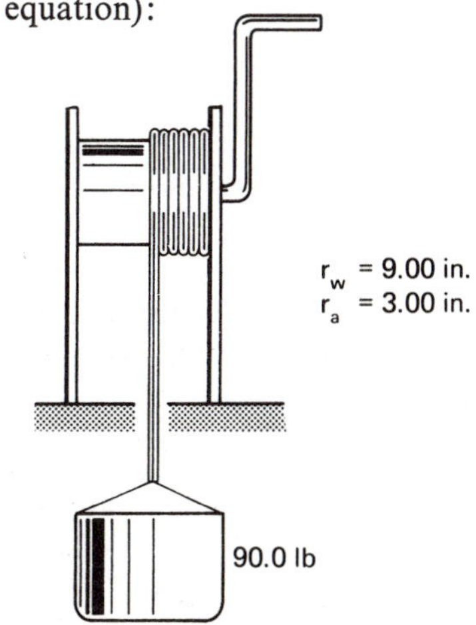

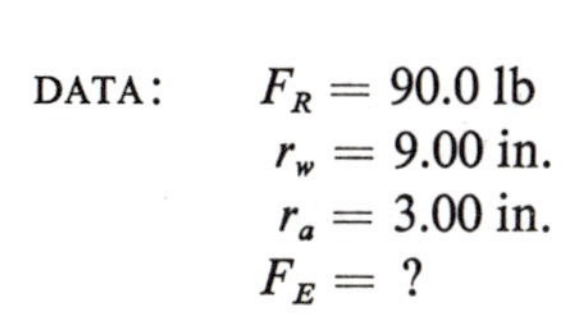

DATA:

$F_R = 90.0$ lb
$r_w = 9.00$ in.
$r_a = 3.00$ in.
$F_E = ?$

BASIC EQUATION: $F_R \cdot r_a = F_E \cdot r_w$

WORKING EQUATION: $F_E = \dfrac{F_R \cdot r_a}{r_w}$

SUBSTITUTION: $F_E = \dfrac{(90.0 \text{ lb})(3.00 \text{ in.})}{(9.00 \text{ in.})} = 30.0 \text{ lb}$

EXAMPLE 2: Calculate the MA of the winch in Example 1.

$$MA = \frac{r_w}{r_a} = \frac{9.00 \text{ in.}}{3.00 \text{ in.}} = \frac{3}{1}$$

PROBLEMS

Given $F_R \cdot r_a = F_E \cdot r_w$, find the missing quantities.

	F_R	F_E	r_a	r_w
1.	20.0	5.30	3.70	____
2.	37.0	12.0	____	112
3.	____	17.5	49.2	76.3
4.	23.4	9.80	____	53.9
5.	119	____	29.7	67.4

Given $MA = r_w/r_a$, find the missing quantities.

	MA	r_w	r_a
6.	7.00	119	____
7.	4.00	____	12.2
8.	____	49.2	31.7
9.	3.00	61.3	____
10.	____	67.4	29.7

11. The radius of the axle of a winch is 3.00 in. The length of the handle (radius of wheel) is 1.50 ft. What weight will be lifted by an effort of 73.0 lb?
12. Calculate the MA of problem 11.
13. A wheel having a radius of 0.700 m is attached to an axle which has a radius of 0.200 m. What force must be applied to the rim of the wheel to raise a weight of 1500 N?

14. Calculate the *MA* of problem 13.
15. What weight can be lifted in problem 13 if a force of 575 N is applied?
16. The diameter of the wheel of a wheel and axle is 10.0 m. If a force of 475 N is to be raised by applying a force of 142 N, find the diameter of the axle.
17. Calculate the *MA* of problem 16.

10-8 *PULLEY*

A pulley is a grooved wheel that turns readily on an axle and is supported in a frame. It can be fastened to a fixed object or it may be fastened to the resistance that is to be moved. If the pulley is fastened to a fixed object, it is called a *fixed pulley*. If the pulley is fastened to the resistance to be moved, it is called a *movable pulley*.

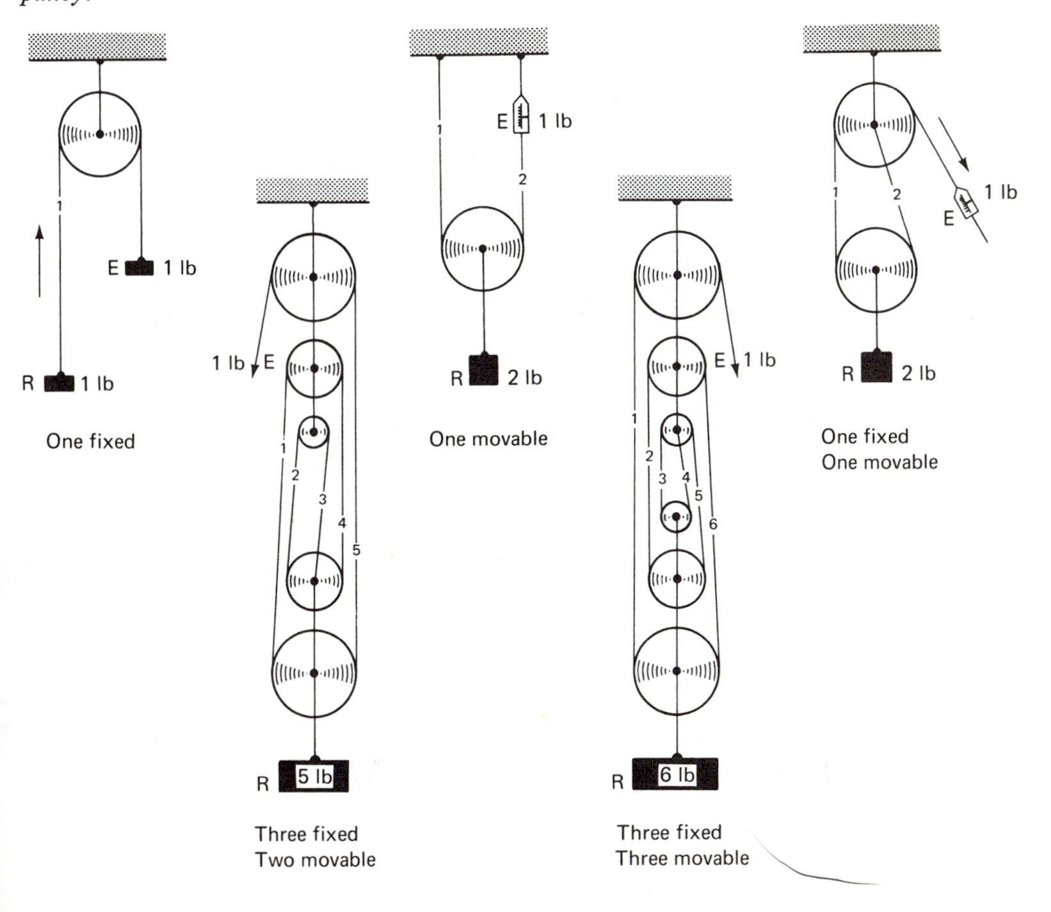

The law of simple machines as applied to pulleys (basic equation):

$$F_R \cdot d_R = F_E \cdot d_E$$

From the last equation,

$$\frac{F_R}{F_E} = \frac{d_E}{d_R} = MA$$

However, when one continuous cord is used, this ratio reduces to be the number of strands holding the resistance in the pulley system. Therefore,

MA = the number of strands holding the resistance.

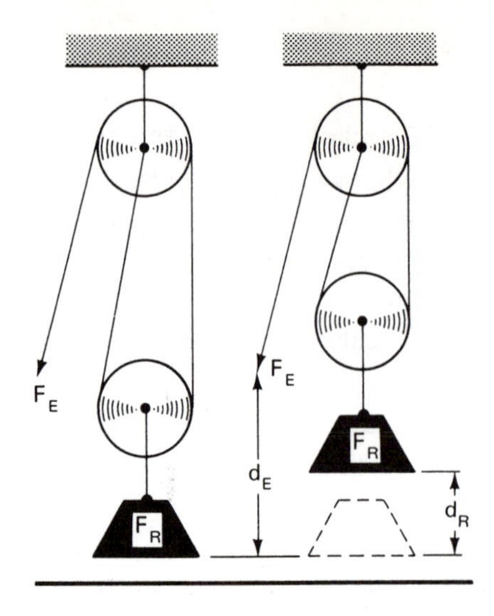

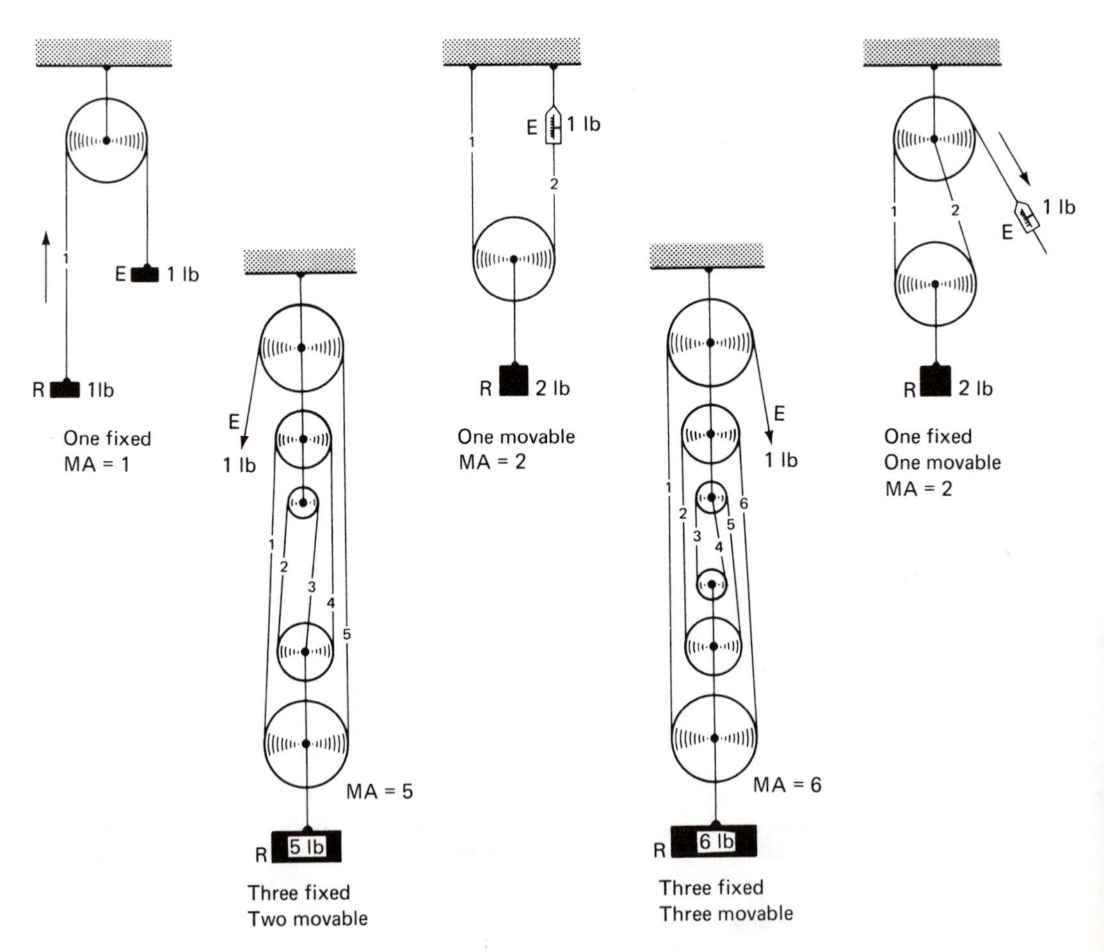

EXAMPLE 1: Draw two different sets of pulleys, each with an MA of 4.

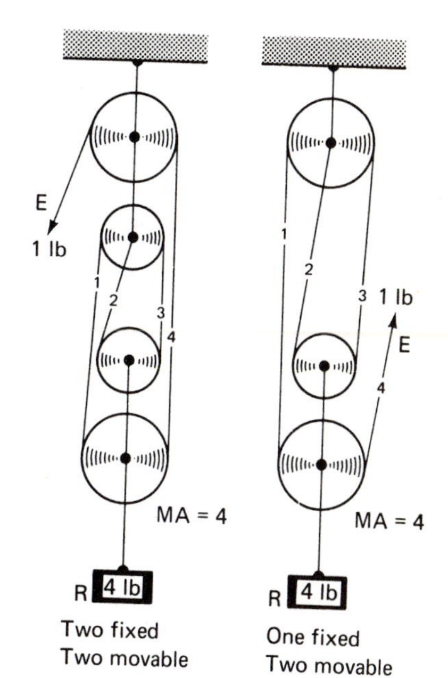

EXAMPLE 2: What effort will lift a resistance of 250 N in the pulley system in Example 1?

DATA: $MA = 4$
$F_R = 250\text{ N}$
$F_E = ?$

BASIC EQUATION: $MA = \dfrac{F_R}{F_E}$

WORKING EQUATION: $F_E = \dfrac{F_R}{MA}$

SUBSTITUTION: $F_E = \dfrac{250\text{ N}}{4} = 62.5\text{ N}$

EXAMPLE 3: If the resistance moves 7.00 ft, what is the effort distance of the pulley system in Example 1?

DATA: $MA = 4$
$d_R = 7.00\text{ ft}$
$d_E = ?$

BASIC EQUATION: $MA = \dfrac{d_E}{d_R}$

WORKING EQUATION: $d_E = d_R\,(MA)$

SUBSTITUTION: $d_E = (7.00\text{ ft})(4)$
$= 28.0\text{ ft}$

EXAMPLE 4: The pulley system shown is used to raise a 650-lb object 25.0 ft. What is the mechanical advantage? What force is exerted?

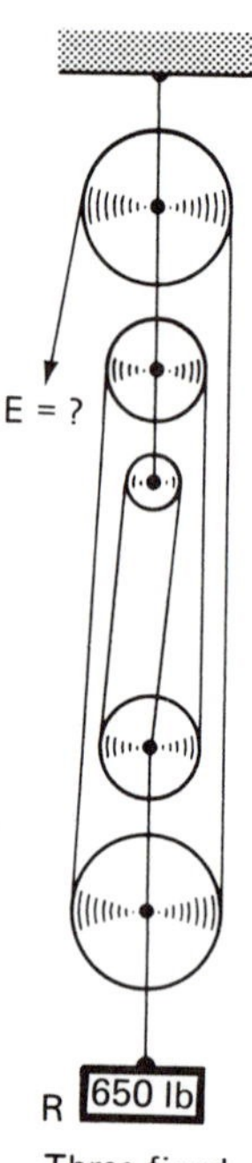

MA = number of strands holding up the resistance
$= 5$

To find the force exerted:

DATA: $d_R = 25.0$ ft
$F_R = 650$ lb
$F_E = ?$

BASIC EQUATION: $MA = \frac{F_R}{F_E}$

WORKING EQUATION: $F_E = \frac{F_R}{MA}$

SUBSTITUTION: $F_E = \frac{650 \text{ lb}}{5}$
$= 130$ lb

PROBLEMS

What is the mechanical advantage of each of the following systems?

1. 2. 3. 4.

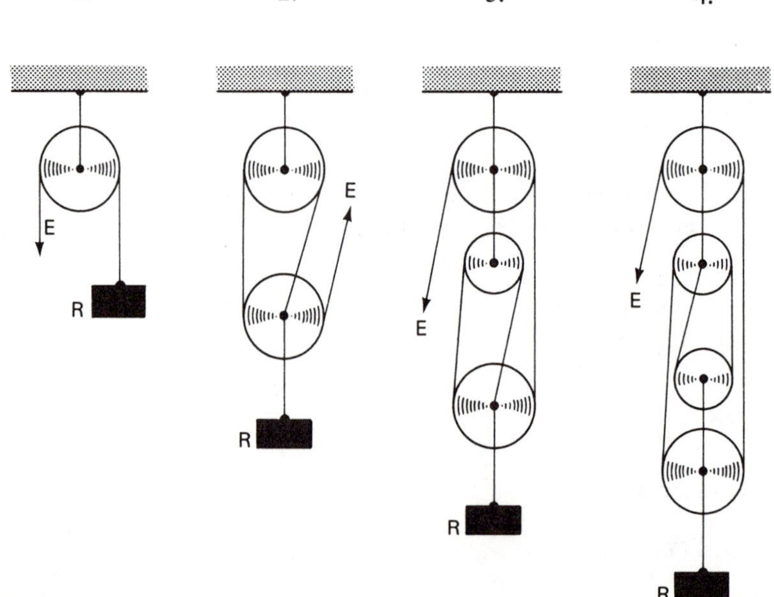

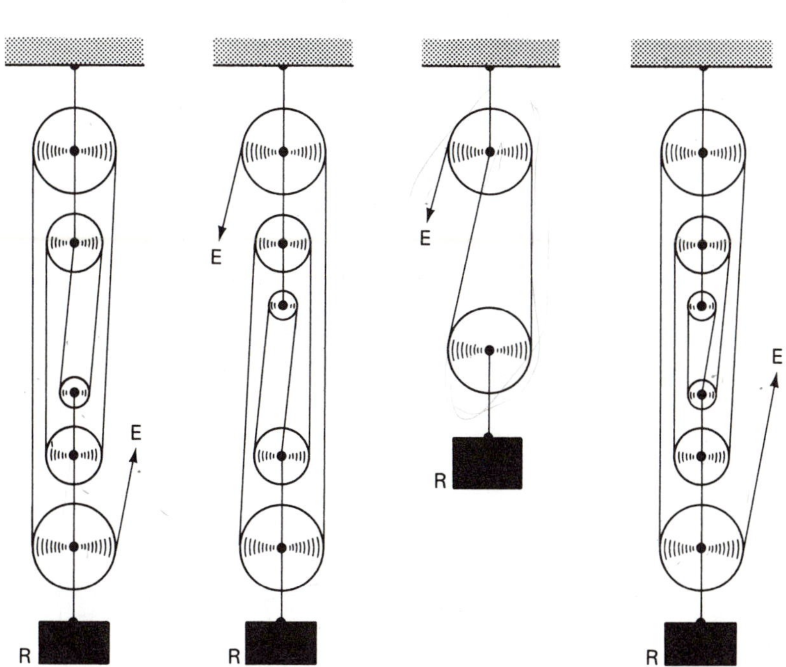

Draw the following pulley systems.

9. One fixed and two movable
10. Two fixed and two movable with a *MA* of 5
11. Three fixed and three movable with a *MA* of 6
12. Four fixed and three movable
13. Four fixed and four movable with a *MA* of 8
14. Three fixed and three movable with a *MA* of 7
15. What is the *MA* of a single movable pulley?
16. What effort will lift a 250-lb weight by using a single movable pulley?
17. If the weight is moved 15.0 ft in problem 16, how many feet of rope must be pulled up by the man exerting the effort?
18. Draw a system consisting of two fixed pulleys and two movable pulleys with a mechanical advantage of 4. If a force of 97.0 N is to be exerted, what weight can be raised?
19. If the weight in problem 18 is raised 20.5 m, what length of rope is pulled?

20. Draw a system of one fixed and two movable pulleys. What is the mechanical advantage of this system?
21. A 400-lb weight is to be lifted 30.0 ft. Using the system in problem 20, find the effort force and effort distance.
22. If an effort force of 65.0 N and an effort distance of 13.0 m is to be applied, find the weight of the resistance and the distance it is moved using the pulley system in problem 20.

10-9 *INCLINED PLANE*

Gangplanks, chutes, and ramps are all examples of the inclined plane. Inclined planes are used to raise heavy objects that are too heavy to lift vertically.

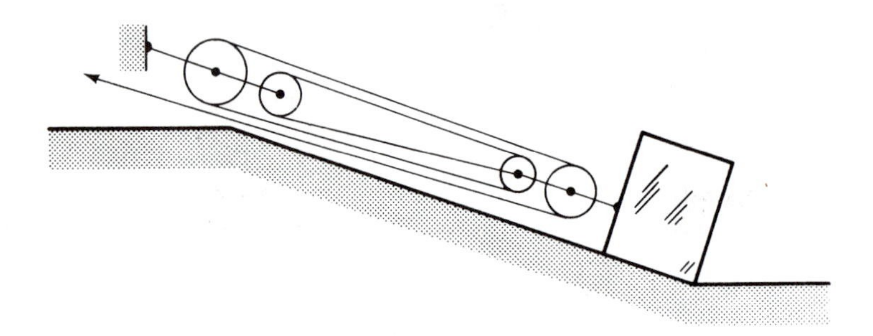

The work done in raising a resistance using the inclined plane equals the resistance times the height. This must also equal the work input which can be found by multiplying the effort times the length of the plane.

$$F_R \cdot d_R = F_E \cdot d_E \qquad \text{(law of machines)}$$

$$F_R \cdot \text{height} = F_E \cdot \text{length}$$

From the above equation:

$$\frac{F_R}{F_E} = \frac{\text{length of plane}}{\text{height of plane}} = MA$$

EXAMPLE 1: A man is pushing a box weighing 375 lb up a ramp 17.0 ft long onto a platform 4.50 ft above the ground. What is the mechanical advantage? What effort is applied?

$$MA = \frac{\text{length of plane}}{\text{height of plane}} = \frac{17.0\,\text{ft}}{4.50\,\text{ft}} = 3.78$$

To find the effort force:

DATA:
$$F_R = 375 \text{ lb}$$
$$MA = 3.78$$
$$F_E = ?$$

BASIC EQUATION: $$MA = \frac{F_R}{F_E}$$

WORKING EQUATION: $$F_E = \frac{F_R}{MA}$$

SUBSTITUTION: $$F_E = \frac{375 \text{ lb}}{3.78} = 99.2 \text{ lb}$$

EXAMPLE 2: What is the shortest ramp that can be used to push a 600-lb resistance onto a platform 3.50 ft high by exerting a force of 72.0 lb?

DATA:
$$F_R = 600 \text{ lb}$$
$$F_E = 72.0 \text{ lb}$$
$$d_R = 3.50 \text{ lb} \qquad d_E = ?$$

BASIC EQUATION: $$F_R \cdot d_R = F_E \cdot d_E$$

WORKING EQUATION: $$d_E = \frac{F_R \cdot d_R}{F_E}$$

SUBSTITUTION: $$d_E = \frac{(600 \cancel{\text{lb}})(3.50 \text{ ft})}{(72.0 \cancel{\text{lb}})}$$
$$d_E = 29.2 \text{ ft}$$

EXAMPLE 3: An inclined plane is 13.0 m long and 7.00 m high. What is the mechanical advantage and what weight can be raised by exerting a force of 22.0 N?

$$MA = \frac{\text{length of plane}}{\text{height of plane}} = \frac{13.0 \cancel{\text{m}}}{7.00 \cancel{\text{m}}} = 1.86$$

To find the weight of the resistance:

DATA:
$$d_R = 7.00 \text{ m}$$
$$d_E = 13.0 \text{ m}$$
$$F_E = 22.0 \text{ N}$$
$$F_R = ?$$

BASIC EQUATION: $$MA = \frac{F_R}{F_E}$$

WORKING EQUATION: $F_R = (F_E)(MA)$

SUBSTITUTION: $F_R = (22.0\ \text{N})(1.86)$
$= 40.9\ \text{N}$

PROBLEMS

Given $F_R \cdot \text{height} = F_E \cdot \text{length}$, find the missing quantities.

	F_R	F_E	Height of Plane	Length of Plane
1.	20.0	5.30	3.40	____
2.	23.4	9.80	____	3.79
3.	119	____	13.2	74.0
4.	____	17.6	0.821	3.79
5.	37.0	12.0	____	112

Given $MA = \dfrac{\text{length of plane}}{\text{height of plane}}$, find the missing quantities.

	MA	Length of Plane	Height of Plane
6.	9.00	3.40	____
7.	____	3.79	0.821
8.	1.30	____	9.72
9.	____	74.0	13.2
10.	17.4	____	13.4

11. An inclined plane is 10.0 m long and 2.50 m high. What is the mechanical advantage?
12. A resistance of 727 N is to be pushed up the plane in problem 11. What effort is needed?
13. An effort of 200 N is applied to push an 815-N resistance up the inclined plane in problem 11. Is the effort enough?
14. A safe is to be lifted onto a truck whose bed is 5.50 ft off the ground. The safe weighs 538 lb. If the effort to be applied is 140 lb, what length of ramp is needed to raise the safe?
15. What is the MA of the inclined plane in problem 14?

16. Another safe weighing only 257 lb is to be raised onto the same truck as in problem 14. If the ramp is 21.1 ft long, what effort is needed?
17. A resistance of 325 N is to be raised by using a ramp 5.76 m long and by applying a force of 75.0 N. How high can it be raised?
18. What is the *MA* of the ramp in problem 17?

10-10 *SCREW*

A screw is an inclined plane wrapped around a cylinder. The jack screw and wood screw are examples of this simple machine.

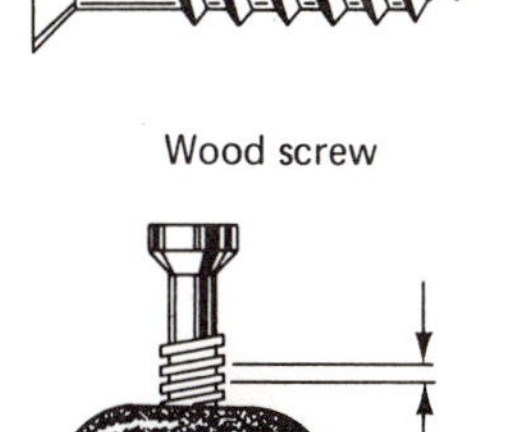

Wood screw

Pitch

Jack screw

The distance a beam rises or the distance the wood screw advances into a piece of wood in one revolution is called the pitch of the screw. Therefore, the pitch of a screw is actually the distance between two successive threads.

From the law of machines:

$$F_R \cdot d_R = F_E \cdot d_E$$

However,

$$d_R = \text{pitch}$$

$$d_E = \text{circumference of the handle of the screwdriver}$$

or,

$$d_E = 2\pi r$$

where r is the radius of the handle of the screwdriver. Therefore,

$$F_R \cdot \text{pitch} = F_E \cdot 2\pi r$$

so,

$$\frac{F_R}{F_E} = \frac{2\pi r}{\text{pitch}} = MA$$

EXAMPLE 1: Find the mechanical advantage of a jack screw having a pitch of 0.125 in. and a handle radius of 12.0 in.

DATA:

Pitch = 0.125 in.
Handle radius = 12.0 in.
$MA = ?$

BASIC EQUATION: $\frac{2\pi r}{\text{pitch}} = MA$

WORKING EQUATION: Same

SUBSTITUTION: $2\pi \frac{(12.0 \text{ in.})}{0.125 \text{ in.}} = 603 = MA$

EXAMPLE 2: What resistance can be lifted using the previous jack screw if an effort of 20.3 lb is exerted?

DATA:

$$MA = 603$$
$$F_E = 20.3$$
$$F_R = ?$$

BASIC EQUATION: $MA = \dfrac{F_R}{F_E}$

WORKING EQUATION: $F_R = (F_E)(MA)$

SUBSTITUTION:

$$F_R = (20.3 \text{ lb})(603)$$
$$F_R = 12{,}200 \text{ lb}$$

EXAMPLE 3: A 4350-lb pickup is to be raised using a jack screw having a pitch of 0.200 in. and a handle radius of 10.0 in. What force must be applied?

DATA:

Pitch = 0.200 in.
Handle radius = 10.0 in.
$F_R = 4350$ lb
$F_E = ?$

BASIC EQUATION: $F_R \cdot \text{pitch} = F_R \cdot 2\pi r$

WORKING EQUATION: $F_E = \dfrac{F_R(\text{pitch})}{2\pi r}$

SUBSTITUTION:

$$F_E = \frac{(4350 \text{ lb})(0.200 \text{ in.})}{2\pi(10.0 \text{ in.})}$$
$$F_E = 13.9 \text{ lb}$$

PROBLEMS

Given $F_R \cdot \text{pitch} = F_E \cdot 2\pi r$, find the missing quantities.

	F_R	F_E	Pitch	r
1.	20.7	5.30	3.70	____
2.	____	17.6	1.30	3.70
3.	23.4	9.80	____	53.9
4.	119	____	2.97	67.4
5.	370	12.0	____	11.2

Given $MA = \frac{2\pi r}{\text{pitch}}$, find the missing quantities.

	MA	r	Pitch
6.	7.00	3.40	_____
7.	_____	3.79	0.812
8.	9.00	_____	0.970
9.	_____	7.40	1.32
10.	13.0	_____	2.10

11. A 3652-lb car is to be raised using a jack screw having eight threads to the inch and a handle 15.0 in. long. What effort must be applied?
12. What is the MA in problem 11?
13. The mechanical advantage of a jack screw is 97.0. If the handle is 13.7 in. long, what is the pitch?
14. How much weight can be raised by applying an effort of 87.0 lb to the jack screw in problem 13?
15. The radius of the handle of a screwdriver represents the effort distance of a wood screw. If a wood screw whose pitch 0.125 in. is advanced into a piece of wood using a screwdriver whose handle is 1.50 in. in diameter, what is the mechanical advantage of the screw?
16. What is the resistance of the wood if 15.0 lb of effort is applied on the wood screw in problem 15?
17. What is the resistance of the wood if 15.0 lb of effort is applied to the wood screw in problem 15 using a screwdriver whose handle is 0.50 in. in diameter?
18. The handle of a jack screw is 0.600 m long. If the mechanical advantage is 78, what is the pitch?
19. How much weight can be raised by applying a force of 430 N to the jack screw handle in problem 18?

SKETCH
DATA
$a = 1, b = 2$
$c = ?$

BASIC
EQUATION
$a = bc$

WORKING
EQUATION
$c = \frac{a}{b}$

SUBSTITUTION
$c = \frac{1}{2}$

10-11 *WEDGE*

A wedge is an inclined plane in which the plane is moved instead of the resistance.

Finding the mechanical advantage of a wedge is not practical because of the large amount of friction. A narrow wedge is easier to drive than a thick wedge. Therefore, the mechanical advantage depends on the ratio of its length to its thickness.

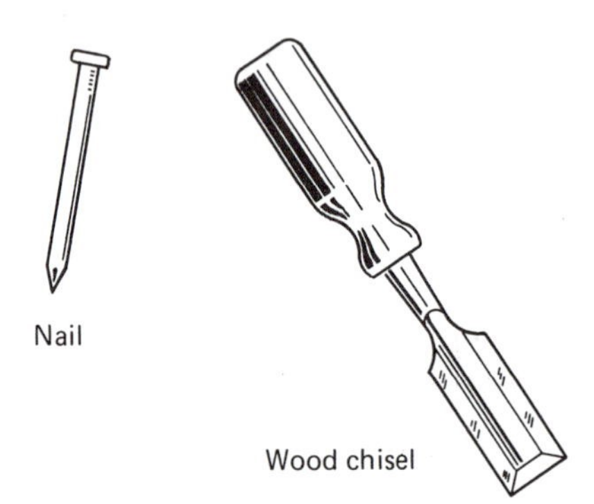

10-12 *COMPOUND MACHINES*

A compound machine is a combination of simple machines. In most all compound machines, the *total mechanical advantage is the product of the mechanical advantages of each simple machine.*

PROBLEMS

1. The solid below being pulled up an inclined plane using the indicated pulley system (called a block and tackle) weighs 400 lb. If the inclined plane is 21.0 ft long and the height of the platform is 7.00 ft, find the mechanical advantage of this compound machine. (Find the mechanical advantage of both simple machines and multiply them.)

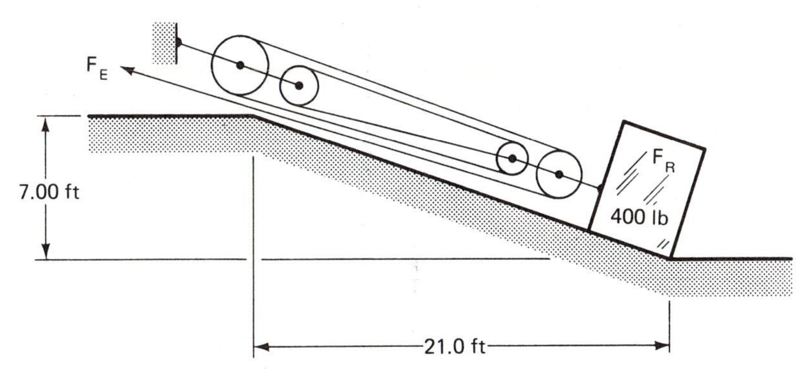

SKETCH
DATA
$a = 1, b = 2,$
$c = ?$

BASIC
EQUATION
$a = bc$

WORKING
EQUATION
$c = \frac{a}{b}$

SUBSTITUTION
$c = \frac{1}{2}$

2. What effort force must be exerted to move the box to the platform in problem 1?
3. Find the mechanical advantage of the compound machine below. The radius of the wheel is 1.00 ft and the radius of the axle is 0.500 ft.

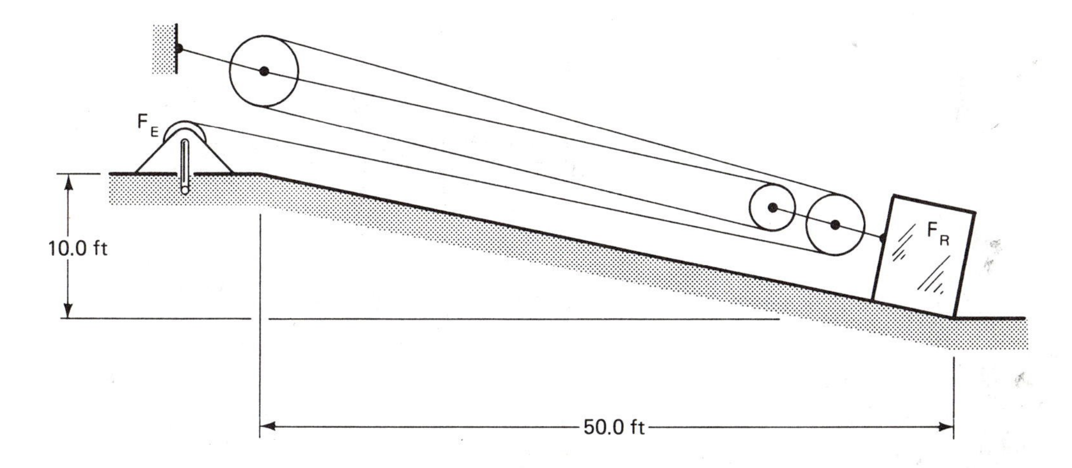

4. If an effort of 300 lb is exerted, what weight can be lifted using the compound machine in problem 3?
5. What effort is required to lift a load of 1.50 tons using the compound machine in problem 3? (1 ton = 2000 lb)

10-13 *DISADVANTAGES OF FRICTION*

Energy is lost in every machine by overcoming friction. This lost energy decreases the efficiency of the machine. That is, more work must be put into a machine than what you get out of the machine. This lost energy becomes heat energy, which results in machine wear or even burning out of certain parts of the machine.

10-14 *ADVANTAGES OF FRICTION*

Although friction causes loss of energy in machines, we could hardly do without it.

Consider:

Walking without friction between your shoes and the ground:

Driving on a road without friction between the road and your tires:

Driving a nail into wood with no friction between the nail and wood:

1. *Friction between rough surfaces is greater than between smooth surfaces.* It is much easier to slide a crate on ice than a wood floor.

2. *Friction increases as the pressure between the surfaces increases.* It is much easier to slide a light crate across the floor than a heavy one.

3. *Starting friction is greater than moving friction.* If you have ever pushed a car by hand, you probably noticed that it took more force to start the car moving than it did to keep it moving.

10-16 *METHODS OF REDUCING FRICTION*

1. Use smoother surfaces.
2. Lubrication provides a thin film between surfaces and thus reduces friction.
3. Teflon greatly reduces friction between surfaces when an oil lubricant is not desirable, such as in electric motors.
4. Substitute rolling friction for sliding friction. Using ball bearings and roller bearings greatly reduces friction.

10-17 *THE EFFECT OF FRICTION ON SIMPLE MACHINES*

Throughout this chapter we have been discussing simple machines, mechanical advantages of simple machines, resistance forces, effort forces, resistance distances, and effort distances.

In the previous calculations, the effect of friction was not considered. Therefore, when we found the effort to lift a resistance of 250 N to be 62.5 N (using the pulley system below), actually an effort of 70.0 N was needed due to the friction of the pulleys on their axles.

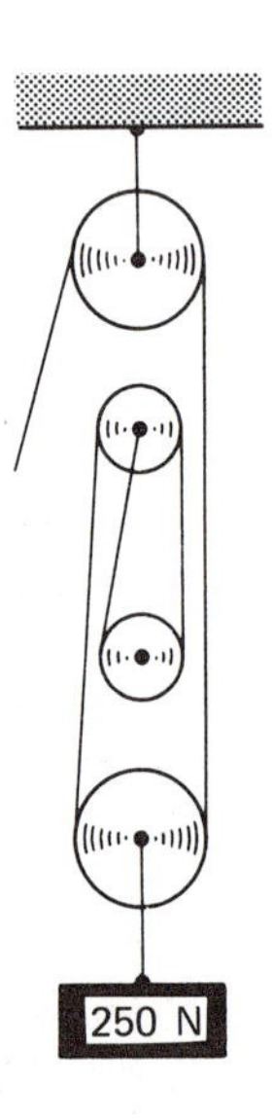

ROTATIONAL MOTION 11

11-1 *MEASUREMENT OF ROTATIONAL MOTION*

Until now we have considered only motion in a straight line—called rectilinear motion. Technicians are faced with many problems, however, that deal with motion along a curved path or objects that are rotating. Though these kinds of motion are similar, we must distinguish them.

Motion along a curved path is called curvilinear motion. A satellite in orbit around the earth is an example of this kind of motion.

Rotary motion occurs when the body itself is spinning. Examples of this kind of motion are the earth spinning on its axis, any turning wheel, a turning driveshaft, and the turning shaft of an electric motor.

We can see a wheel turn but, again, to turn an interesting happening into useful information, we need a system of measurement. The basic unit of measurement in rotary motion is the number of rotations—how often the object goes around. We usually need to know not only the number of rotations but also the time for each rotation. The unit of rotation is the revolution (rev). This is the unit most often used in industry. Another system of measurement divides the circle of rotation into 360 degrees ($360° =$

1 rev). A radian is another angular measure, used by scientists, which is approximately 57.3° or exactly, 360°/2π. Therefore:

$$1 \text{ rev} = 360° = 2\pi \text{ radians}$$

You will find it necessary to refer to the box above when a conversion between systems of measurement is required.

In the automobile industry, technicians are concerned with the *rate* of rotary motion. Recall that in the linear system velocity is the rate of motion. Similarly, we have what is called *angular* velocity in the rotary system. Angular velocity (designated ω—the Greek letter omega) is usually measured in rev/min for relatively slow rotations (e.g., automobile engines) and rev/s for high-speed instruments. It is defined as the rate of *angular* displacement.

$$\omega = \text{angular velocity} = \frac{\text{number of revolutions}}{\text{time}}$$

EXAMPLE 1: A motorcycle wheel turns 3600 times while being ridden for 6.40 min. What is its angular velocity in rev/min?

SKETCH: None

DATA:

$$t = 6.40 \text{ min}$$
$$\text{No. of turns} = 3600$$
$$\omega = ?$$

BASIC EQUATION:

$$\omega = \frac{\text{no. of turns}}{\text{time}}$$

WORKING EQUATION: Same

SUBSTITUTION:

$$= \frac{3600 \text{ rev}}{6.40 \text{ min}}$$
$$= 563 \text{ rev/min}$$

SKETCH
DATA
$a = 1, b = 2$
$c = ?$

BASIC EQUATION
$a = bc$

WORKING EQUATION
$c = \frac{a}{b}$

SUBSTITUTION
$c = \frac{1}{2}$

PROBLEMS

Find the angular velocity in each of the following.

1. Number of turns = 525
   $t = 3.42$ min
   ω in rev/min = ?

2. Number of turns = 7360
   $t = 37.0$ s
   ω in rev/s = ?
3. A motor turns at a rate of 11.0 rev/s. What is its angular velocity in rev/min?
4. Number of turns = 4.00
   $t = 3.00$ s
   ω in rad/s = ?
5. Number of turns = 6370
   $t = 4.00$ min
   ω in rev/s = ?

11-2 *TORQUE*

We have already discussed force in the linear system and defined it as a push or a pull. In rotational systems we have a "twist" which we will call torque. Torque is the tendency to produce change in rotary motion. It is, in rotational systems, similar to force in the linear system. You may already have studied torque in connection with automobile engines. We shall consider the simpler example of pedaling a bicycle.

In pedaling we apply a torque to the sprocket causing it to rotate. The torque developed depends on two factors:

1. The size of the force applied.
2. How far from the axle (shaft) the force is applied.

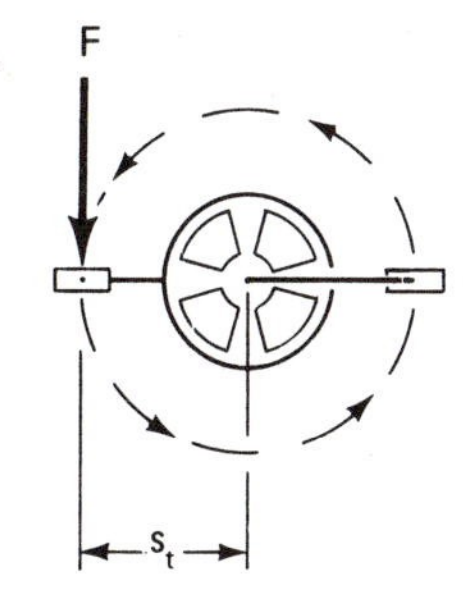

We can express torque with an equation:

$$T = Fs_t$$

where: T = torque (lb ft or N m)
F = applied force (lb or N)
s_t = length of torque arm (ft or m)

Note that s_t, the length of the torque arm, is different from s in the work equation. Recall that s in work is the linear distance over which the force acts.

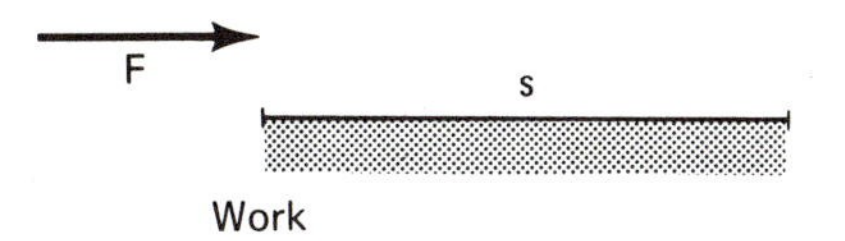

In all torque problems we are concerned with motion about a point or axis of rotation. The torque arm is the *perpendicular* distance from the point of rotation to the direction of the applied force.

Point of rotation
F
Torque
s_t

s_t is always perpendicular to the force in torque problems. Note that in the first figure on page 147, s_t is the distance from the axle to the pedal. The units of torque look similar to those of work, but don't forget the difference between s and s_t.

EXAMPLE 1: A force of 10.0 lb is applied to a bicycle pedal as shown. If the length of the pedal arm is 0.850 ft, what torque is applied to the shaft?

SKETCH:

10.0 lb
0.850 ft

DATA: $F = 10.0$ lb
$s_t = 0.850$ ft $\quad T = ?$

BASIC EQUATION: $T = Fs_t$

WORKING EQUATION: Same

SUBSTITUTION: $T = (10.0 \text{ lb})(0.850 \text{ ft})$
$= 8.50$ lb ft

If the force is not exerted tangent to the circle made by the pedal, the length of the torque arm is *not* the length of the pedal arm. s_t is measured as the perpendicular distance to the force.

Since s_t is therefore shorter, the product $F \cdot s_t$ is smaller and the turning effect, the torque, is less.

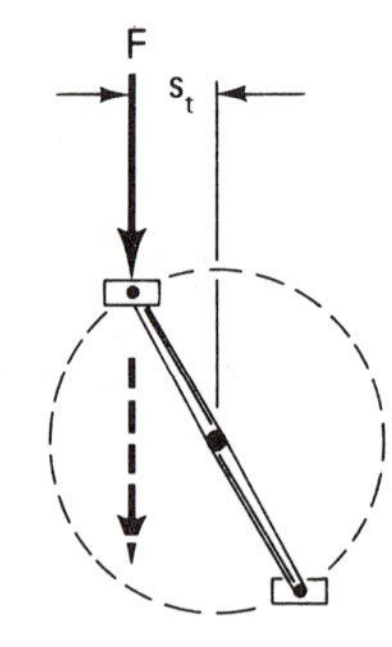

SKETCH
DATA
$a = 1, b = 2,$
$c = ?$
BASIC EQUATION
$a = bc$
WORKING EQUATION
$c = \frac{a}{b}$
SUBSTITUTION
$c = \frac{1}{2}$

PROBLEMS

1. Given: $F = 16.0$ lb
   $s_t = 6.00$ ft
   $T = ?$
2. Given: $F = 100$ N
   $s_t = 0.420$ m
   $T = ?$
3. Given: $T = 35.7$ lb ft
   $s_t = 0.0240$ ft
   $F = ?$

4. If the torque on a shaft of radius 0.0237 m is 38.0 N m, what force is applied to the shaft?

5. If the force applied to a torque wrench 1.50 ft long is 56.2 lb, what torque is indicated by the wrench?

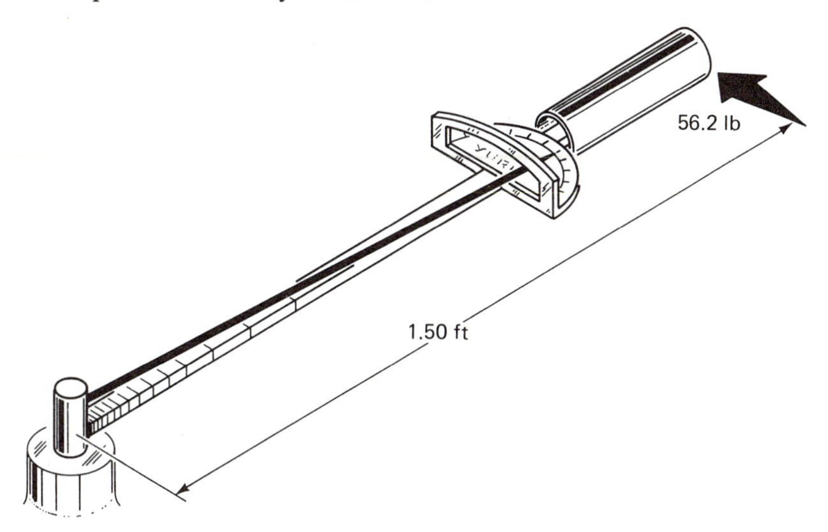

6. If a motorcycle head bolt is torqued to 20.0 lb ft, how long a shaft do we need on a wrench on which we can exert a maximum force of 35.0 lb?
7. If a force of 24.0 lb is applied to a shaft of radius 0.140 ft, what is the torque on the shaft?
8. If a torque of 175 lb ft is needed to free a large rusted-on nut and the length of the wrench is 1.10 ft, what force must be applied to free it?

11-3 *MOTION IN A CURVED PATH*

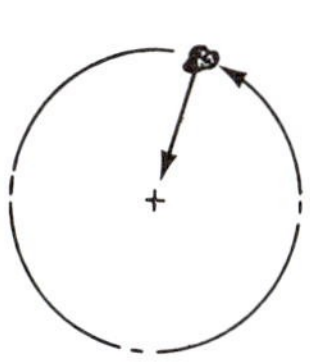

Newton's laws of motion apply to motion along a curved path as well as in a straight line. Recall that a moving body tends to continue in a straight line because of inertia. If we are to cause the body to move in a circle, we must constantly apply a force perpendicular to the line of motion of the body. The simplest example is a rock being swung in a circle on the end of a string. By Newton's first law, the rock tends to go in a straight line but the string exerts a constant force on the rock perpendicular to this line of travel. The resulting path of the rock is a circle.

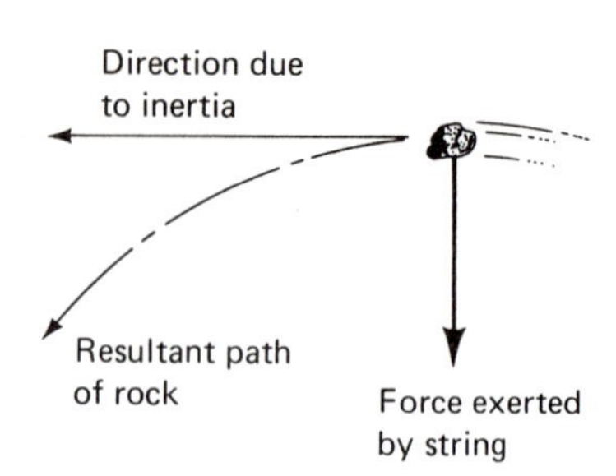

The force of the string on the rock is the *centripetal* (toward the center) force.

If the string should break, however, there would no longer be a centripetal force acting on the rock, and it would fly off tangent to the circle.

The equation for calculating the *centripetal force* on any body moving along a curved path is:

$$F = \frac{mv^2}{r}$$

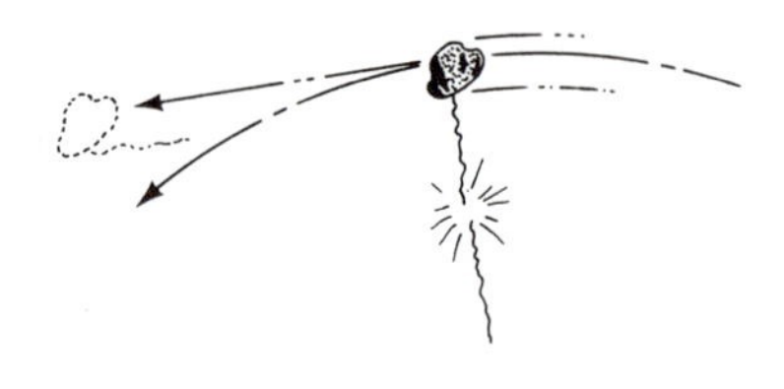

where: F = the centripetal force,
m = mass of the body,
v = velocity of the body,
r = radius of curvature of the path of the body.

What then is the difference between centripetal and centrifugal forces? The term centrifugal force is widely misused to mean centripetal force. Recall Newton's third law of motion that for every action there is an equal and opposite reaction. Centrifugal force is the reaction force. It is present only when there is a centripetal force. It is equal in magnitude to the centripetal force but is in the opposite direction. It is exerted by the rock on the string.

EXAMPLE: An automobile of mass 113 slugs rounds a curve of radius 75.0 ft with a velocity of 44.0 ft/s (30.0 miles per hour). What centripetal force is exerted on the automobile while rounding the curve?

SKETCH:

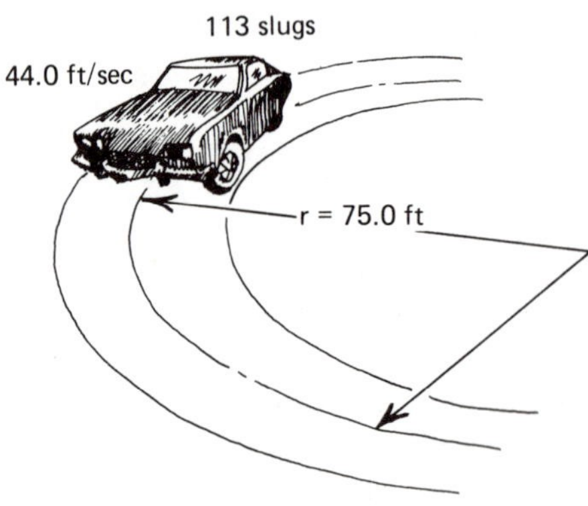

DATA: $m = 113$ slugs
$v = 44.0$ ft/s
$r = 75.0$ ft $\quad F = ?$

BASIC EQUATION: $F = \dfrac{mv^2}{r}$

WORKING EQUATION: Same

SUBSTITUTION: $F = \dfrac{(113 \text{ slugs})(44.0 \text{ ft/s})^2}{75.0 \text{ ft}}$

$= 2920 \text{ slug ft/s}^2 \quad \left(\text{Recall } 1 \text{ lb} = \dfrac{1 \text{ slug ft}}{\text{s}^2}\right)$

$= 2920$ lb

PROBLEMS

1. Given: $m = 64.0$ kg
   $v = 34.0$ m/s
   $r = 17.0$ m
   F in N = ?

2. Given: $m = 11.3$ slugs
   $v = 3.00$ ft/s
   $r = 3.24$ ft
   F in lb = ?

3. Given: $F = 2500$ lb
   $v = 47.6$ ft/s
   $r = 72.0$ ft
   m in slugs = ?

4. Given: $F = 587$ N
   $v = 0.780$ m/s
   $m = 67.0$ kg
   r in m = ?

5. Given: $F = 602$ N
   $m = 63.0$ kg
   $r = 3.20$ m
   v in m/s = ?

6. An automobile of mass 117 slugs follows a curve of radius 79.0 ft with a velocity of 49.3 ft/s. What centripetal force is exerted on the automobile while it is rounding the curve?
7. What is the centripetal force exerted on a 7.12-kg mass moving at a speed of 2.98 m/s in a circle of radius 2.72 m?
8. The centripetal force on a car of mass 97.0 slugs rounding a curve is 3250 lb. If its velocity is 40.1 ft/s, what is the radius of the curve?
9. The centripetal force on a runner is 17.0 lb. If the runner weighs 175 lb and his velocity is 14.0 miles per hour, what is the radius of the curve?

SKETCH
DATA
$a = 1, b = 2,$
$c = ?$

BASIC EQUATION
$a = bc$

WORKING EQUATION
$c = \frac{a}{b}$

SUBSTITUTION
$c = \frac{1}{2}$

11-4 *POWER IN ROTARY SYSTEMS*

Probably the most important aspect of rotational motion to the technician is the power developed in rotary systems. Recall that torque in rotary systems was discussed in Section 11-2. Power, however, must be considered whenever an engine or motor is used to turn a shaft. Some common examples are the use of winches and automobile drive trains.

Earlier we learned that:

$$\text{Power} = \frac{\text{force} \times \text{displacement (work)}}{\text{time}}$$

in the linear system. In the rotary system:

$$\text{Power} = \frac{\text{(torque)(angular displacement)}}{\text{time}}$$
$$= \text{(torque)(angular velocity)} = T\omega$$

Angular displacement is the angle through which the shaft is turned.

In the English system we measure angular displacement by multiplying the number of revolutions by 2π.

$$\text{Angular displacement} = \text{number of revolutions} \times 2\pi$$

For the rotary system:

$$\text{Power} = \frac{\text{torque} \times 2\pi \text{ revolutions}}{\text{time}}$$

When time is in minutes:

$$\text{Power} = \text{torque} \times 2\pi \times \frac{\text{rev}}{\text{min}} \times \frac{1 \text{ min}}{60 \text{ s}}$$

In ft lb/s:

$$\text{Power} = \text{torque} \times \frac{\text{no. of rev}}{\text{min}} \times 0.105 \frac{\text{min}}{\text{rev s}}$$

In hp:

$$\text{Power} = \frac{\text{torque} \times \dfrac{\text{rev}}{\text{min}}}{\dfrac{5250 \text{ ft lb rev}}{\text{min}}}$$

EXAMPLE 1: What power (in ft lb/s) is developed by an electric motor with torque of 5.70 ft lb and speed (ω) of 425 rev/min?

SKETCH: None

DATA: $T = 5.70$ ft lb

$\omega = 425 \dfrac{\text{rev}}{\text{min}} \quad P = ?$

BASIC EQUATION: $P = \text{torque} \times \dfrac{\text{rev}}{\text{min}} \times 0.105 \dfrac{\text{min}}{\text{rev s}}$

WORKING EQUATION: Same

SUBSTITUTION: $P = (5.70 \text{ ft lb})\left(425 \dfrac{\cancel{\text{rev}}}{\cancel{\text{min}}}\right)\left(0.105 \dfrac{\cancel{\text{min}}}{\cancel{\text{rev}}\text{ s}}\right)$

$= 254 \dfrac{\text{ft lb}}{\text{s}}$

EXAMPLE 2: What horsepower is developed by a racing engine with torque of 545 ft lb at 6500 rev/min?

SKETCH: None

DATA: $T = 545$ ft lb

$\omega = 6500 \dfrac{\text{rev}}{\text{min}} \qquad P = ?$

BASIC EQUATION: $\text{Power} = \dfrac{\text{torque} \times \dfrac{\text{rev}}{\text{min}}}{5250 \dfrac{\text{ft lb rev}}{\text{min}}}$

WORKING EQUATION: Same

SUBSTITUTION: $P = \dfrac{(545 \cancel{\text{ft}} \cancel{\text{lb}})\left(6500 \dfrac{\cancel{\text{rev}}}{\cancel{\text{min}}}\right)}{5250 \dfrac{\cancel{\text{ft}} \cancel{\text{lb}} \cancel{\text{rev}}}{\cancel{\text{min}}}}$

$= 675$ hp

In the metric system, angular displacement is measured directly in radians. (Recall 1 rev $= 2\pi$ radians.) The power formula is then:

$$\text{Power} = \text{torque} \times \frac{\text{angular displacement}}{\text{time}}$$
$$= \text{torque} \times \text{angular velocity}$$

Substituting symbols and units,

$$\begin{aligned}\text{Power} &= T\omega \\ &= (\text{N m})\left(\frac{1}{\text{s}}\right) \\ &= \frac{\text{N m}}{\text{s}} \\ &= \text{watt}\end{aligned}$$

Note: When used in problem solving, the radian unit is not used, and ω is expressed with the unit 1/s.

EXAMPLE 3: How many watts of power are developed by a mechanic tightening bolts using 50.0 N m of torque at a rate of 2.10 rad/s?

SKETCH: None

DATA:

$T = 50.0$ N m
$\omega = 2.10/\text{s}$
$P = ?$

BASIC EQUATION: $P = T\omega$

WORKING EQUATION: Same

SUBSTITUTION:

$$\begin{aligned}P &= (50.0 \text{ N m})(2.10/\text{s}) \\ &= 105 \text{ N m/s} \\ &= 105 \text{ W}\end{aligned}$$

PROBLEMS

1. Given: $T = 372$ ft lb
   $\omega = 264 \dfrac{\text{rev}}{\text{min}}$
   $P = ?$ hp
2. Given: $T = 39.4$ N m
   $\omega = 6.70/\text{s}$
   $P = ?$ W
3. Given: $T = 125$ ft lb
   $\omega = 555 \dfrac{\text{rev}}{\text{min}}$
   $P = ? \dfrac{\text{ft lb}}{\text{s}}$
4. What horsepower is developed by an engine with torque of 400 ft lb at 4500 rev/min?
5. What torque must be applied to develop 175 (ft lb)/s of power in a motor if $\omega = 394$ rev/min?

SKETCH
DATA
$a = 1, b = 2,$
$c = ?$

BASIC EQUATION
$a = bc$

WORKING EQUATION
$c = \dfrac{a}{b}$

SUBSTITUTION
$c = \dfrac{1}{2}$

6. What is the angular velocity of a motor developing 649 watts of power with torque of 131 N m?
7. A high-speed industrial drill develops $\frac{1}{2}$ hp at 1600 rev/min. What torque is applied to the drill bit?
8. An engine has torque of 505 ft lb at 3250 rev/min. What power in (ft lb)/s does it develop?
9. What is the angular velocity of a motor developing 350 (ft lb)/s of power with a torque of 96.0 ft lb?
10. What power (in hp) is developed by an engine with torque of 524 ft lb?
    (a) at 3000 rev/min?
    (b) at 6000 rev/min?

GEARS AND PULLEYS 12

12-1 *INTRODUCTION*

Suppose we have two discs touching each other as indicated below. Disc A is driven by a motor and turns disc B by making use of the friction between them.

The relationship between the diameters of the two discs and their number of revolutions is:

$$D \cdot N = d \cdot n$$

where: $D =$ the diameter of the driver disc,
$d =$ the diameter of the driven disc,
$N =$ the number of revolutions of the driver disc,
$n =$ the number of revolutions of the driven disc.

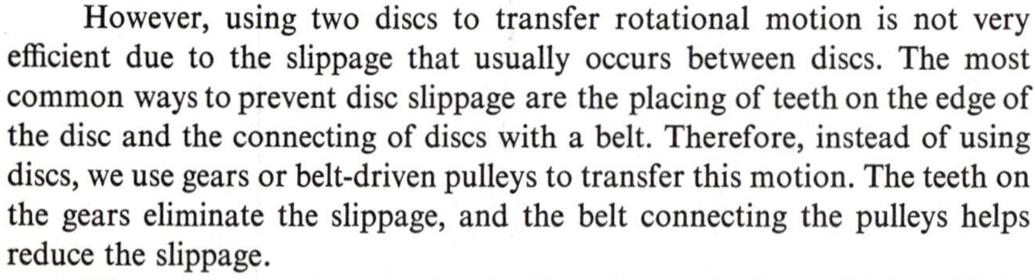

However, using two discs to transfer rotational motion is not very efficient due to the slippage that usually occurs between discs. The most common ways to prevent disc slippage are the placing of teeth on the edge of the disc and the connecting of discs with a belt. Therefore, instead of using discs, we use gears or belt-driven pulleys to transfer this motion. The teeth on the gears eliminate the slippage, and the belt connecting the pulleys helps reduce the slippage.

We can change the equation $D \cdot N = d \cdot n$ to the form $D/d = n/N$ by dividing both sides by dN. The left side indicates the ratio of the diameters of the discs. If the ratio is 2, this means that the larger disc must have a diameter two times the diameter of the smaller disc. The same would apply to gears

and pulleys. The ratio of the diameter of the gears must be 2 to 1, and the ratio of the diameters of the pulleys must be 2 to 1. In fact, the ratio of the number of teeth on the gears must be 2 to 1.

The right side of the equation indicates the ratio of the number of revolutions of the two discs. If the ratio is 2, this means that the smaller disc makes two revolutions while the larger disc makes one revolution. The same would be true for gears and for pulleys connected by a belt.

12-2 *GEARS*

Gears are used to transfer rotational motion from one gear to another. The gear that causes the motion is called the *driver* gear. The gear to which the motion is transferred is called the *driven* gear.

There are many different sizes, shapes, and types of gears. A few examples are shown on the next two pages.

Some of these gears and more are shown in the transmission below.

Precision herringbone gears and racks.

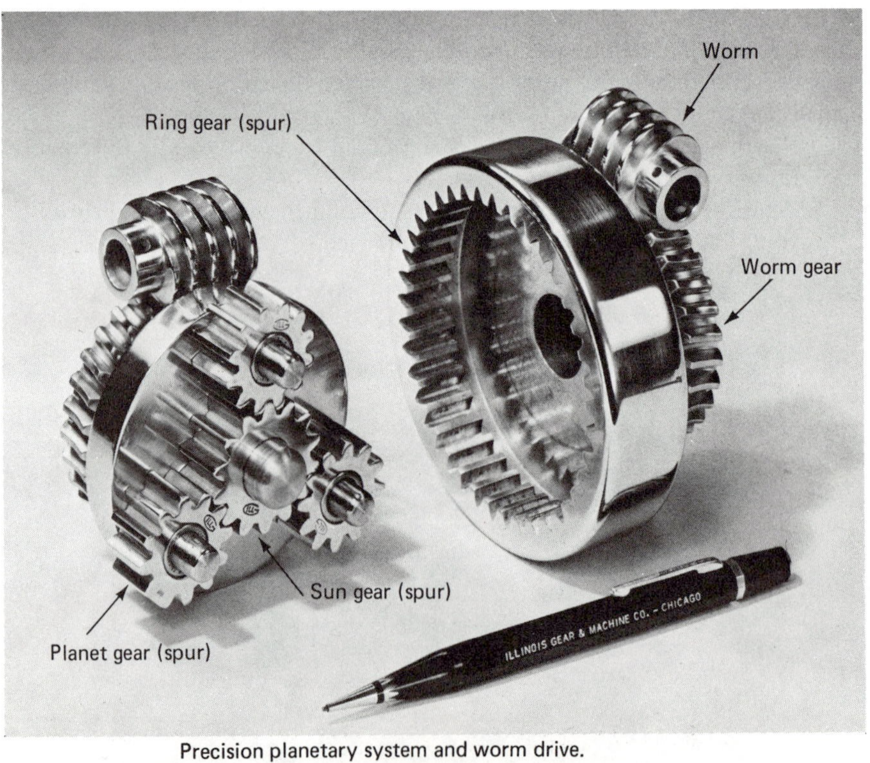

Precision planetary system and worm drive.

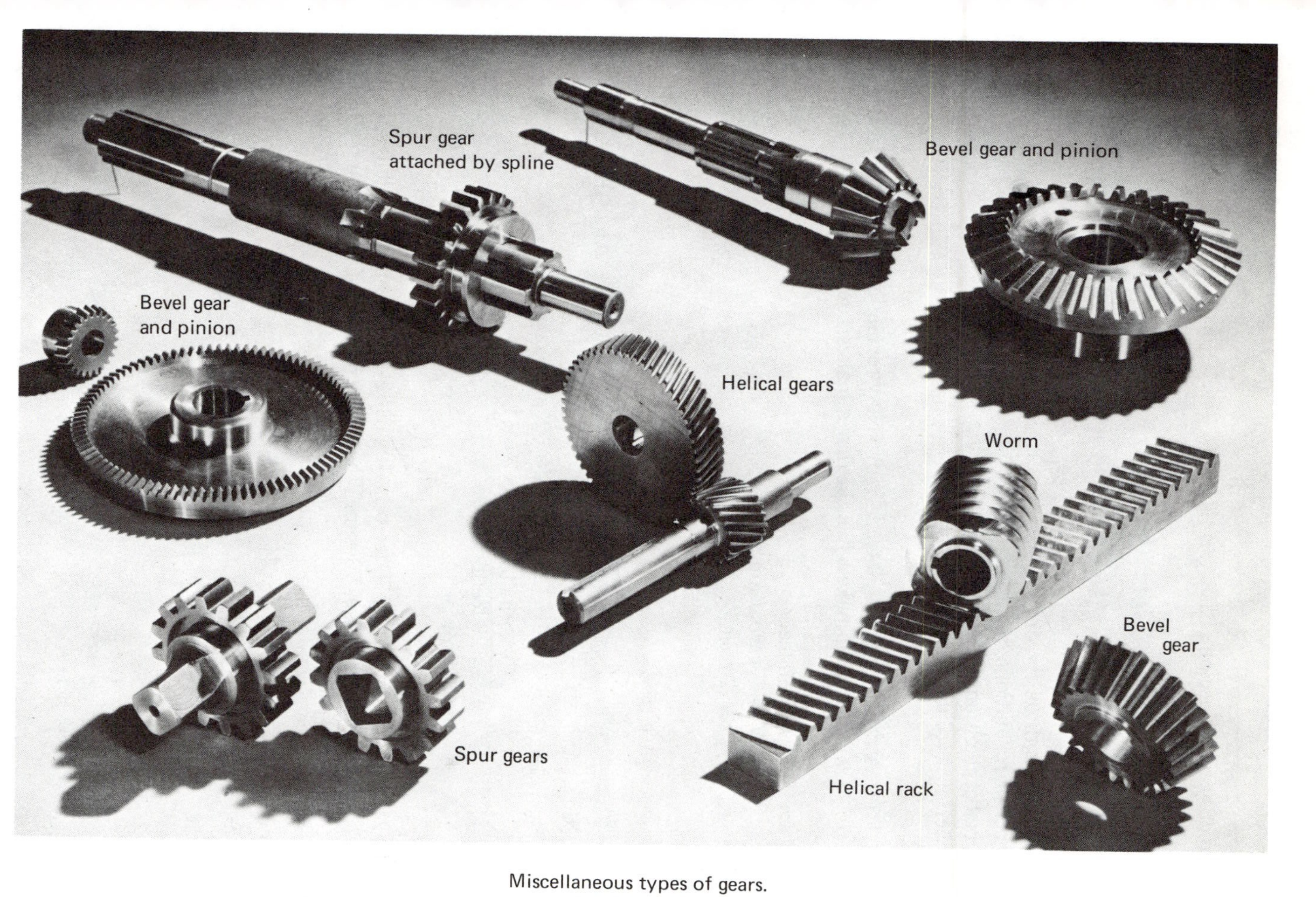

Miscellaneous types of gears.

Photographs courtesy Illinois Gear Division, Wallace-Murray Corporation.

For any type of gear, we use one basic formula:

$$T \cdot N = t \cdot n$$

where: T = number of teeth on the driver,
N = number of revolutions of the driver,
t = number of teeth on the driven,
n = number of revolutions of the driven.

EXAMPLE 1: A driver gear has 30 teeth. How many revolutions does the driven gear with 20 teeth make while the driver makes 1 revolution?

DATA:
$T = 30$ teeth
$N = 1$ revolution
$t = 20$ teeth
$n = ?$

BASIC EQUATION: $T \cdot N = t \cdot n$

WORKING EQUATION: $$n = \frac{T \cdot N}{t}$$

SUBSTITUTION: $$n = \frac{(30 \cancel{\text{teeth}})(1 \text{ rev})}{20 \cancel{\text{teeth}}}$$

$n = 1.5$ rev

EXAMPLE 2: A driven gear of 70 teeth makes 63 revolutions per minute (rpm). The driver gear makes 90 rpm. What is the number of teeth required for the driver gear?

DATA:
$N = 90$ rpm
$t = 70$ teeth
$n = 63$ rpm
$T = ?$

BASIC EQUATION: $T \cdot N = t \cdot n$

WORKING EQUATION: $$T = \frac{t \cdot n}{N}$$

SUBSTITUTION: $$T = \frac{(70 \text{ teeth})(63 \text{ rpm})}{90 \text{ rpm}}$$

$T = 49$ teeth

PROBLEMS

Fill in the blanks using the equation $T \cdot N = t \cdot n$.

	T	N	t	n
1.	17	160	37	____
2.	36	____	14	370
3.	60	1600	____	480
4.	____	100	20	150
5.	12	____	44	234
6.	73	169	99	____
7.	39	23.4	____	70.2
8.	____	39	23	156
9.	19	____	34	93.4
10.	7	324	19	____

11. A driver gear has 38 teeth and makes 85 rpm. What is the rpm of the driven gear with 72 teeth?

12. A motor turning at 1250 rpm is fitted with a gear having 56 teeth. Find the speed of the driven gear if it has 66 teeth.

13. A gear running at 250 rpm meshes with another revolving at 100 rpm. If the smaller gear has 30 teeth, how many teeth does the larger gear have?

Fill in the blanks:

	No. of teeth		rpm	
	Driver	Driven	Driver	Driven
14.	58	____	230	145
15.	____	150	240	120
16.	190	240	____	162
17.	70	____	420	700
18.	80	65	260	____
19.	____	80	480	780

20. A driven gear with 40 teeth makes 154 rpm. How many teeth must the driver have if it makes 220 rpm?
21. Two gears have a speed ratio of 4.6 to 1. If the smaller gear has 15 teeth, what must be the number of teeth on the larger gear?
22. What size gear should be mated with a 15-tooth pinion to achieve a speed reduction of 10 to 3?

12-3 *GEAR TRAINS*

When two gears mesh as shown below,* they turn in opposite directions. If gear *A* turns clockwise, gear *B* turns counterclockwise. If gear *A* turns counterclockwise, gear *B* turns clockwise.

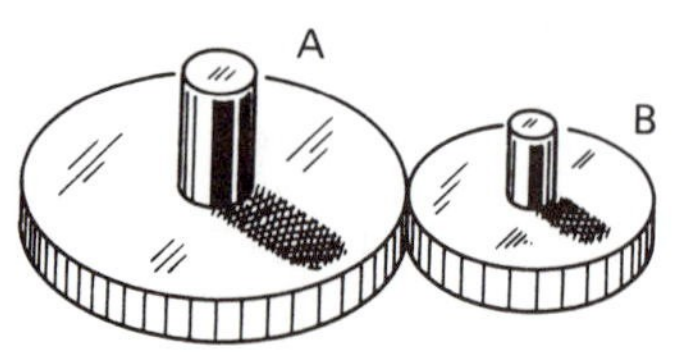

If a third gear is inserted between the two, as below, then gears *A* and *B* are rotating in the same direction. This third gear is called an *idler*, and such an arrangement of gears is called a *gear train.*

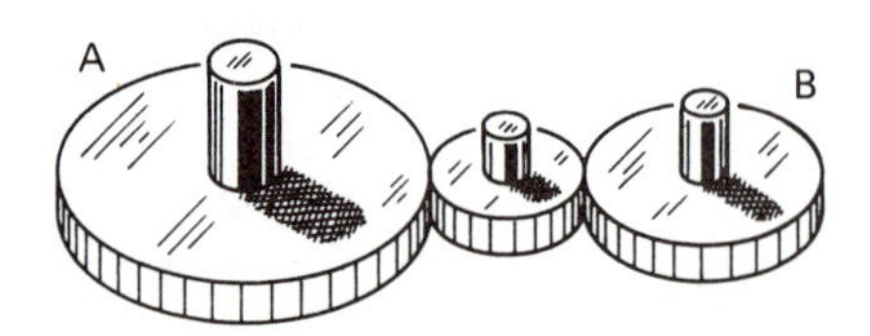

RULE

When the number of shafts in a gear train is odd (such as 1, 3, 5, . . .), *the first gear and last gear rotate in the same direction. When the number of shafts is even, the gears rotate in opposite directions.*

---

*Although gears have teeth, in technical work they are usually shown as cylinders.

Example 1:

When a very complicated gear train is considered, the relationship between revolutions and number of teeth is still present. This relationship is: The number of revolutions of the first driver times the product of the numbers of teeth of all the drivers equals the number of revolutions of the final driven times the product of the number of teeth on all the driven gears.

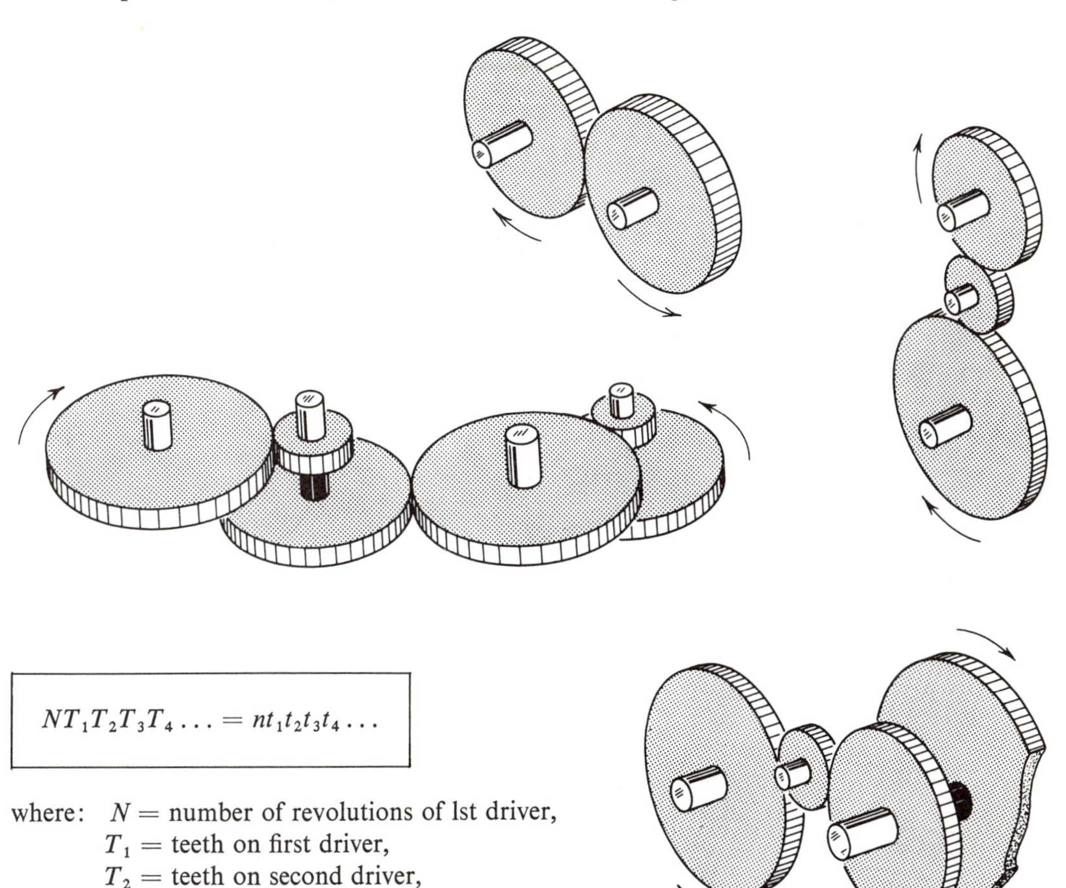

$$NT_1T_2T_3T_4 \ldots = nt_1t_2t_3t_4 \ldots$$

where: N = number of revolutions of lst driver,
T_1 = teeth on first driver,
T_2 = teeth on second driver,
T_3 = teeth on third driver,
T_4 = teeth on fourth driver,
n = number of revolutions of last driven,
t_1 = teeth on first driven,
t_2 = teeth on second driven,
t_3 = teeth on third driven,
t_4 = teeth on fourth driven.

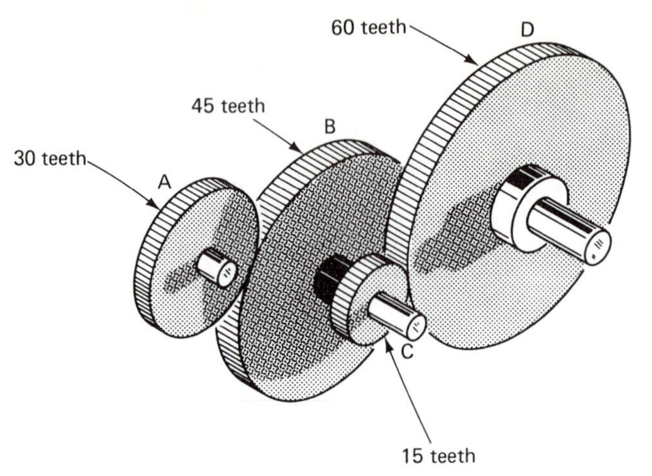

EXAMPLE 2: Find the number of revolutions per minute of gear D if gear A rotates at 20 revolutions per minute.

Gears A and C are drivers and gears B and D are driven.

DATA:

$$N = 20 \text{ rpm}$$
$$T_1 = 30 \text{ teeth}$$
$$T_2 = 15 \text{ teeth}$$
$$t_1 = 45 \text{ teeth}$$
$$t_2 = 60 \text{ teeth}$$
$$n = ?$$

BASIC EQUATION: $NT_1T_2 = nt_1t_2$

WORKING EQUATION:

$$n = \frac{NT_1T_2}{t_1t_2}$$

SUBSTITUTION:

$$n = \frac{(20 \text{ rpm})(30 \text{ teeth})(15 \text{ teeth})}{(45 \text{ teeth})(60 \text{ teeth})}$$
$$n = 3.33 \text{ rpm}$$

EXAMPLE 3: Find the revolutions per minute of gear D in the train below.

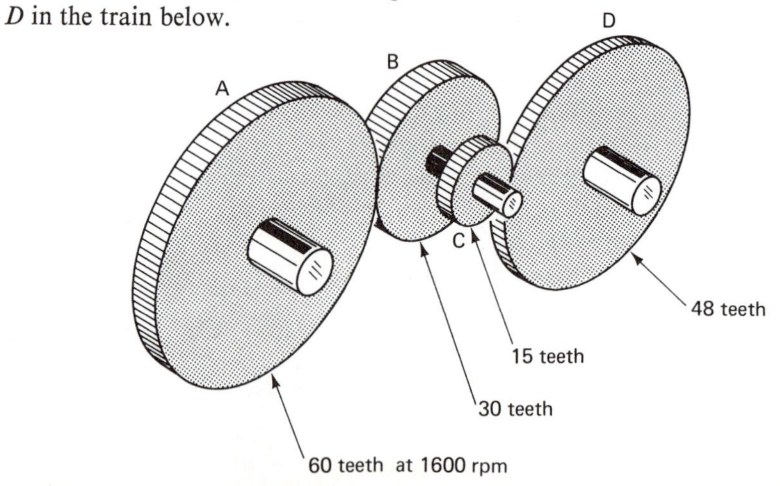

Gears A and C are drivers and gears B and D are driven.

DATA:

$N = 1600$ rpm
$T_1 = 60$ teeth
$T_2 = 15$ teeth
$t_1 = 30$ teeth
$t_2 = 48$ teeth
$n = ?$

BASIC EQUATION: $NT_1T_2 = nt_1t_2$

WORKING EQUATION: $$n = \frac{NT_1T_2}{t_1t_2}$$

SUBSTITUTION: $$n = \frac{(1600 \text{ rpm})(60 \cancel{\text{teeth}})(15 \cancel{\text{teeth}})}{(30 \cancel{\text{teeth}})(48 \cancel{\text{teeth}})}$$

$n = 1000$ rpm

EXAMPLE 4: In the gear train below, find the speed in rpm of gear A.

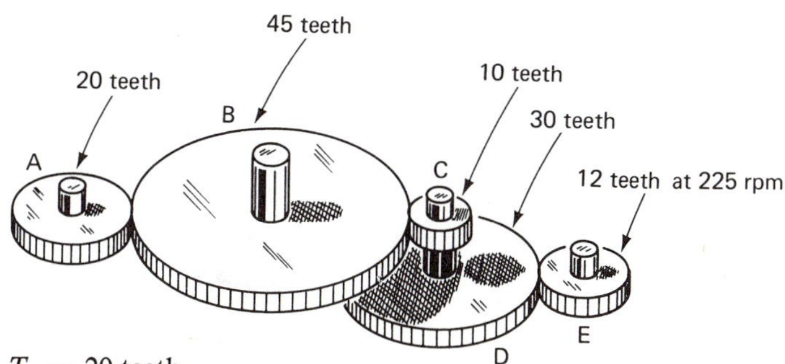

DATA:

$T_1 = 20$ teeth
$T_2 = 45$ teeth
$T_3 = 30$ teeth
$t_1 = 45$ teeth
$t_2 = 10$ teeth
$t_3 = 12$ teeth
$n = 225$ rpm
$N = ?$

Gear B is both a driver and a driven gear.

BASIC EQUATION: $NT_1T_2T_3 = nt_1t_2t_3$

WORKING EQUATION: $$N = \frac{nt_1t_2t_3}{T_1T_2T_3}$$

SUBSTITUTION: $$N = \frac{(225 \text{ rpm})(45 \cancel{\text{teeth}})(10 \cancel{\text{teeth}})(12 \cancel{\text{teeth}})}{(20 \cancel{\text{teeth}})(45 \cancel{\text{teeth}})(30 \cancel{\text{teeth}})}$$

$N = 45$ rpm

In a gear train, when a gear is both a driver gear and a driven gear, it may be left out of the computation.

EXAMPLE 5: The problem in Example 4 could have been worked as follows since gear *B* is both a driver and a driven.

BASIC EQUATION: $NT_1T_3 = nt_2t_3$

WORKING EQUATION: $$N = \frac{nt_2t_3}{T_1T_3}$$

SUBSTITUTION: $$N = \frac{(225\text{ rpm})(10\ \cancel{\text{teeth}})(12\ \cancel{\text{teeth}})}{(20\ \cancel{\text{teeth}})(30\ \cancel{\text{teeth}})}$$

$$N = 45\text{ rpm}$$

PROBLEMS

If gear *A* turns in a clockwise motion, determine the motion of gear *B* in each of the following:

1.

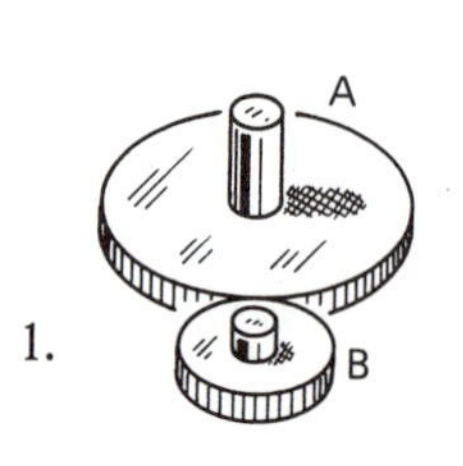

2.

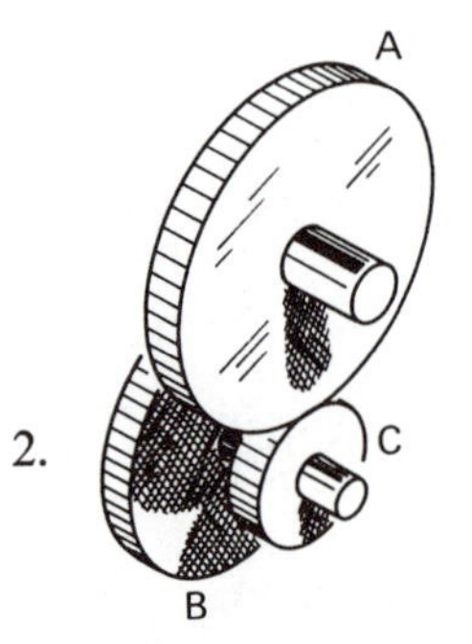

3.

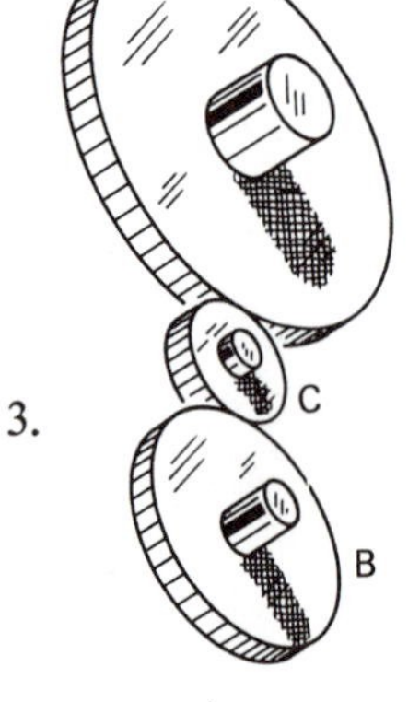

4.

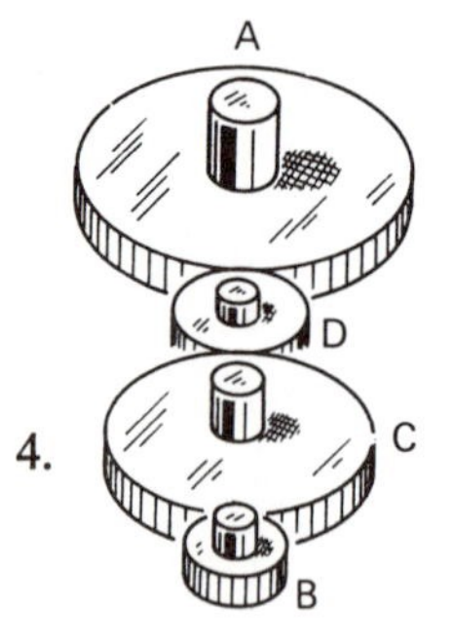

5.

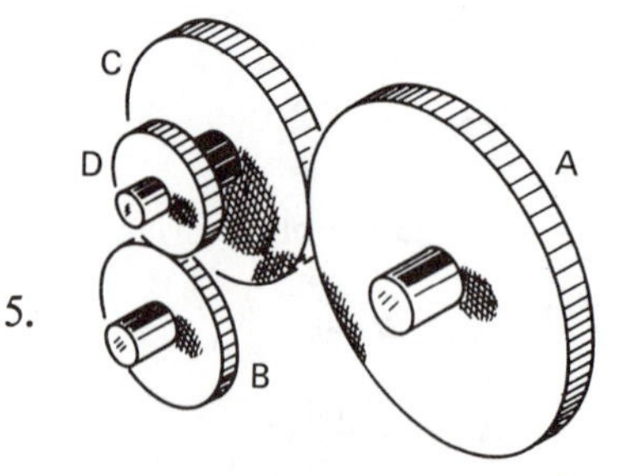

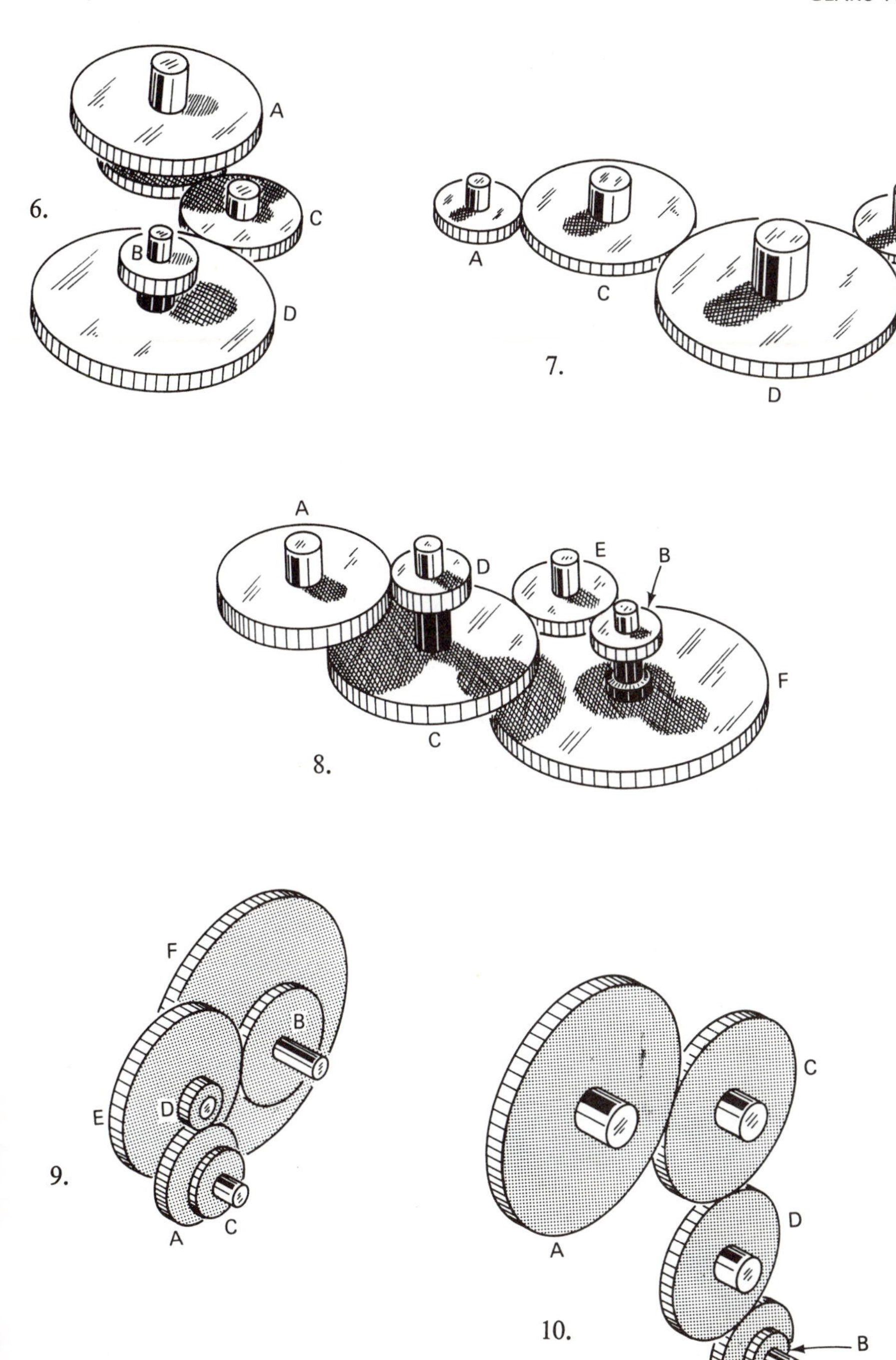
6.
A
B
C
D
7.
A
C
D
E
B
8.
A
D
E
B
F
C
9.
F
B
E
D
A
C
10.
A
C
D
B
E

Find the speed in rpm of gear D in the following figures (11–15):

11.

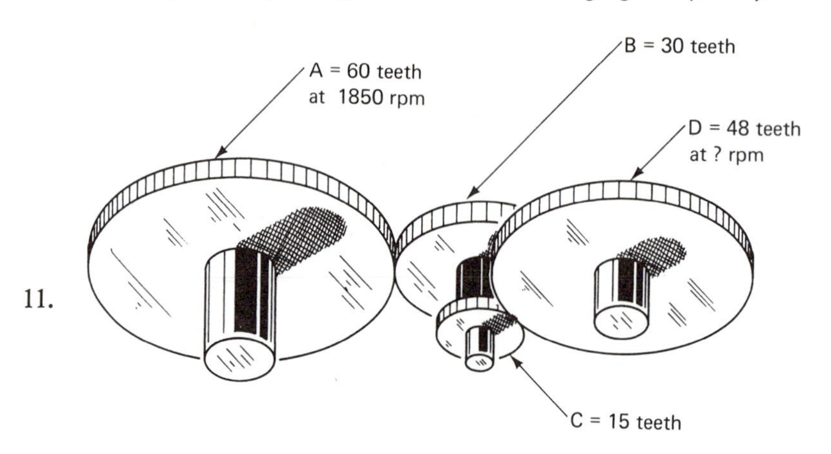

12.

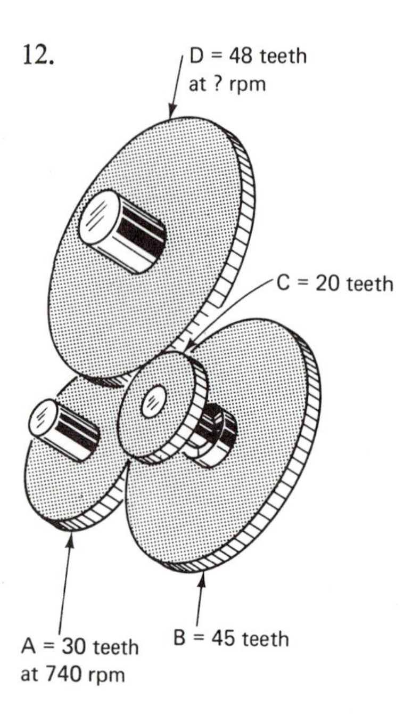

13.

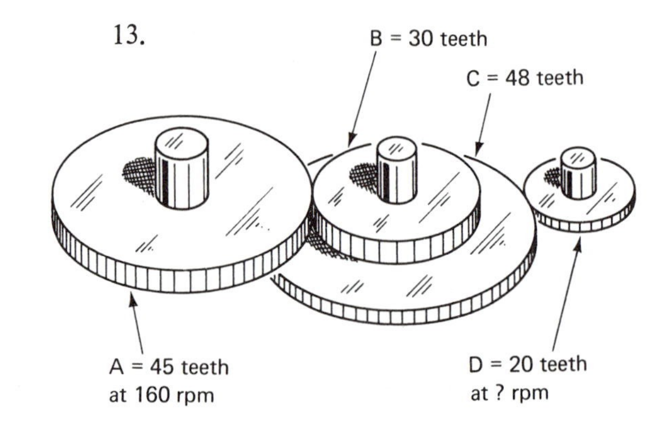

14.

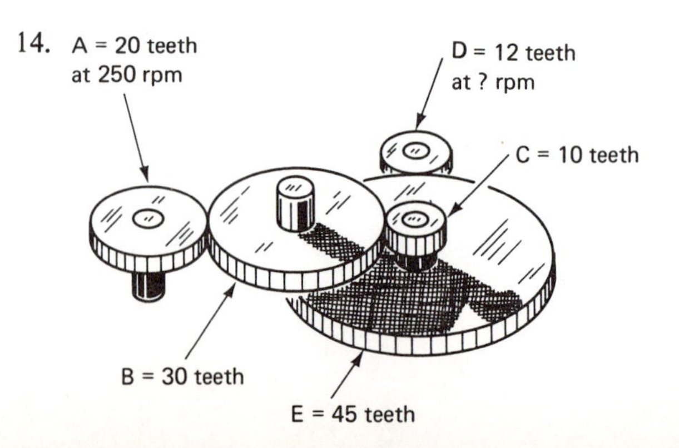

15.

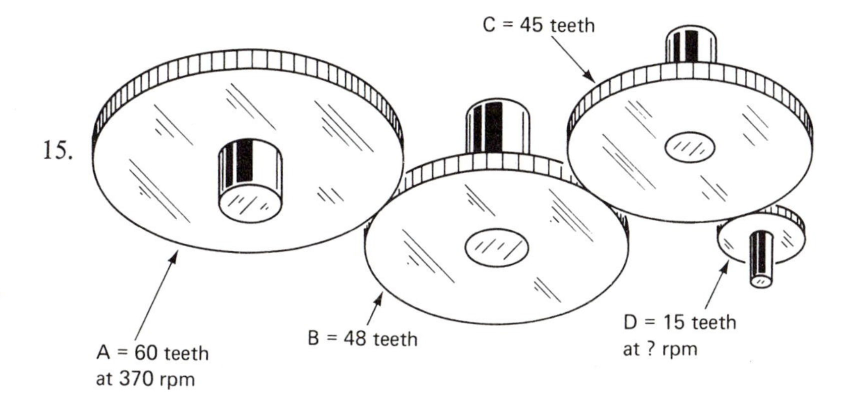

Find the number of teeth for gear D in each of the following:

16.

17.

18.

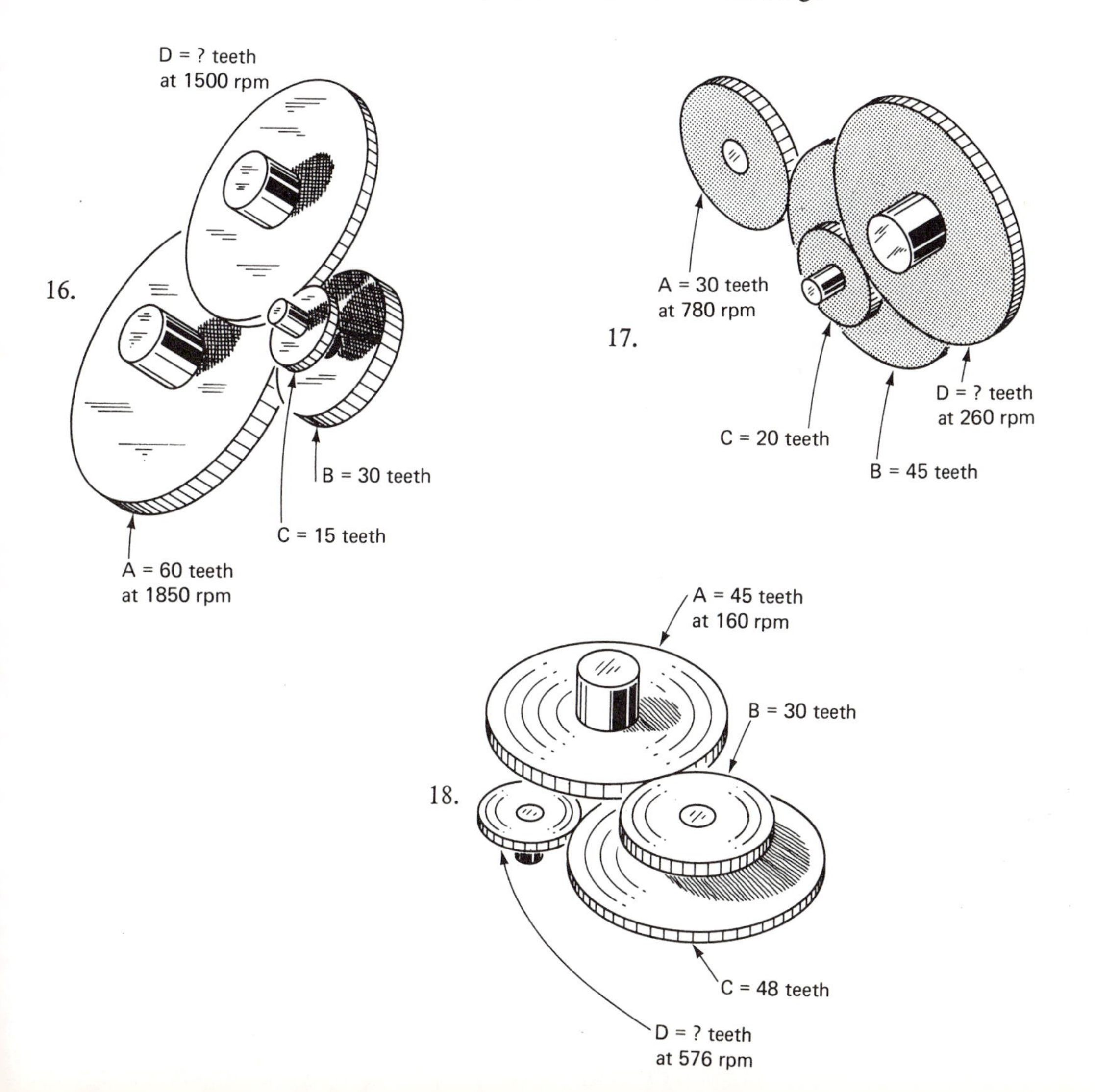

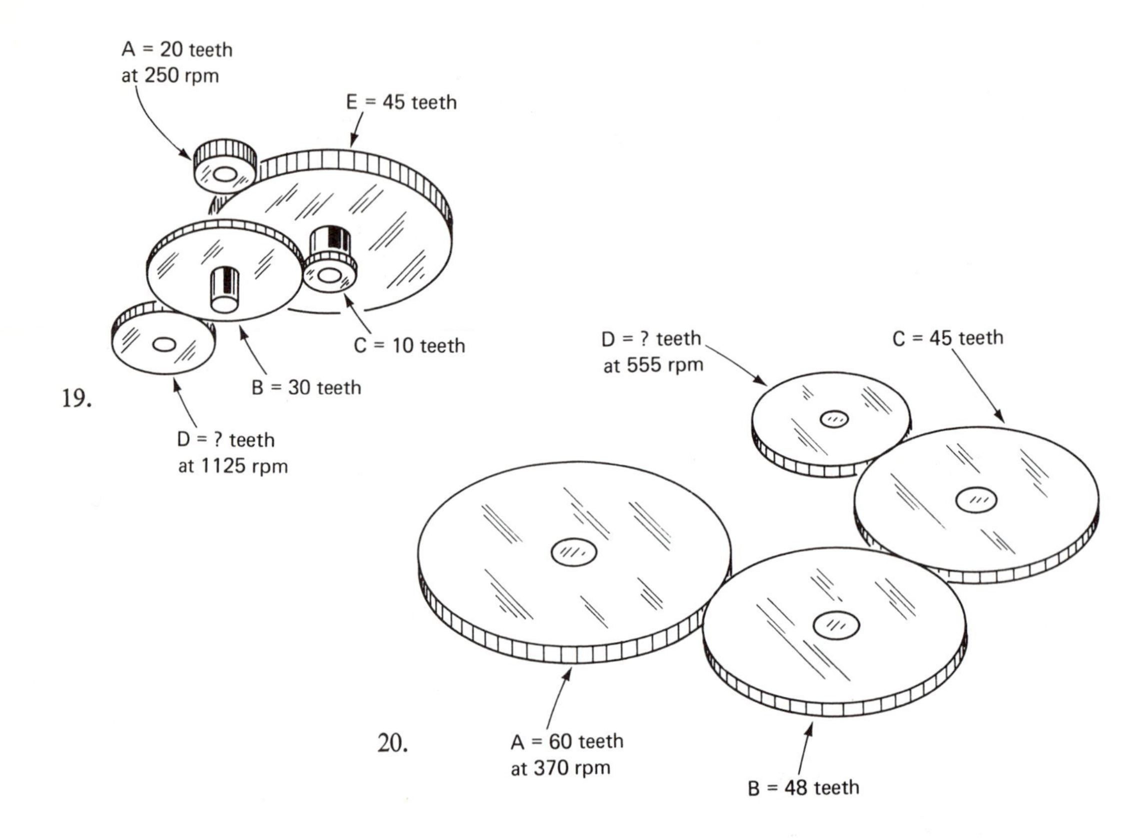

12-4 *PULLEYS CONNECTED WITH A BELT*

Pulleys connected with a belt are used to transfer rotational motion from one shaft to another.

Two pulleys connected with a belt have a relationship similar to gears.

Assuming no slippage, when two pulleys are connected

$$D \cdot N = d \cdot n$$

where: $D =$ the diameter of the driver pulley,
$N =$ the number of revolutions per minute of the driver pulley,
$d =$ the diameter of the driven pulley,
$n =$ the number of revolutions per minute of the driven pulley.

The preceding equation may be generalized in the same manner as for gear trains to get

$$ND_1D_2D_3 \ldots = nd_1d_2d_3 \ldots$$

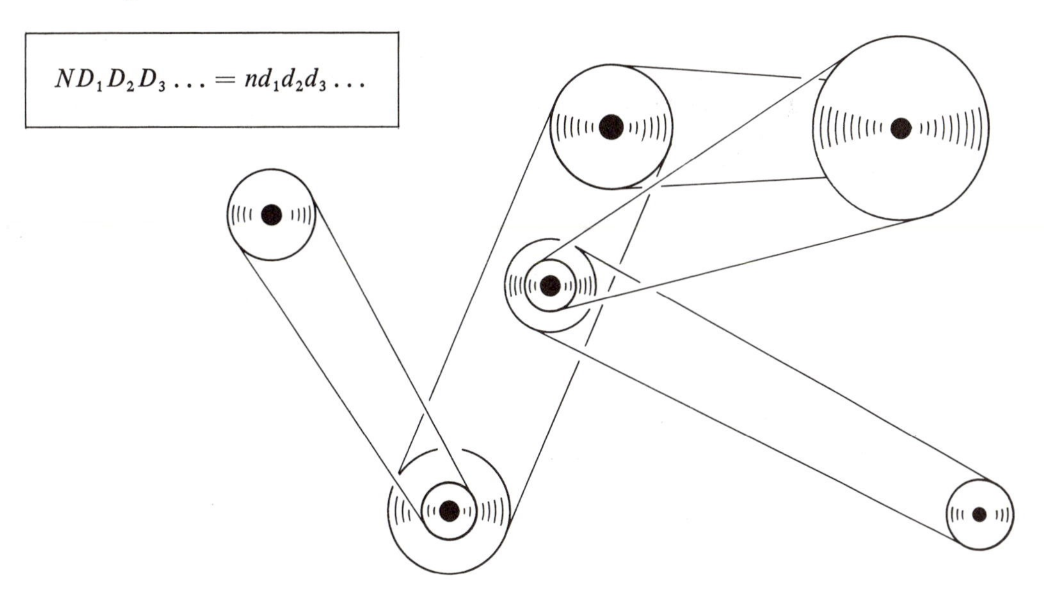

Power Transmission

EXAMPLE: Find the speed in rpm of pulley A below.

DATA:

$D = 6$ in.
$d = 30$ in.
$n = 350$ rpm
$N = ?$

BASIC EQUATION: $D \cdot N = d \cdot n$

WORKING EQUATION: $N = \dfrac{dn}{D}$

SUBSTITUTION: $N = \dfrac{(30 \text{ in.})(350 \text{ rpm})}{6 \text{ in.}}$

$N = 1750$ rpm

When two pulleys are connected with an open-type belt, the pulleys turn in the same direction. When two pulleys are connected with a cross-type belt, the pulleys turn in opposite directions. This is illustrated at the top of page 172.

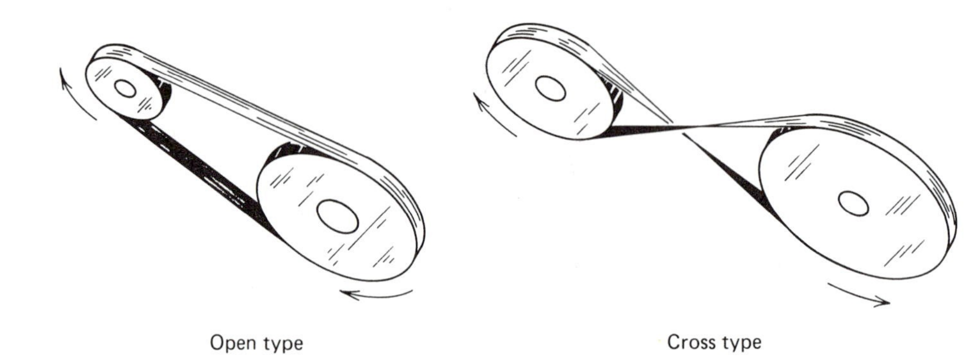

Open type   Cross type

PROBLEMS

Find the missing quantities using $D \cdot N = d \cdot n$.

	D	N	d	n
1.	18	1500	14	____
2.	8	____	9	972
3.	12	1800	6	____
4.	____	2250	9	1125
5.	49	1860	____	620

6. The diameter of a driving pulley is 6.5 in. and revolves at 1650 rpm. At what speed will the driven pulley revolve if it is 26 in. in diameter?
7. The diameter of a driving pulley is 25 cm and makes 120 rpm. At what speed will the driven pulley turn if it is 42 cm in diameter?
8. The diameter of a driving pulley is 18 in. and makes 600 rpm. What is the diameter of the driven pulley if it rotates at 360 rpm?
9. A driving pulley rotates at 440 rpm. The diameter of the driven pulley is 15 in. and makes 680 rpm. What is the diameter of the driving pulley?
10. The radius of a driving pulley is 4 in. and rotates at 120 rpm. The radius of the driven pulley is 6 in. Find the rpm of the driven pulley.

Determine the direction of pulley B in each of the following:

11. 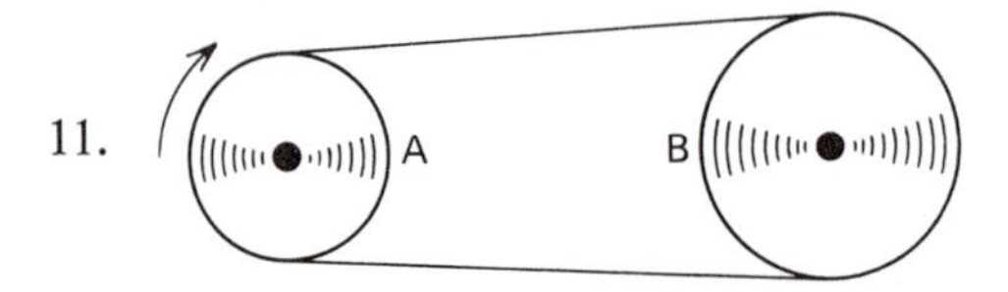

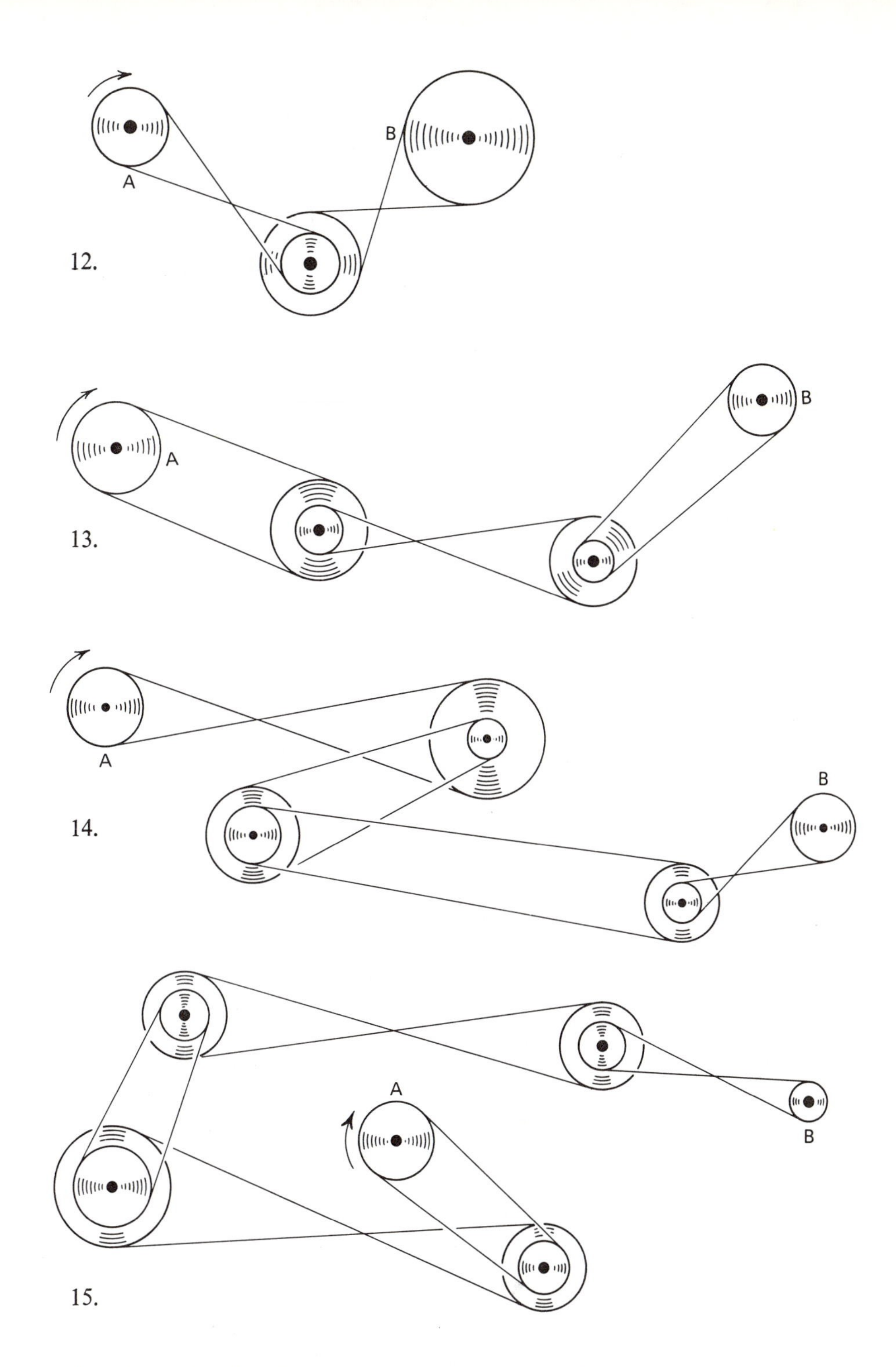

16. What size pulley should be placed on a countershaft turning 150 rpm to drive a grinder with a 4-in. pulley which is to turn at 1200 rpm?

NONCONCURRENT FORCES 13

13-1 *PARALLEL FORCE PROBLEMS*

A painter stands 2.00 ft from one end of a 6.00-ft plank which is supported at each end by a ladder (see next page). How much of the painter's weight must each ladder support?

Problems of this kind are often faced in the construction industry, particularly in the design of bridges and buildings. Using some things we learned about torques and equilibrium, we can now solve problems of this type.

Let's look at the painter problem above. A force diagram is on the next page. The arrow pointing down represents the weight of the man. The arrows pointing up represent the forces exerted by each of the ladders in holding up the plank and painter. (For now, we will neglect the weight of the plank.)

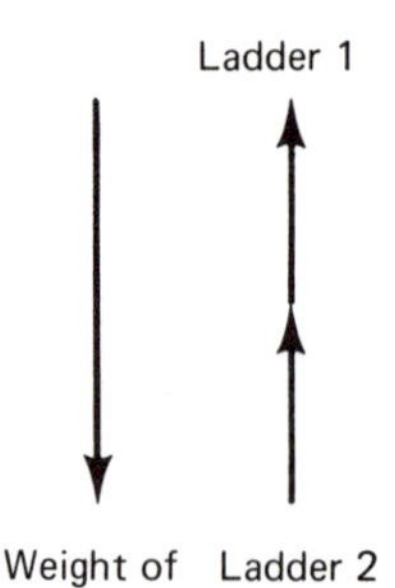

We have a condition of equilibrium. The plank and painter are not moving. The sum of the forces exerted by the ladders is equal to the weight of the painter. Since these forces are vectors and are parallel, we can show that their sum is zero. This is our *first condition of equilibrium*: *The sum of all parallel forces must be zero*. If the vector sum was not zero (forces up unequal to forces down), we would have an unbalanced force tending to cause motion.

Now consider this situation: one ladder is strong enough to support the man, and the other is removed. What happens to the painter? The plank, supported only on one end collapses, and the painter has a mess to clean up!

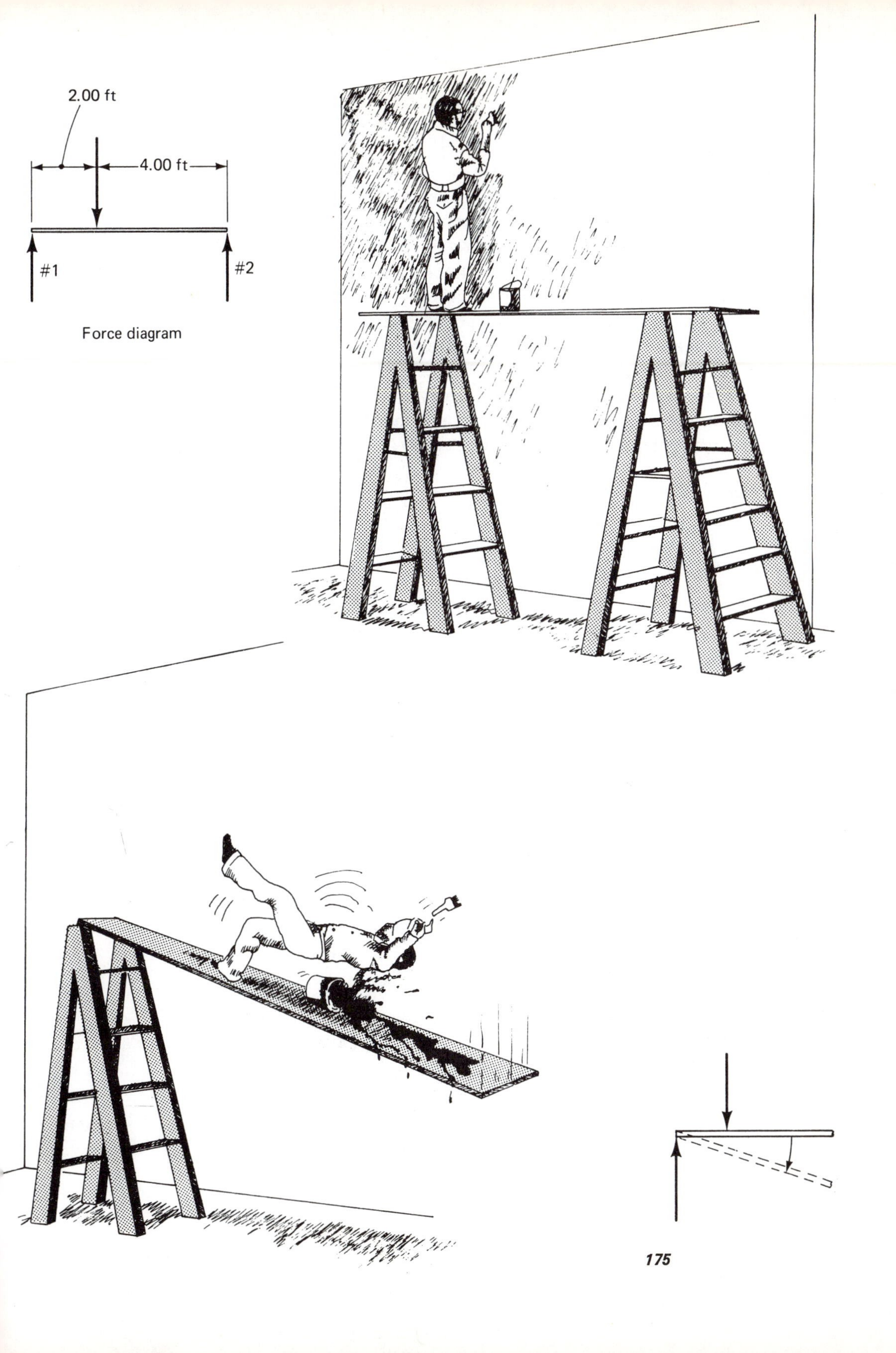
2.00 ft
4.00 ft
#1
#2
Force diagram

Not only must the forces balance each other (vector sum = 0), but they must also be positioned so there is no rotation in the system (here, rotation occurred about the supported end of the plank).

Do you see how torques are important in this problem? To avoid rotation, we can have no unbalanced torques. Sometimes there will be a natural point of rotation as in our painter problem. We can, however, choose any point as our center of rotation as we consider the torques present. We will soon see that one of a number of points could be selected. What is necessary, though, is that there be no rotation (no unbalanced torques). Stated as our *second condition of equilibrium: The sum of the clockwise torques must equal the sum of the counterclockwise torques.*

To illustrate the above principles, we will find how much weight each stepladder must support if our painter weighs 150 lb.

Sketch:

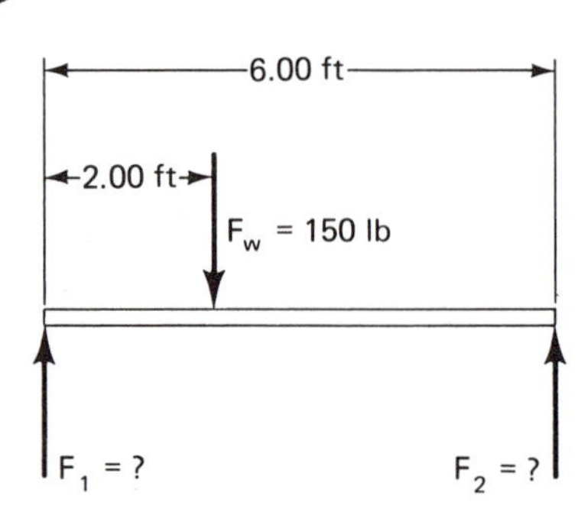

Data:

$F_w = 150$ lb

plank = 6.00 ft

F_w is 2.00 ft from one end

Basic Equation: (1) sum of forces = 0

$$F_1 + F_2 - F_w = 0$$

$$\text{or,} \quad F_1 + F_2 = F_w$$

$$F_1 + F_2 = 150\text{ lb}$$

(*Note*: F_w is negative here because its direction is opposite that of F_1 and F_2.)

(2) $$T_{\text{clockwise}} = T_{\text{counterclockwise}}$$

We must first select a point of rotation. Picking an end is usually helpful. We will pick the left end in this problem where F_1 acts. What are the clockwise torques about this point?

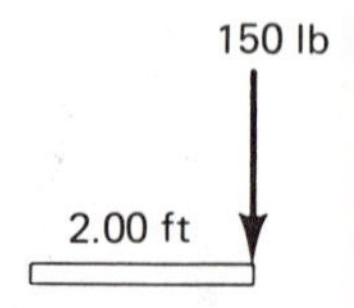

The force due to the weight of the painter tends to cause clockwise motion. The torque arm is 2.00 ft. Therefore, $T = (150\text{ lb})(2.00\text{ ft})$. This is the only clockwise torque in this problem.

The only counterclockwise torque is F_2 times its torque arm, 6.00 ft. $T = (F_2)(6.00 \text{ ft})$.

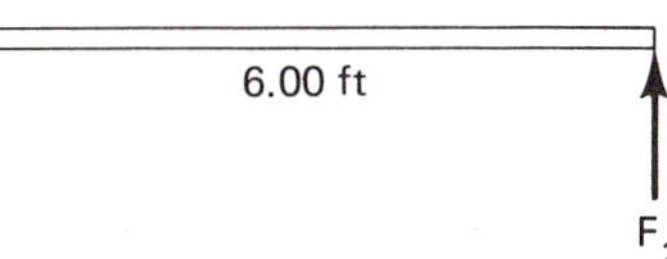

There is no torque involving F_1 because its torque arm is zero. Setting $T_{\text{clockwise}} = T_{\text{counterclockwise}}$ we have an equation:

F1

$$(150 \text{ lb})(2.00 \text{ ft}) = (F_2)(6.00 \text{ ft})$$

Note that by selecting an end as the point of rotation, we were able to set up our equation with just one variable (F_2). Solving for F_2 gives the working equation.

Working Equation: $F_2 = \dfrac{(150 \text{ lb})(2.00 \text{ ft})}{6.00 \text{ ft}}$

Substitution: The substitution was necessarily made earlier.

Simplifying: $F_2 = 50.0 \text{ lb}$

Since $F_1 + F_2 = F_w$, we can now substitute for F_w and F_2 to find F_1.

$$F_1 + 50.0 \text{ lb} = 150 \text{ lb}$$
$$F_1 = 150 \text{ lb} - 50.0 \text{ lb}$$
$$F_1 = 100 \text{ lb}$$

Outline of Procedure to Solve Parallel Force Problems

1. Sketch the problem.
2. Write an equation setting the sums of the opposite forces equal to each other.
3. Pick a point of rotation, eliminating a variable, if possible.
4. Write the sum of all clockwise torques.
5. Write the sum of all counterclockwise torques.
6. Set $T_{\text{clockwise}} = T_{\text{counterclockwise}}$
7. Solve the equation $T_{\text{clockwise}} = T_{\text{counterclockwise}}$ for the unknown quantity.
8. Substitute the value found in step 7 back into the equation in step 2 to find the other unknown quantity.

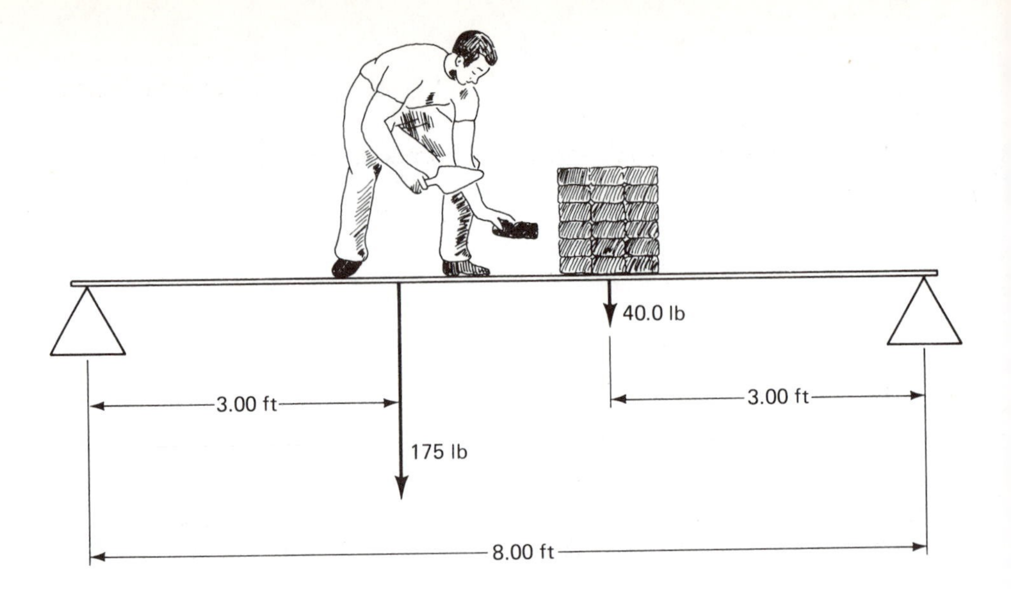

EXAMPLE: A bricklayer weighing 175 lb stands on an 8.00-ft scaffold 3.00 ft from one end. He has a pile of bricks, which weighs 40.0 lb, 3.00 ft from the other end. How much weight must each end support?

1. Sketch:

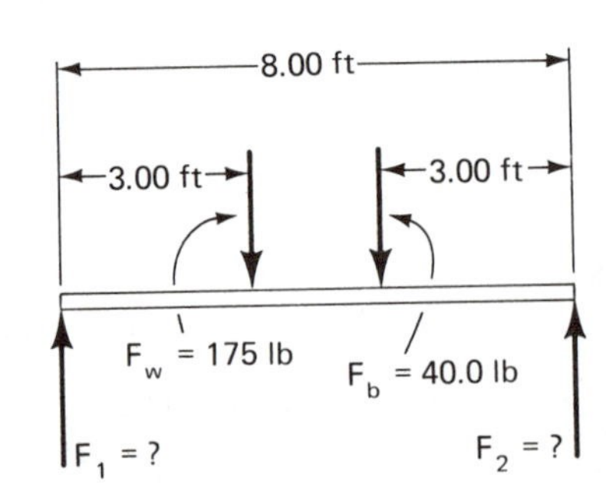

2. $F_1 + F_2 = 175 \text{ lb} + 40.0 \text{ lb}$
3. Pick a point of rotation. We should pick either end here to eliminate one of the variables F_1 or F_2. Let's pick the right end.
4. $T_{\text{clockwise}} = (F_1)(8.00 \text{ ft})$
5. $T_{\text{counterclockwise}} = (40.0 \text{ lb})(3.00 \text{ ft}) + (175 \text{ lb})(5.00 \text{ ft})$
   Note there are 2 counterclockwise torques.
6. $T_{\text{clockwise}} = T_{\text{counterclockwise}}$
   $F_1(8.00 \text{ ft}) = (40.0 \text{ lb})(3.00 \text{ ft}) + (175 \text{ lb})(5.00 \text{ ft})$
7. $$F_1 = \frac{(40.0 \text{ lb})(3.00 \text{ ft}) + (175 \text{ lb})(5.00 \text{ ft})}{8.00 \text{ ft}}$$
   $$= \frac{120 \text{ lb ft} + 875 \text{ lb ft}}{8.00 \text{ ft}} = \frac{995 \text{ lb ft}}{8.00 \text{ ft}}$$
   $$F_1 = 124 \text{ lb}$$

8. $F_1 + F_2 = 175\text{ lb} + 40.0\text{ lb}$

Substituting:

$$124\text{ lb} + F_2 = 215\text{ lb}$$
$$F_2 = 215\text{ lb} - 124\text{ lb}$$
$$F_2 = 91.0\text{ lb}$$

PROBLEMS

1. A 165-lb painter stands 3.00 ft from one end of an 8.00-ft scaffold. If the scaffold is supported at each end by a stepladder, how much weight must each ladder support?
2. A 5000-lb truck is 20.0 ft from one end of a 50.0-ft bridge. A 4000-lb car is 40.0 ft from the same end. How much weight must the other end of the bridge support? (Neglect the weight of the bridge.)
3. A 2400-kg truck is 6.00 m from one end of a 27.0 m long bridge. A 1500-kg car is 10.0 m from the same end. How much weight must each end of the bridge support?

13-2 *CENTER OF GRAVITY*

In the preceding section we neglected the weight of the plank in the painter example. In practice, the weight of the plank or bridge is extremely important. An engineer must know the weight of the bridge he is designing so that he uses methods and materials of sufficient strength to hold up the traffic and not collapse.

An important idea in this kind of problem is center of gravity. *The center of gravity of any object is that point at which all of its weight can be considered to be concentrated.*

Objects like a brick or a uniform rod have their centers of gravity at their middles or centers. The center of gravity of something like an automobile, however, is not at its center or middle because its weight is not evenly distributed throughout. Its center of gravity is located nearer the heavy engine.

You have probably had the experience of carrying a long board by yourself. If the board wasn't too heavy, you could carry it yourself by suspending it in about the middle. You didn't have to hold up both ends. You applied the principle of center of gravity and suspended the board at that point.

We shall represent the weight of an object by a vector through its center of gravity. We use a vector to show the weight (force due to gravity) of the object. It is placed through the center of gravity to show that all the weight of the object may be considered concentrated at that point. If the center of gravity is not at the middle of the object, its location will be given.

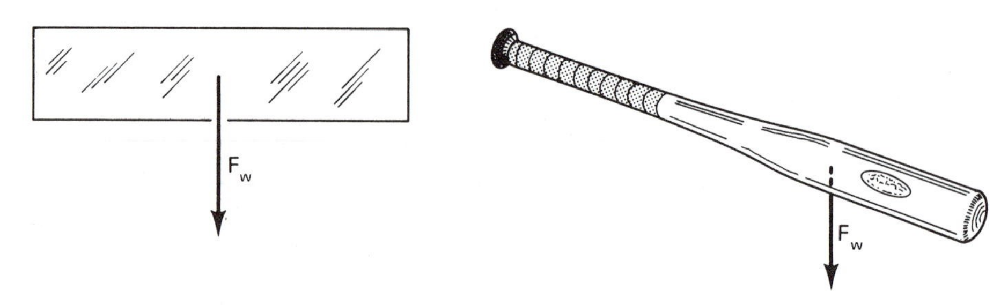

In solving problems, the weight of the plank or bridge is represented like the other forces by a vector at the center of gravity of the object.

EXAMPLE: A carpenter stands 2.00 ft from one end of a 6.00-ft scaffold which is uniform and weighs 20.0 lb. If the carpenter weighs 165 lb, how much weight must each end support?

1. Sketch:

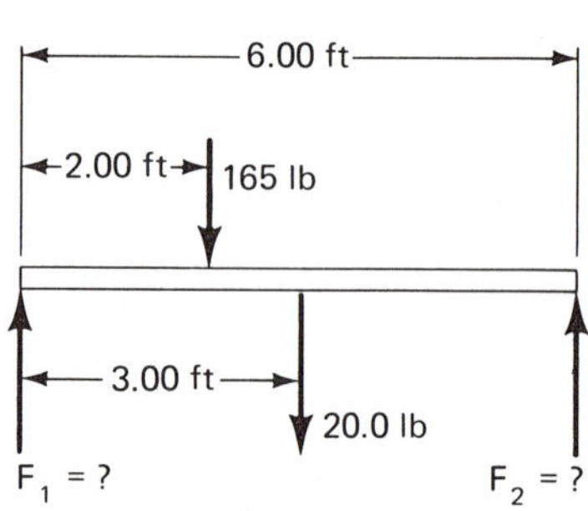

Since the plank is uniform, its center of gravity is at the middle.

2. $F_1 + F_2 = 165 \text{ lb} + 20.0 \text{ lb}$
3. Pick the left end as the point of rotation.
4. $T_{\text{clockwise}} = (165 \text{ lb})(2.00 \text{ ft}) + (20.0 \text{ lb})(3.00 \text{ ft})$
5. $T_{\text{counterclockwise}} = (F_2)(6.00 \text{ ft})$
6. $(165 \text{ lb})(2.00 \text{ ft}) + (20.0 \text{ lb})(3.00 \text{ ft}) = (F_2)(6.00 \text{ ft})$
7. $F_2 = \dfrac{330 \text{ lb ft} + 60.0 \text{ lb ft}}{6.00 \text{ ft}} = \dfrac{390 \text{ lb ft}}{6.00 \text{ ft}}$

   $F_2 = 65.0 \text{ lb}$
8. $F_1 + 65.0 \text{ lb} = 165 \text{ lb} + 20.0 \text{ lb}$

   $F_1 = 165 \text{ lb} + 20.0 \text{ lb} - 65.0 \text{ lb}$

   $F_1 = 120 \text{ lb}$

PROBLEMS

Solve the following problems using the method outlined in this chapter.

1. Solve for F_1:

$$30F_1 = (14)(18) + (25)(17)$$

2. Solve for F_2:

$$39F_2 + (60)(55) = (200)(40) + (52)(27)$$

3. Solve for F_w:

$$(12)(15) + 45F_\omega = (21)(65) + (22)(32)$$

4. Two men carry a uniform 15.0-ft plank which weighs 22.0 lb. There is a load of blocks which weighs 165 lb and is located 7.00 ft from the first man. What force must each man exert to hold up the plank and load?

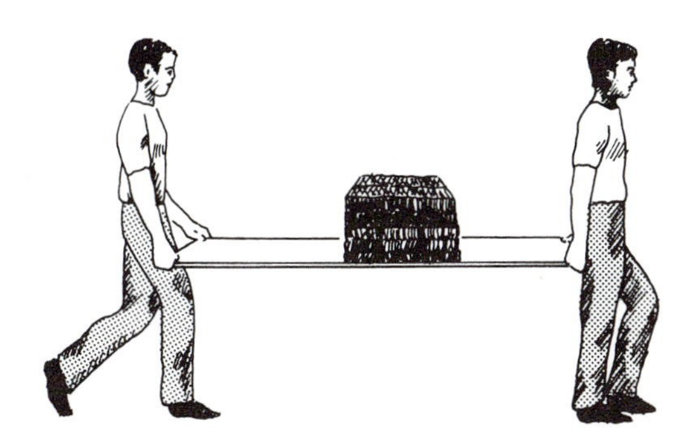

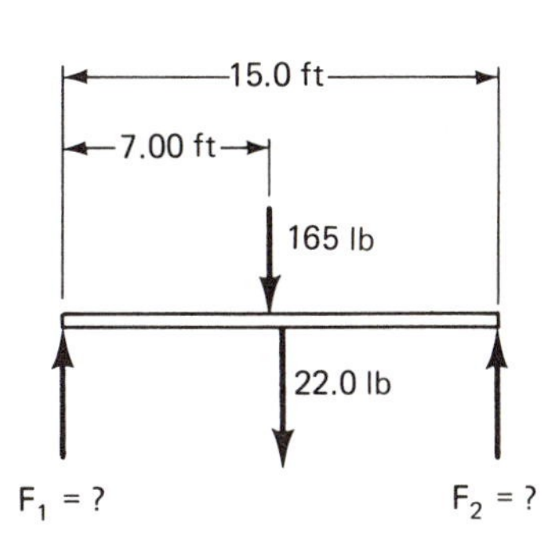

5. A wooden beam is 3.30 m long and has its center of gravity 1.30 m from one end. If the beam weighs 2.50×10^4 N, what force is needed to support each end?
6. An auto engine weighs 650 lb and is located 4.00 ft from one end of a 10.0-ft workbench. If the bench is uniform and weighs 75.0 lb, what weight must each end of the bench support?
7. An old covered bridge across a country stream weighs 20,000 lb. A large truck stalls 12.0 ft from one end of the 32.0-ft bridge. What weight must each of the piers support if the truck weighs 22,000 lb?
8. A window washer's scaffold 12.0 ft long and weighing 75.0 lb is suspended from each end. One washer weighs 155 lb and is 3.00 ft from one end. The other washer is 4.00 ft from the other end. If the force supported by the end near the first washer is 200 lb, how much does the second washer weigh?
9. A porch swing weighs 29.0 lb. It is 4.40 ft long and has a dog weighing 14.0 lb sleeping on it 1.90 ft from one end. What weight must the support ropes on each end hold up?

10. A wooden plank is 5.00 m long and supports a 75.0-kg block 2.00 m from one end. If the plank is uniform and has a mass of 30.0 kg, how much force is needed to support each end?
11. A bridge has a mass of 1.60×10^4 kg, is 21.0 m long, and has a 3500-kg truck 7.00 m from one end. What force must each end of the bridge support?

MATTER 14

14-1 *INTRODUCTION*

In our study of mechanics we did not need to discuss the structure of matter. When we studied pulley systems, we did not need any information about the kind of matter it was. However, when we study gases, heat, temperature, and electricity, we will find it helpful to understand a few basic facts about the structure of matter. The pressure of a gas on its container (such as an expanding gas in a combustion chamber) is the result of the bombardment of the walls by very small bits of matter, called atoms and molecules, which make up the gas. The higher the temperature of a gas, the greater is the velocity of these small bits of matter. The flow of electricity is a result of the movement of certain parts of atoms, namely electrons.

The unit building block of matter is the atom. There are over 100 different types of atoms which occur in nature. The smallest atom, hydrogen, is about 2.9×10^{-9} in. or 7.4×10^{-9} cm in diameter which is much too small to see with even the most powerful microscope. We can, for our purposes, consider atoms to be hard spheres of different sizes. When many atoms of the *same* type are collected together, we have a simple substance called an element (aluminum, copper, and oxygen are elements). When a small number of different atoms are bonded together in a chemical process, a molecule is formed. Large numbers of molecules can form together to make a compound (examples are water and carbon dioxide).

Matter exists in three states: solids, liquids, and gases. A solid is a substance that has a definite shape and volume. A liquid is a substance that takes the shape of its container and has a definite volume. A gas is a substance that takes the shape of its container and has the same volume as its container.

14-2 *CHARACTERISTICS OF MATTER*

The molecules of a solid are fixed in relation to each other. They vibrate in a back and forth motion. They are so close that a solid can be compressed only slightly. Solids are usually crystalline substances, meaning their molecules are arranged in a definite pattern. This is why a solid tends to hold its shape and has a definite volume.

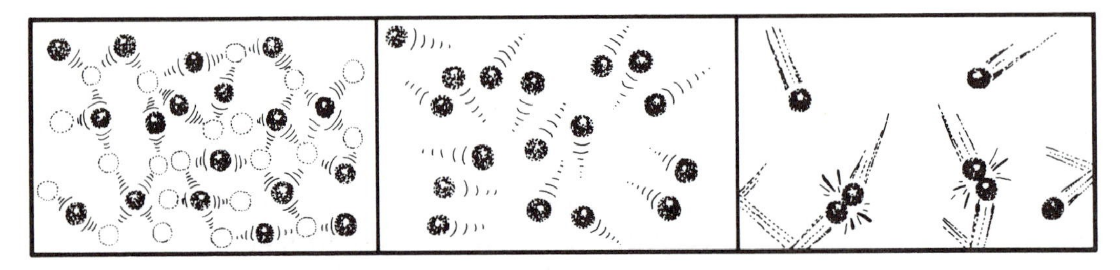

Solid molecules vibrate in fixed positions

Liquid molecules flow over each other

Gas molecules move rapidly in all directions and collide

The molecules of a liquid are not fixed in relation to each other. They normally move in a flowing type of motion but yet are so close together that they are practically incompressible, thus having a definite volume. Because the molecules move in a smooth flowing motion and not in any fixed manner, a liquid takes the shape of its container.

The molecules of a gas are not fixed in relation to each other and move rapidly in all directions, colliding with each other. They are much farther apart than molecules in a liquid, and they are extremely far apart when compared to the distance between molecules in solids. The movement of the molecules is limited only by its container. Therefore, a gas takes the shape of its container. Because the molecules are far apart, a gas can easily be compressed and it has the same volume as its container.

14-3 *PROPERTIES OF SOLIDS*

Solids, as previously noted, are composed of molecules. Sometimes these molecules attract each other and sometimes they repel each other. For instance, take a rubber ball and try to pull it apart. You notice that the ball stretches out of shape.

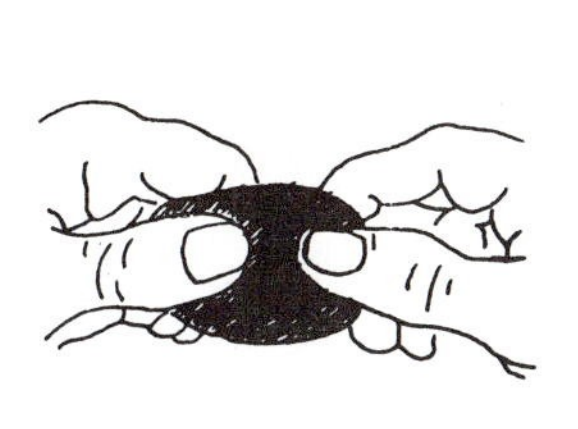 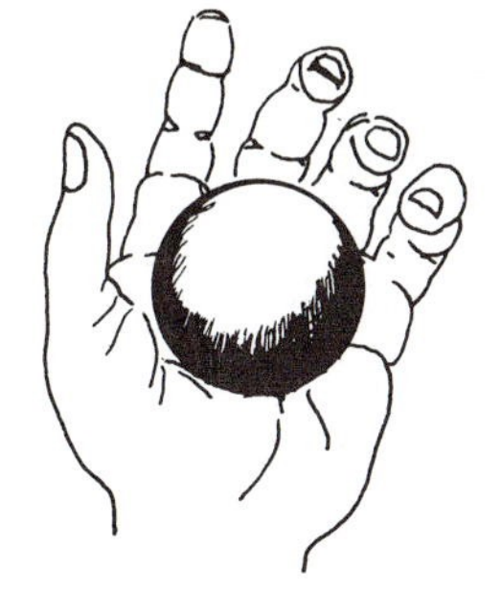 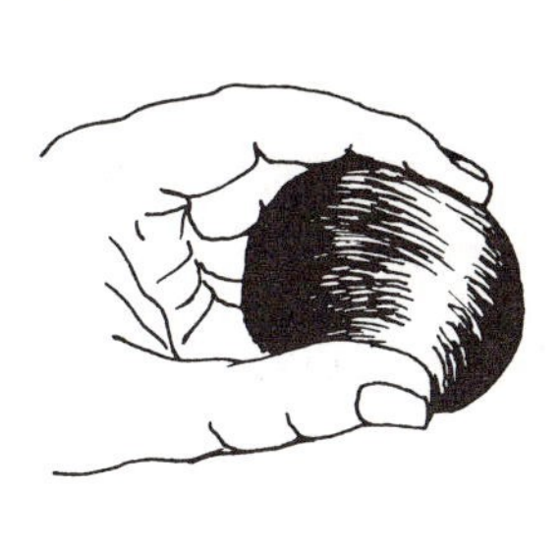

However, when you release the pulling force, the ball returns to its original shape because the molecules, being farther apart than normal, attract each other. If you squeeze the ball together, the ball will again become out of shape. Now release the pressure and the ball will again return to its original shape because the molecules, being too close together, repel each other. Therefore, we can see that when molecules are slightly pulled out of position, they attract each other. When they are pressed too close together, they repel each other.

This combination of attraction and repulsion is called *elasticity.* Elasticity is the ability of a solid to regain its shape after a deforming force has been applied. Most solids have the property of elasticity; however, some are only slightly elastic. For example, wood and styrofoam are two solids whose elasticity is small.

Not every elastic object returns to its original shape. If too large a deforming force is applied, it will become permanently deformed. Take a door spring and pull it apart far as you can. When you let go, it will probably not return to its exact original shape. When a solid is deformed as such, it is said to have been deformed past its elastic limit. If the deforming force is enough greater, the body breaks apart.

Spring before stretching.

Stretched near elastic limit.

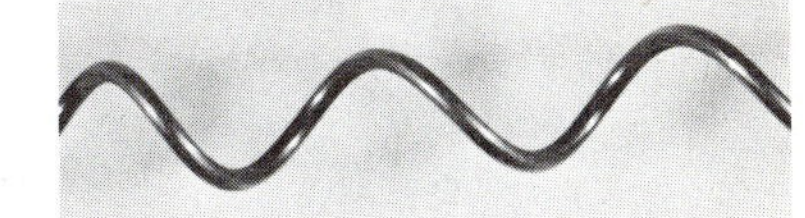

Stretched beyond elastic limit ... break occurs!

Hooke's Law relates to elasticity of solids.

RULE

Hooke's Law: The ratio of the change in length of an object that is stretched or compressed to the force causing this change is constant as long as the elastic limit has not been exceeded.

In equation form:

$$\frac{F}{\Delta l} = k$$

where: F = applied force
Δl = change in length
k = elastic constant

(*Note*: Δ is often used in mathematics and science to mean "change in.")

EXAMPLE 1: A force of 5.00 N is applied to a spring whose elastic constant is 0.250 N/cm. Find its change in length.

SKETCH:

5.00 N

DATA: $F = 5.00\text{ N}$

$k = 0.250\,\frac{\text{N}}{\text{cm}}$

BASIC EQUATION: $\frac{F}{\Delta l} = k$

WORKING EQUATION: $\Delta l = \frac{F}{k}$

SUBSTITUTION: $\Delta l = \frac{5.00\text{ N}}{0.250\text{ N/cm}} = \frac{5.00\text{ N (cm)}}{0.250\text{ N}}$

$= 20.0\text{ cm}$

EXAMPLE 2: A force of 3.00 lb stretches a spring 12.0 in. What force is required to stretch the spring 15.0 in?

SKETCH:

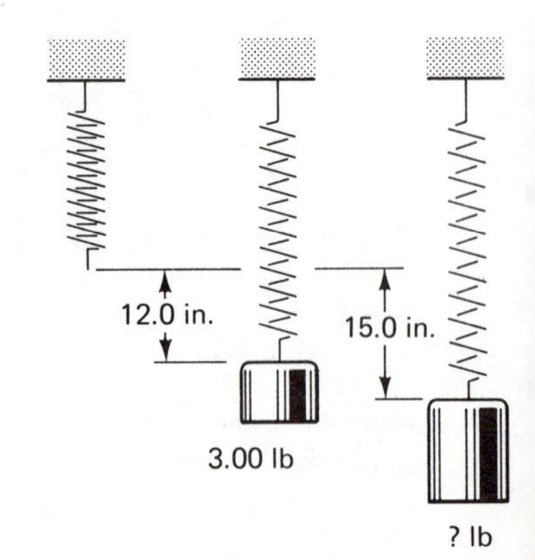

DATA: $F_1 = 3.00\text{ lb}$

$l_1 = 12.0\text{ in.}$

$l_2 = 15.0\text{ in.}$

$F_2 = ?$

BASIC EQUATION: $\frac{F}{\Delta l} = k$

WORKING EQUATIONS: $\frac{F}{\Delta l} = k$ and $F = k(\Delta l)$

SUBSTITUTION: There are two substitutions in this problem, one to find k, and one to find the second force (F_2).

$$\frac{3.00 \text{ lb}}{12 \text{ in.}} = k$$

$$0.250 \frac{\text{lb}}{\text{in.}} = k$$

$$F_2 = 0.250 \frac{\text{lb}}{\text{in.}} (15.0 \text{ in.})$$

$$F_2 = 3.75 \text{ lb}$$

When the molecular forces in a solid are very great, it is difficult to scratch the solid. This property is called *hardness*. Diamond is the hardest solid.

The *tensile strength* of a solid is a measure of its resistance to being pulled apart.

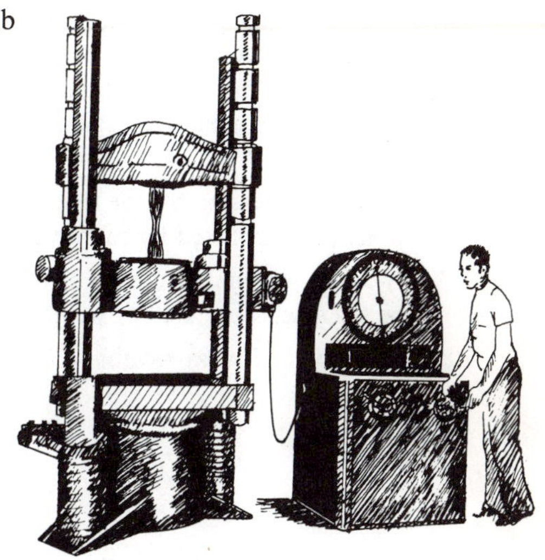

Tensile strength testing machine

Elasticity, hardness, and tensile strength are determined by the composition of the metal and by heat processes. For instance, the tensile strength of 0.3% carbon steel is 75,000 psi (lb/in^2); the tensile strength of 0.7% carbon steel is 100,000 psi; and the tensile strength for 3% carbon cast iron is 20,000 psi. So you see that, by increasing the amount of carbon in steel, the tensile strength is increased up to a point, and then it is decreased.

Metals have two other special properties, *ductility* and *malleability*. Ductility is the ability to be drawn through a die to produce a wire. Malleability is the ability to be hammered and rolled into sheets.

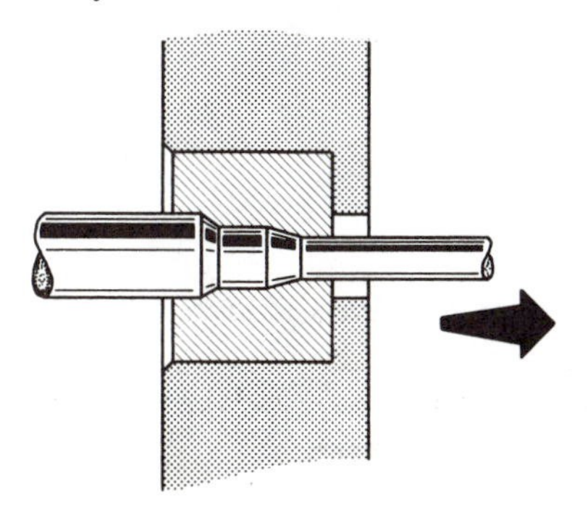

Ductility: A metal being drawn into a wire

Rollers

Metal

Malleability: a metal being rolled into a sheet

PROBLEMS

1. A spring is stretched 24.0 in. by a force of 50.0 lb. How far will it stretch if a force of 104 lb is applied?
2. What weight will stretch the spring 9.00 in. in problem 1?
3. If a 17.0-N force stretches a wire 0.650 cm, what force will stretch a similar piece of wire 1.87 cm?
4. If a force of 21.3 N is applied on a similar piece of wire as in problem 3, how far will it stretch?
5. The vertical steel columns of an office building each support a weight of 30,000 lb at the second floor, with each being compressed 0.00234 in. What will be the compression of each column if a weight of 125,000 lb is supported?
6. If the compression for the steel columns of problem 5 is 0.0279 in., what weight is supported by each column?
7. A coiled spring is stretched 30.0 cm by a 5.00-N weight. How far will it be stretched by a 15.0-N weight?
8. In problem 7 what weight will stretch the spring 40.0 cm?

SKETCH
DATA
$a = 1, b = 2$
$c = ?$

BASIC EQUATION
$a = bc$

WORKING EQUATION
$c = \frac{a}{b}$

SUBSTITUTION
$c = \frac{1}{2}$

14-4 *PROPERTIES OF LIQUIDS*

Liquids have three basic properties: *capillary action*, *surface tension*, and *viscosity*. Capillary action is the ability of a liquid to rise or fall in a narrow tube.

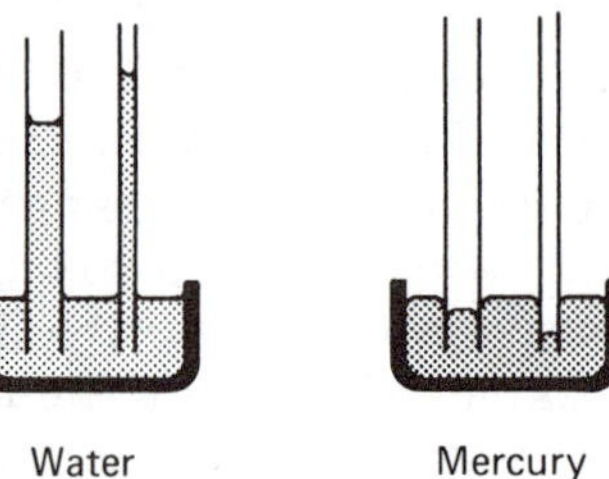

The ability of the surface of water to support a needle is an example of surface tension. However, surface tension can be reduced by adding soap to the water, and the needle sinks.

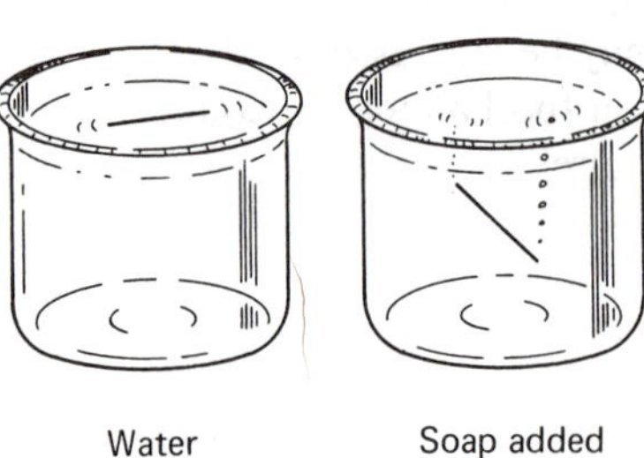

One characteristic of liquids is called flow. In liquids there is friction, which is called viscosity. The more the friction, the greater is the viscosity. For example, it takes more force to move a block of wood through oil than through gasoline. This is because oil is more viscous than gasoline.

If a liquid's temperature is increased, its viscosity decreases. For example, the viscosity of oil in a car engine before it is started on a winter morning at 0°F is greater than after the engine has been running for an hour.

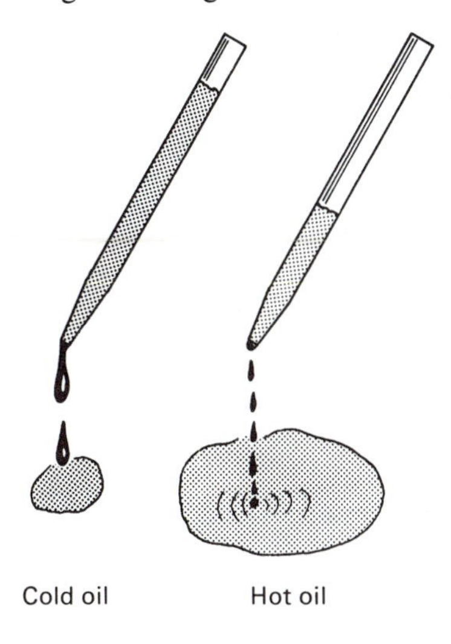

Cold oil Hot oil

14-5 *PROPERTIES OF GASES*

Due to the rapid random movement of its molecules, a gas spreads to completely occupy the volume of its container. This property is called *expansion.*

Diffusion of a gas is the process by which molecules of the gas mix with the molecules of a solid, liquid, or another gas. If you remove the cap from a bottle of rubbing alcohol, you soon smell the fumes of the alcohol. The air molecules and the alcohol molecules mix throughout the room because of diffusion.

A ballon inflates due to the pressure of the air molecules on the inside surface of the balloon. This pressure is caused by the bombardment on the walls by the moving molecules. The pressure may be increased by increasing the number of molecules by blowing more air in the balloon. Pressure may also be increased by heating the air molecules already in the balloon. Heat increases the velocity of the molecules.

14-6 *DENSITY*

Density is a property of all three states of matter. *Mass density*, D_m, is defined as mass per unit volume. *Weight density*, D_w, is defined as weight per unit volume, or,

$$D_m = \frac{M}{V} \qquad\qquad D_w = \frac{W}{V}$$

where: D_m = mass density $\quad D_w$ = weight density
M = mass $\quad W$ = weight
V = volume $\quad V$ = volume

Although either mass density or weight density can be expressed in both the metric system and the English system, mass density is usually given in the metric units g/cm^3 and weight density is usually given in the English units lb/ft^3.

The weight density of water is 62.4 lb/ft^3, that is, one cubic foot of water weighs 62.4 lb. (A suggested project is to make a container one cubic foot in volume, pour it full of water, and find the weight of the water. If you fill the container with a gallon container, you will also find that 1 ft^3 is approximately $7\frac{1}{2}$ gal.)

Substance	Mass Density (g/cm^3)	Weight Density (lb/ft^3)
Solids		
Copper	8.89	555
Iron	7.8	490
Lead	11.3	708
Aluminum	2.70	169
Ice	0.917	57
Wood, white pine	0.42	26
Concrete	2.3	145
Cork	0.24	15
Liquids		
Water	1.00	62.4
Sea Water	1.03	64.0
Oil	0.87	54.2
Mercury	13.6	846
Alcohol	0.79	49.4
Gasoline	0.68	42.0
Gases*	at 0°C and 1 atm pressure	at 32°F and 1 atm pressure
Air	1.29×10^{-3}	0.081
Carbon Dioxide	1.96×10^{-3}	0.123
Carbon Monoxide	1.25×10^{-3}	0.078
Helium	1.78×10^{-4}	0.011
Hydrogen	8.99×10^{-5}	0.0056
Oxygen	1.43×10^{-3}	0.089
Nitrogen	1.25×10^{-3}	0.078
Ammonia	7.6×10^{-4}	0.047
Propane	2.02×10^{-3}	0.126

*The density of a gas is found by pumping the gas into a container, by measuring its volume and mass or weight, and then by using the appropriate density formula.

The mass density of water is 1 g/cm³, that is, one cubic centimeter of water has a mass of one gram.

In all forms of matter, the density usually decreases as the temperature increases and increases as the temperature decreases. The densities of various substances are listed in the table on page 190.

EXAMPLE 1: What is the weight density of a block of wood having dimensions 3.00 in. × 4.00 in. × 5.00 in. and a weight of 0.700 lb?

SKETCH:

5.00 in.

3.00 in.

4.00 in.

DATA:

$l = 4.00$ in.
$w = 3.00$ in.
$h = 5.00$ in.
$W = 0.700$ lb
$D_w = ?$

BASIC EQUATIONS: $V = lwh$ and $D_w = \frac{W}{V}$

WORKING EQUATIONS: Same

SUBSTITUTIONS:

$V = (4.00 \text{ in.})(3.00 \text{ in.})(5.00 \text{ in.})$
$V = 60.0 \text{ in}^3$

$$D_w = \frac{0.700 \text{ lb}}{60.0 \text{ in}^3}$$

$$D_w = 0.0117 \frac{\text{lb}}{\cancel{\text{in}^3}} \times \frac{1728 \cancel{\text{in}^3}}{1 \text{ ft}^3}$$

$$D_w = 20.2 \frac{\text{lb}}{\text{ft}^3}$$

EXAMPLE 2: Find the mass density of a ball bearing with mass 125 g and radius 0.875 cm.

DATA:

$r = 0.875$ cm
$M = 125$ g
$D_m = ?$

BASIC EQUATIONS: $V = \frac{4}{3}\pi r^3$ and $D_m = \frac{M}{V}$

WORKING EQUATIONS: Same

SUBSTITUTIONS: $V = \frac{4}{3}(3.14)(0.875\text{ cm})^3$

$V = 2.80\text{ cm}^3$

$$D_m = \frac{125\text{ g}}{2.80\text{ cm}^3}$$

$$D_m = 44.6\frac{\text{g}}{\text{cm}^3}$$

EXAMPLE 3: Find the weight density of a gallon of water weighing 8.34 lb.

DATA: $W = 8.34\text{ lb}$

$V = 1\text{ gal} = 231\text{ in}^3$

$D_w = ?$

BASIC EQUATION: $D_w = \frac{W}{V}$

WORKING EQUATION: Same

SUBSTITUTION: $D_w = \frac{8.34\text{ lb}}{231\text{ in}^3}$

$D_w = 0.0361\frac{\text{lb}}{\cancel{\text{in}^3}} \times \frac{1728\cancel{\text{in}^3}}{1\text{ ft}^3}$

$D_w = 62.4\frac{\text{lb}}{\text{ft}^3}$

EXAMPLE 4: Find the weight density of a can of oil (1 quart) weighing 1.90 lb.

DATA: $1\text{ qt} = \frac{1}{4}\text{ gal} = \frac{1}{4}(231\text{ in}^3) = 57.8\text{ in}^3 = V$

$W = 1.90\text{ lb}$

$D_w = ?$

BASIC EQUATION: $D_w = \frac{W}{V}$

WORKING EQUATION: Same

SUBSTITUTION: $D_w = \frac{1.90\text{ lb}}{57.8\text{ in}^3}$

$D_w = 0.0329\frac{\text{lb}}{\cancel{\text{in}^3}} \times \frac{1728\cancel{\text{in}^3}}{1\text{ ft}^3}$

$D_w = 56.9\frac{\text{lb}}{\text{ft}^3}$

EXAMPLE 5: A quantity of gasoline weighs 5.50 lb with weight density of 42.0 lb/ft³. What is its volume?

DATA: $D_w = 42.0 \frac{\text{lb}}{\text{ft}^3}$

$W = 5.50 \text{ lb}$

$V = ?$

BASIC EQUATION: $D_w = \frac{W}{V}$

WORKING EQUATION: $V = \frac{W}{D_w}$

SUBSTITUTION: $V = \frac{5.50 \text{ lb}}{42.0 \text{ lb/ft}^3}$

$V = \frac{(5.50 \cancel{\text{lb}})(\text{ft}^3)}{42.0 \cancel{\text{lb}}}$

$V = 0.131 \text{ ft}^3$

The density of an irregular solid (rock) cannot be found directly because of the difficulty of finding its volume. However, we could find the amount of water the solid displaces, which is the same as the volume of the irregular solid. Volume of water in beaker = volume of rock.

EXAMPLE 6: A rock of mass 30.8 kg displaces 320 cm³ of water. What is the mass density of the rock?

DATA: $M = 30.8 \text{ kg}$

$V = 320 \text{ cm}^3$

$D_m = ?$

BASIC EQUATION: $D_m = \frac{M}{V}$

WORKING EQUATION: Same

SUBSTITUTION: $D_m = \frac{30.8 \cancel{\text{kg}}}{320 \text{ cm}^3} \times \frac{1000 \text{ g}}{1 \cancel{\text{kg}}}$

$D_m = 96.3 \frac{\text{g}}{\text{cm}^3}$

EXAMPLE 7: A rock displaces 3.00 gallons of water and has weight density of 76.0 lb/ft³. What is its weight?

DATA: $D = 76.0 \frac{\text{lb}}{\text{ft}^3}$

1 gallon water = 231 in³

3 gallon water = 693 in³ = V

$W = ?$

BASIC EQUATION: $D_w = \frac{W}{V}$

WORKING EQUATION: $W = D_w V$

SUBSTITUTION: $W = \frac{76.0 \text{ lb}}{\cancel{\text{ft}^3}} \times 693 \cancel{\text{in}^3} \times \frac{1 \cancel{\text{ft}^3}}{1728 \cancel{\text{in}^3}}$

$W = 30.5 \text{ lb}$

PROBLEMS

Express mass density in g/cm³ and weight density in lb/ft³.

1. What is the mass density of a chunk of rock of mass 210 g and which displaces a volume of 75 cm³ of water?
2. A block of wood is 55.9 in. × 71.1 in. × 25.4 in. and weighs 1814 lb. What is its weight density?
3. If the block of wood in problem 2 has weight density of 30.0 lb/ft³, what does it weigh?
4. What volume does 1300 g of mercury occupy?
5. What volume does 1300 g of cork occupy?
6. What volume does 1300 g of nitrogen occupy at 0°C and one atmosphere pressure?
7. A block of gold 9.00 in. × 8.00 in. × 6.00 in. weighs 301 lb. What is its weight density?
8. If a cylindrical piece of copper is 9.00 in. tall and 1.40 in. in radius, how much does it weigh?

SKETCH
DATA
$a = 1, b = 2, c = ?$
BASIC EQUATION
$a = bc$
WORKING EQUATION
$c = \frac{a}{b}$
SUBSTITUTION
$c = \frac{1}{2}$

9. A piece of aluminum of mass 6.22 kg. If it displaces water that fills a container 12.0 cm $\times$ 12.0 cm $\times$ 16.0 cm, find the mass density of the aluminum.
10. If one pint of turpentine weighs 0.907 lb, what is its weight density?
11. Find the mass density of gasoline if 102 g occupy 150 cm^3.
12. How much does a gallon of gasoline weigh?
13. Determine the volume in gallons of 400 lb of oil.
14. How many ft^3 will 573 lb of water occupy?
15. If 20.4 in^3 of linseed oil weighs 0.694 lb, what is its weight density?
16. If 108 in^3 of ammonia gas weighs 0.00301 lb, what is its weight density?
17. What is the volume of 3.00 kg of propane at 0°C and one atmosphere pressure?
18. Granite has a mass density of 2.65 g/cm^3. Find its weight density in lb/ft^3.

FLUIDS 15

15-1 *INTRODUCTION*

In many respects liquids and gases behave in much the same manner. For this reason they are often studied together as fluids. The gas and water piped to your home are fluids having several common characteristics. We will now study some of these characteristics.

15-2 *PRESSURE*

If you press your hand against the table, you are applying a force to the table. You are also applying pressure to the table. Is there a difference? The difference is an important one. Pressure is the force applied to a unit area. It is the concentration of the force.

$$P = \frac{F}{A}$$

where: P = pressure (usually in N/m^2 or lb/in^2–psi)
F = force applied (N or lb)
A = area (m^2 or in^2)

Imagine a brick first lying on its side on a table and then standing on one end. The weight of the brick is the same no matter what its position. So, the total force (the weight of the brick) on the table is the same in both cases.

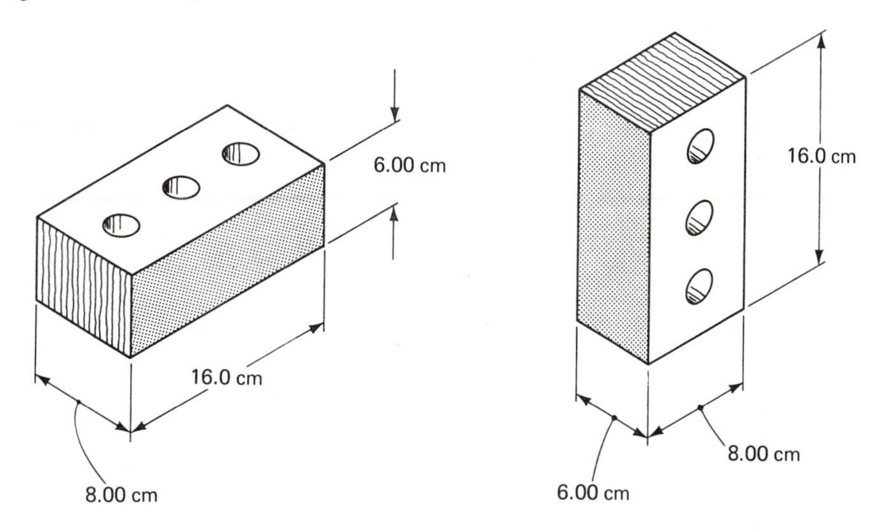

However, the position of the brick does make a difference on the pressure exerted on the table. In which case is the pressure greater? When standing on end, the brick exerts a greater pressure on the table. The reason is that the area of contact on end is *smaller* than on the side. Using $P = F/A$, let's find the pressure exerted in each case.

Case 1:

$$F = 12.0 \text{ N}$$
$$A = 8.00 \text{ cm} \times 16.0 \text{ cm}$$
$$A = 128 \text{ cm}^2$$
$$P = \frac{F}{A} = \frac{12.0 \text{ N}}{128 \cancel{\text{cm}^2}} \times \frac{10^4 \cancel{\text{cm}^2}}{1 \text{ m}^2}$$
$$P = 938 \text{ N/m}^2$$

Case 2:

$$F = 12.0 \text{ N}$$
$$A = 6.00 \text{ cm} \times 8.00 \text{ cm}$$
$$A = 48.0 \text{ cm}^2$$
$$P = \frac{F}{A} = \frac{12.0 \text{ N}}{48.0 \cancel{\text{cm}^2}} \times \frac{10^4 \cancel{\text{cm}^2}}{1 \text{ m}^2}$$
$$P = 2500 \text{ N/m}^2$$

This shows that when the same force is applied to a smaller area the pressure is greater. From the discussion so far, would you rather the same woman step on your foot with a pointed-heel shoe or a flat shoe? Before you

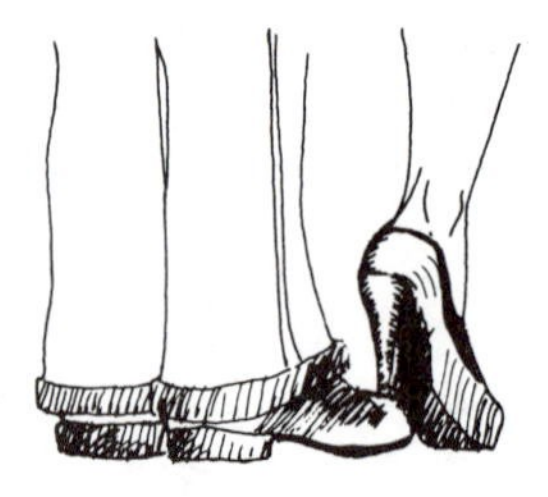
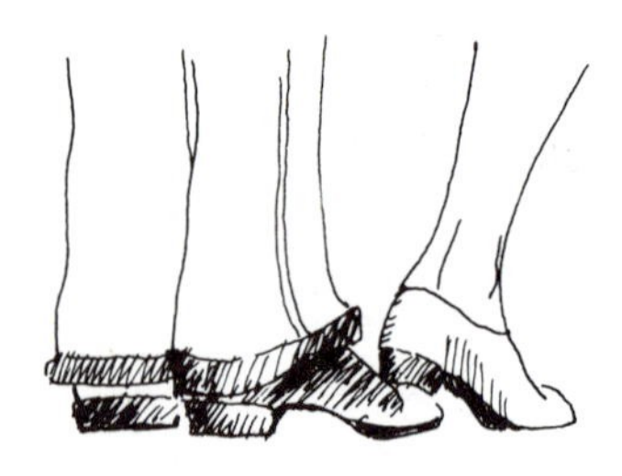

snicker, thinking this question is silly, the aircraft industry doesn't think it is. They must design and construct floors light in weight but strong enough to stand the pressure of women wearing pointed-heel shoes. This was a serious problem for them.

For example, if a 160-lb woman rests her weight on a 4-in² heel, the pressure is

$$P = \frac{F}{A} = \frac{160\ \text{lb}}{4\ \text{in}^2} = 40\,\frac{\text{lb}}{\text{in}^2}$$

But if she rests her weight on a pointed heel of $\frac{1}{8}$ in², which is a common area of a pointed heel, the pressure is

$$P = \frac{F}{A} = \frac{160\ \text{lb}}{\frac{1}{8}\ \text{in}^2} = 1280\,\frac{\text{lb}}{\text{in}^2}$$

Liquids present a slightly different situation. As you probably know already, the pressure increases as you go deeper in water. Liquids are different in this respect from solids in that, where solids exert only a downward force due to gravity, the force exerted by liquids is in all directions.

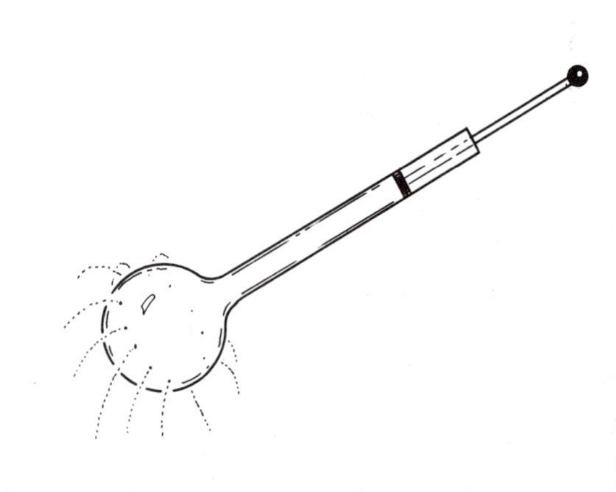

The pressure in a liquid depends only on the depth and weight density of the liquid. Because the pressure exerted by water increases with depth, dams are built much thicker at the base than at the top.

To find the pressure at a given depth in a liquid use the formula:

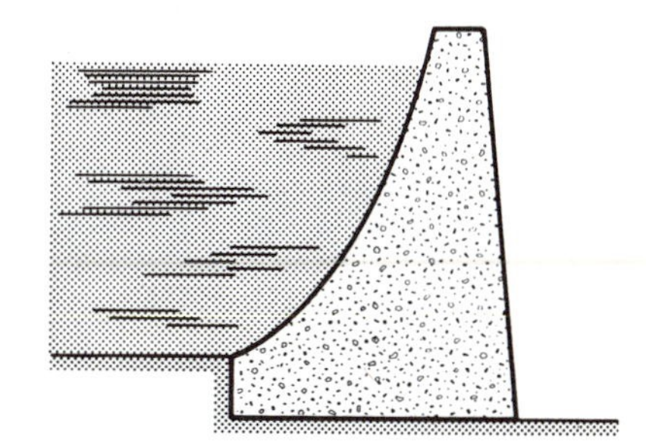

$$P = hD_w$$

where: P = pressure
h = height (or depth)
D_w = weight density of the liquid

EXAMPLE 1: Find the pressure at the bottom of a water-filled drum 4.00 ft high.

SKETCH:

DATA:
$h = 4.00$ ft
$D_w = 62.4$ lb/ft³
$P = ?$

BASIC EQUATION: $P = hD_w$

WORKING EQUATION: Same

SUBSTITUTION:
$$P = (4.00 \text{ ft})\left(62.4 \frac{\text{lb}}{\text{ft}^3}\right)$$
$$P = 250 \frac{\text{lb}}{\cancel{\text{ft}^2}} \cdot \frac{1 \cancel{\text{ft}^2}}{144 \text{ in}^2}$$
$$P = 1.74 \frac{\text{lb}}{\text{in}^2}$$

EXAMPLE 2: Find the depth in a lake at which the pressure is 100 lb/in.²

SKETCH: None

DATA:
$P = 100$ lb/in²
$D_w = 62.4$ lb/ft³
$h = ?$

BASIC EQUATION: $P = hD_w$

WORKING EQUATION: $h = \dfrac{P}{D_w}$

SUBSTITUTION:

$$h = \frac{100 \text{ lb/in}^2}{62.4 \text{ lb/ft}^3}$$

$$h = 1.60 \frac{\cancel{\text{lb}}}{\text{in}^2} \cdot \frac{\text{ft}^3}{\cancel{\text{lb}}}$$

$$h = 1.60 \frac{\overset{\text{ft}}{\cancel{\text{ft}^3}}}{\text{in}^2} \cdot \frac{144 \text{ in}^2}{1 \cancel{\text{ft}^2}}$$

$$h = 230 \text{ ft}$$

15-3 *TOTAL FORCE EXERTED BY LIQUIDS*

The *total* force exerted by a liquid on a horizontal surface (such as the bottom of a barrel) depends on the area of the surface, the depth of the liquid, and the weight density of the liquid. By formula:

$$F_t = AhD_w$$

where: F_t = total force
A = area
h = height or depth of the liquid
D_w = weight density

The total force on a vertical surface F_s (such as the *side* of a barrel) is found by using half the vertical height (average height):

$$F_s = \tfrac{1}{2}AhD_w$$

EXAMPLE 1: What is the total force on the bottom of a rectangular tank 10.0 ft by 5.00 ft by 4.00 ft deep?

SKETCH:

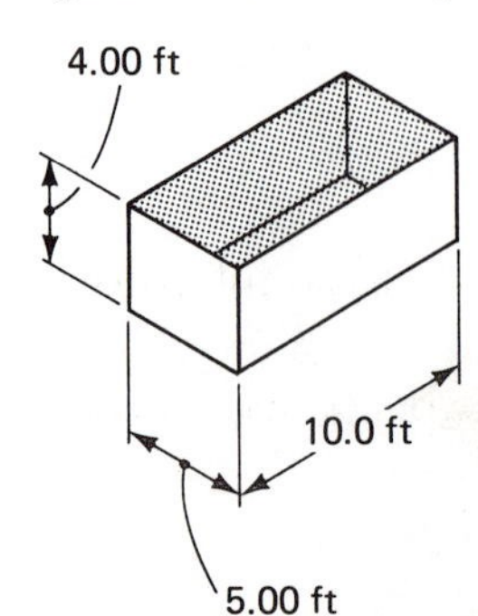

DATA: $A = lw = (10.0 \text{ ft})(5.00 \text{ ft}) = 50.0 \text{ ft}^2$
$h = 4.00 \text{ ft}$
$d = 62.4 \text{ lb/ft}^3$
$F_t = ?$

BASIC EQUATION: $F_t = AhD_w$

WORKING EQUATION: Same

SUBSTITUTION: $F_t = (50.0 \cancel{ft^2})(4.00 \cancel{ft})\left(62.4 \frac{lb}{\cancel{ft^3}}\right)$

$F_t =$ 12,500 lb

PROBLEMS

1. What is the water pressure at the bottom of a water tower 50.0 ft high?
2. If the water pressure is 20.0 lb/in², how high can a column of water be raised?
3. What is the density of a liquid which exerts a pressure of 0.400 lb/in² at a depth of 42.0 in.?
4. What is the total force on the bottom of a water-filled circular cattle tank 2.50 ft high with a radius of 4.00 ft?
5. What is the total force on the side of the tank in problem 4?
6. What must the water pressure be to supply water to the third floor of a building (35.0 ft up) with a pressure of 40 lb/in² at that level?
7. A small tank 3 in. by 5 in. is filled with mercury. If the total force on the bottom of the tank is 14.5 lb, how deep is the mercury? (weight density of mercury = 0.490 lb/in³.)
8. What is the total force on the largest side of the tank?

SKETCH
DATA
$a = 1, b = 2,$
$c = ?$

BASIC EQUATION
$a = bc$

WORKING EQUATION
$c = \frac{a}{b}$

SUBSTITUTION
$c = \frac{1}{2}$

15-4 *PASCAL'S PRINCIPLE*

The hydraulic jack or press illustrates an important principle that a liquid can be used as a simple machine to multiply force.

RULE

Pascal's Principle: Pressure applied to a confined liquid is transmitted undiminished throughout the entire liquid.

If we apply a force to the small piston of the hydraulic jack below, the pressure is transmitted undiminished in all directions. The reason for this is the virtual noncompressibility of liquids.

The *pressure* on the large piston is the same as on the small piston; however, the *total force* on the large piston is greater because of its larger surface area.

EXAMPLE: In the diagram of the hydraulic jack:
(a) What is the pressure on the small piston?
(b) What is the pressure on the large piston?
(c) What is the total force on the large piston?
(d) What is the mechanical advantage of the jack?

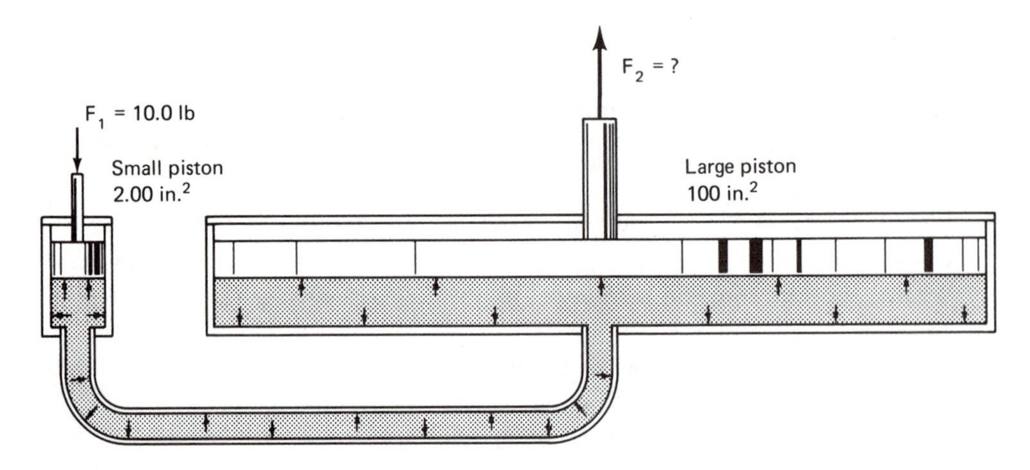

DATA:

$$F_1 = 10.0 \text{ lb}$$
$$A_{\text{small piston}} = 2.00 \text{ in}^2$$
$$A_{\text{large piston}} = 100 \text{ in}^2$$
$$P_1 = ? \qquad P_2 = ?$$
$$F_2 = ?$$
$$MA = ?$$

(a) BASIC EQUATION: $P_1 = \dfrac{F_1}{A}$

WORKING EQUATION: Same

SUBSTITUTION: $P_1 = \dfrac{10.0 \text{ lb}}{2.00 \text{ in}^2} = 5.00 \dfrac{\text{lb}}{\text{in}^2}$

(b) APPLYING PASCAL'S PRINCIPLE: $P_2 = 5.00 \dfrac{\text{lb}}{\text{in}^2}$

(c) BASIC EQUATION: $P_2 = \dfrac{F_2}{A}$

WORKING EQUATION: $F_2 = P_2 A$

SUBSTITUTION: $F_2 = \left(5.00 \dfrac{\text{lb}}{\cancel{\text{in}^2}}\right)(100 \cancel{\text{in}^2})$

$F_2 = 500 \text{ lb}$

(d) BASIC EQUATION: $MA = \dfrac{F_R}{F_E}$

WORKING EQUATION: Same

SUBSTITUTION: $MA = \dfrac{500 \cancel{\text{lb}}}{10.0 \cancel{\text{lb}}}$

$MA = 50$

PROBLEMS

1. The area of the small piston in a hydraulic jack is 0.750 in^2. The area of the large piston is 3.00 in^2. If a force of 15.0 lb is applied to the small piston, what weight can be lifted by the large one?
2. If the mechanical advantage of a hydraulic press is 25, what applied force is necessary to produce a pressing force of 1200 N?
3. What is the mechanical advantage of a hydraulic press which produces a pressing force of 7500 N when the applied force is 600 N?
4. If the mechanical advantage of a hydraulic press is 20, what applied force is necessary to produce a pressing force of 325 lb?
5. What is the mechanical advantage of a hydraulic press which produces a pressing force of 1000 lb when the applied force is 150 lb?
6. The small piston of a hydraulic press has an area of 8.00 cm^2. If the applied force is 25.0 N, what must the area of large piston be to exert a pressing force of 3600 N?
7. The *MA* of a hydraulic jack is 360. What force must be applied to lift an automobile weighing 12,000 N?
8. The small piston of a hydraulic press has an area of 4.00 in^2. If the applied force is 10.0 lb, what must the area of the large piston be to exert a pressing force of 1000 lb?
9. The *MA* of a hydraulic jack is 450. What is the weight of the heaviest automobile that can be lifted by an applied force of 8.00 lb?

SKETCH
DATA
$a = 1, b = 2, c = ?$

BASIC EQUATION
$a = bc$

WORKING EQUATION
$c = \frac{a}{b}$

SUBSTITUTION
$c = \frac{1}{2}$

15-5 *AIR PRESSURE*

We saw in Chapter 14 that air has weight and, as any fluid, it must exert pressure. For example, when a straw is used to drink, the air pressure inside the straw is reduced. As a result, the outside air pressure is higher than in the straw which forces the fluid up the straw.

Air pressure pushes down on fluid with one atmosphere pressure.

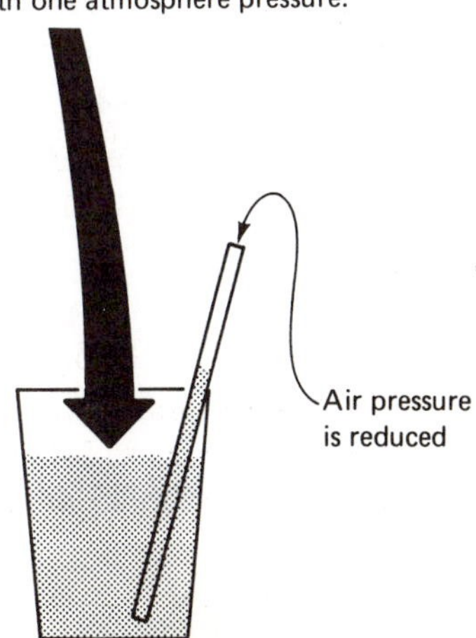

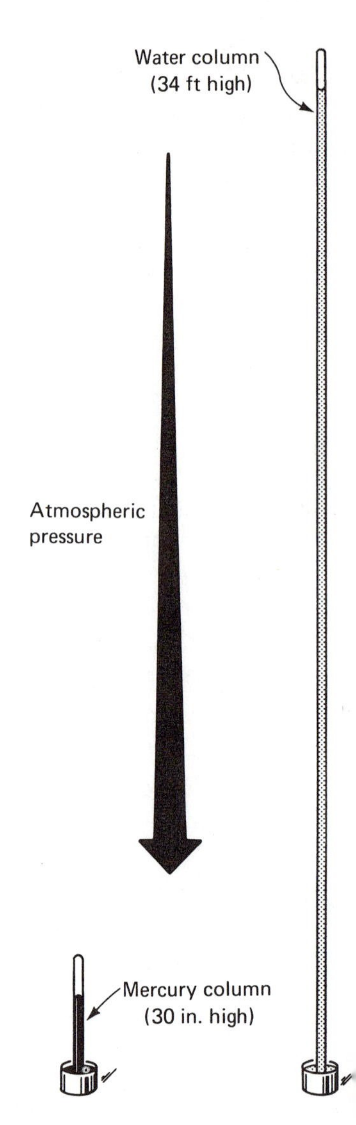

Earlier in this chapter we saw that the pressure on a submerged body increases as the body goes deeper into the liquid. Some creatures live near the bottom of the ocean where the pressure of the water above is so great that it would collapse any human body and most submarines. But through the process of evolution, such creatures have adapted to this tremendous pressure.

Similarly, we on earth live at the bottom of a fluid, air, which is several miles deep. The pressure from this fluid is normally 14.7 lb/in² at sea level. We do not feel this pressure because it normally is almost the same from all directions and also because our bodies have become accustomed to it. When the air pressure becomes unequal, its force becomes quite evident in the form of wind. This wind may be a cool summer breeze or the tremendous concentrated force of a tornado.

What is the pressure of our atmosphere equivalent to? Experiments have shown that the atmosphere supports a column of water 34 ft high in a tube in which the air has been removed. The atmosphere supports 30 in. of mercury in a similar tube. This is not surprising as mercury is 13.6 times as dense as water and

$$\frac{1}{13.6} \times 34 \text{ ft} \times \frac{12 \text{ in.}}{1 \text{ ft}} = 30 \text{ in.}$$

This 30-in. measurement should be familiar since the barometer reading on the TV weather programs is usually very close to this reading.

The pressure of the atmosphere can be and is usually expressed in terms of the pressure of an equivalent column of mercury. Air pressure at sea level is normally 29.9 in. or 76.0 cm or 760 mm of mercury.

How do we arrive at the 14.7 lb/in² measurement? In terms of mercury, its height is 29.9 in. or 2.49 ft. Its density is 13.6 × 62.4 lb/ft³. Therefore

$$\begin{aligned} P &= hD_w \\ &= 2.49 \cancel{\text{ft}} \times 13.6 \times 62.4 \frac{\text{lb}}{\cancel{\text{ft}^3}} \times \frac{1 \cancel{\text{ft}^2}}{144 \text{ in}^2} \\ &= 14.7 \frac{\text{lb}}{\text{in}^2} \end{aligned}$$

The pressure of two atmospheres would be 29.4 lb/in² or 59.8 in. or 152.0 cm of mercury. If the pressure is $\frac{1}{2}$ atmosphere at one point in the sky, it would be 7.35 lb/in² or approximately 15.0 in. or 38.0 cm of mercury.

15-6 *MEASURING GAS PRESSURE*

When we purchase bottled gas, the amount of gas and its density varies with the pressure. If the pressure is low, the amount of gas in the bottle is low. If the pressure is high, the bottle is "nearly full."

The gage that is usually used for checking the pressure in bottles and tires shows a reading of zero at normal atmospheric pressure. The pressure of the atmosphere is not included in this reading. The actual pressure, called *absolute pressure*, is the gage pressure reading plus the normal atmospheric pressure (14.7 lb/in^2).

15-7 *BUOYANCY*

A floating boat displaces its own weight of the water in which it floats. If two or more people get in the boat, the boat rides lower due to the increased weight in the boat.

Archimedes, a Greek philosopher, was one of the first to study fluids and formulated what is now called Archimedes' principle:

RULE

Any object apparently loses weight equal to the weight of the displaced liquid.

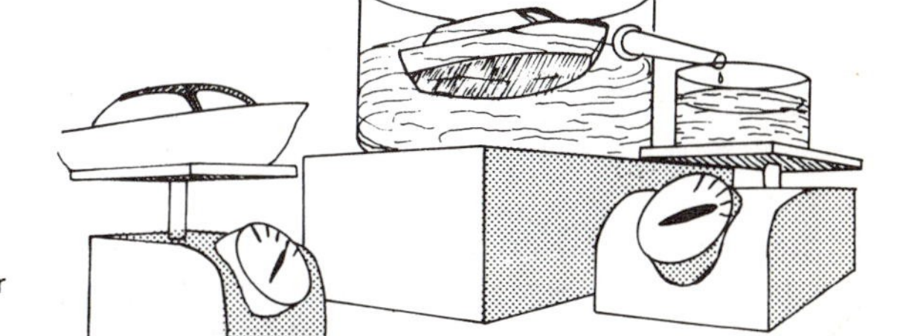

Weight of boat = Weight of displaced water

The object doesn't actually lose weight, but because of the depth difference, the force pushing up on the bottom of the object is greater than the force pushing down on the top of the object. This net upward force is called the *buoyant force.*

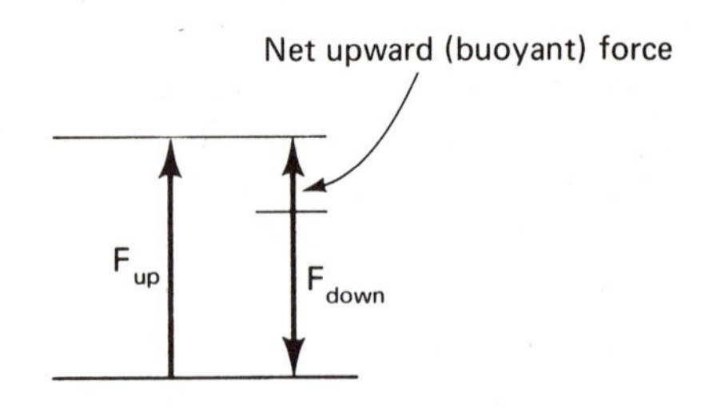

Archimedes' principle applies to gases as well as liquids. Lighter-than-air craft (for example, Goodyear's blimps) operate on this principle. Since they are filled with a gas lighter than air (helium), the buoyant force on them causes them to be supported by the air. Being "submerged" in the air, a blimp is buoyed up by the weight of the air it displaces.

PROBLEMS

Use Archimedes' principle to do the following.

1. If a metal alloy weighs 81 lb in air and 68 lb when under water, what is the buoyant force of the water?
2. If a rock weighs 25.7 N in air and 21.8 N in water, what is the buoyant force of the water?
3. If a rock displaces 1.21 ft³ of water, what is the buoyant force of the water?
4. If a metal casting displaces 327 cm³ of water, what is the buoyant force of the water?
5. A flat-bottom river barge is 30.0 ft wide, 85.0 ft long, and 15.0 ft deep. How many ft³ of water will it displace while the top stays 3.00 ft above the water? And what load in tons will the barge contain under these conditions if the barge weighs 160 tons in dry dock?
6. A flat-bottom river barge is 12.0 m wide, 30.0 m long, and 6.00 m deep. How many m³ of water will it displace while the top stays 1.00 m above the water? And what load in newtons will the barge contain under these conditions if the barge weighs 3.55×10^6 newtons in dry dock? The metric weight density of water is 9800 N/m³.

15-8 *SPECIFIC GRAVITY*

When we check the antifreeze in a radiator in winter, we are really determining the specific gravity of the liquid. *Specific gravity is a comparison of the density of a substance to that of water.* Because the density of antifreeze is different than the density of water, we can find the concentration of antifreeze (and thus the amount of protection from freezing) by measuring the specific gravity of the solution in the radiator.

The instrument used to measure specific gravity is called the hydrometer. It is a hollow glass tube weighted on one end so it will float upright. Its scale is calibrated so that in water it floats at the 1.000 mark. In less dense liquids the hydrometer will sink lower. In more dense liquids it does not sink as far. One other factor must be considered in the use of the hydrometer—that of temperature. Significant differences in readings will occur over a range of temperatures. Specific gravities of some common liquids:

Liquid	Specific Gravity
Ethyl alcohol	0.79
Benzene	0.9
Gasoline	0.68
Kerosene	0.8
Mercury	13.6
Sulfuric acid	1.84
Turpentine	0.87
Water	1.00
Seawater	1.025

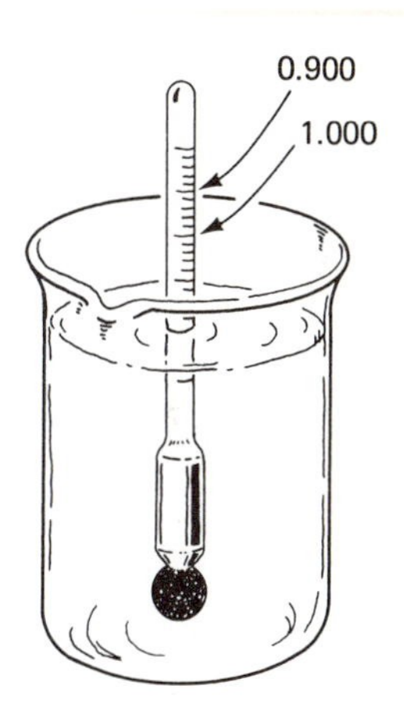

TEMPERATURE AND HEAT 16

16-1 *INTRODUCTION*

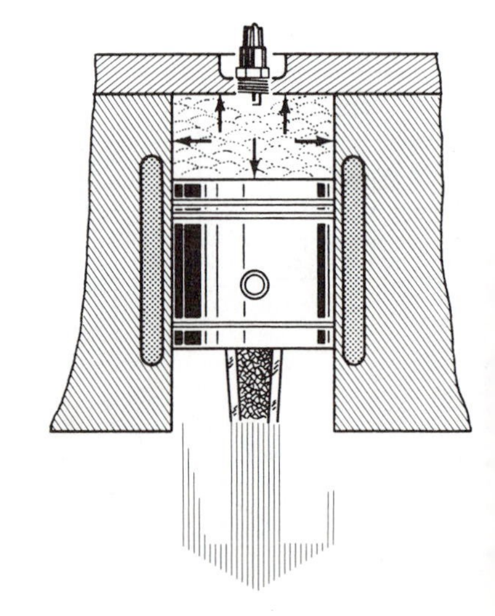

An understanding of temperature and heat is very important to all technicians. An automotive technician is concerned with the heat energy released by a fuel mixture in a combustion chamber. The excess heat produced by an engine must be transferred to the atmosphere. The highway technician is concerned with the expansion of bridges and roads when the temperature changes.

16-2 *TEMPERATURE*

Basically, temperature is a measure of the hotness or coldness of an object. Temperature could be measured in a simple way by using your hand to sense the hotness or coldness of an object. However, the range of temperatures which your hand can stand is too small, and your hand is not accurate enough to measure temperature. Therefore, other methods are used for measuring temperature.

*Temperatures, Steel Colors and Related Processes**

	Colors	Fahrenheit	Centigrade	Processes
Heat colors	White	2500°	1371°	Welding
		2400°	1315°	High speed steel hardening (2150–2450°F)
	Yellow white	2300°	1259°	
		2200°	1204°	
		2100°	1149°	
	Yellow	2000°	1093°	
		1900°	1036°	Alloy tool steel hardening (1500–1950°F)
	Orange red	1800°	981°	
		1700°	926°	
		1600°	871°	
	Light cherry red	1500°	815°	Carbon tool steel hardening (1350–1550°F)
	Cherry red	1400°	760°	
		1300°	704°	
	Dark red	1200°	648°	
		1100°	593°	High speed steel tempering (1000–1100°F)
	Very dark red	1000°	538°	Carbon tool steel tempering (300–1050°F)
		900°	482°	
	Black red in dull light or darkness	800°	426°	
		700°	371°	
Temper colors	Pale blue (590°F)	600°	315°	
	Violet (545°F)			
	Purple (525°F)	500°	260°	
	Yellowish brown (490°F)			
	Dark straw (465°F)	400°	204°	
	Light straw (425°F)			
		300°	149°	
		200°	93°	
		100°	38°	
		0°	18°	

*Allegheny Ludlum Steel Corp. Reprinted by permission.

We will use the fact that certain properties of matter vary with their temperature. For example, we could use the fact that when objects are heated they give off light of different colors. When an object is heated, in the absence of chemical reactions, it first gives off red light. As it is heated more, it appears white (see the table on page 209).

Chemical reactions sometimes cause different colors to be given off. When carbon steel is heated and exposed to air, several colors are given off before it appears red. This is due to a chemical reaction involving the carbon. If we could measure the color of the light given off, we could then determine the temperature. Although this works only for high temperatures, it is used in the production of metal alloys. The temperature of hot molten metals is determined this way.

Another property of matter that can be used is that the volume of a liquid or a solid changes as its temperature changes. The liquid in glass thermometers is an example. This type of thermometer consists of a hollow glass bulb and a hollow glass tube joined together as shown. A small amount of liquid mercury or alcohol is placed in the bulb. The air is removed from the tube. When the liquid is heated, it expands and rises up the glass tube. The height to which the liquid rises indicates the temperature.

The thermometer is standardized by marking two points on the glass which indicate the liquid level at two known temperatures. The temperatures used are the melting point of ice (called the ice point) and the boiling point of water at sea level. The distance between these marks is then divided up into equal segments called degrees.

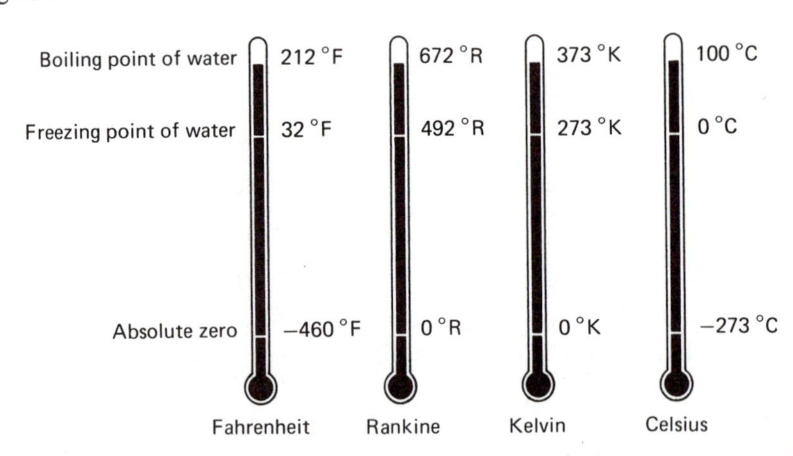

Four Basic Temperature Scales

There are four temperature scales which we will study. You are most familiar with the Fahrenheit scale on which the ice point is 32°F and the boiling point is 212°F. When we write a certain temperature, we give a number then the degree symbol (°) and a capital letter which indicates the scale. The letter F following the degree symbol is not to be considered as a unit when working problems. Temperatures below zero on a scale are written as negative numbers. Thus 30° below zero on the Fahrenheit scale is written as −30°F.

Another common temperature scale is the Celsius scale, which is replacing the centigrade scale. The ice point is 0°C and the boiling point is 100°C. The relationship between Fahrenheit temperatures (T_F) and Celsius temperatures (T_C) is shown below.

$$T_F = \frac{9T_C}{5} + 32^\circ$$

$$T_C = \frac{5(T_F - 32^\circ)}{9}$$

where: T_F = the Fahrenheit temperature
T_C = the Celsius temperature

EXAMPLE 1: The average human body temperature is 98.6°F. What is it in Celsius?

DATA: $T_F = 98.6^\circ$
$T_C = ?$

BASIC EQUATION: $T_C = \frac{5(T_F - 32^\circ)}{9}$

WORKING EQUATION: Same

SUBSTITUTION: $T_C = \frac{5(98.6^\circ - 32^\circ)}{9}$

$T_C = \frac{5(66.6^\circ)}{9}$

$T_C = 37.0^\circ$

In some technical work it is necessary to use the absolute temperature scales, which are the Rankine scale and the Kelvin scale. These are called absolute scales because 0° on either scale refers to the lowest limit of temperature, called absolute zero. Absolute zero and lower temperatures can never be attained. The reasons are discussed in the next section. There is no such limit on high temperatures.

The Rankine scale is closely related to the Fahrenheit scale. The relationship is:

$$T_R = T_F + 460^\circ$$

The Kelvin scale is closely related to the Celsius scale. The relationship is:

$$T_K = T_C + 273°$$

EXAMPLE 2: The lower limit on temperatures is 0°R. Find this limit on the Fahrenheit scale.

DATA: $T_R = 0°$
$T_F = ?$

BASIC EQUATION: $T_R = T_F + 460°$

WORKING EQUATION: $T_F = T_R - 460°$

SUBSTITUTION: $T_F = 0° - 460°$
$T_F = -460°$

PROBLEMS

Convert the following temperatures as indicated.

1. $T_F = 70°$, $T_C =$ ______
2. $T_F = 120°$, $T_C =$ ______
3. $T_F = 250°$, $T_C =$ ______
4. $T_C = 17°$, $T_F =$ ______
5. $T_C = 125°$, $T_F =$ ______
6. $T_C = 5°$, $T_F =$ ______
7. $T_F = 10°$, $T_C =$ ______
8. $T_F = 20°$, $T_C =$ ______
9. $T_C = -5°$, $T_F =$ ______
10. $T_F = -50°$, $T_C =$ ______
11. $T_F = 150°$, $T_R =$ ______
12. $T_F = 55°$, $T_R =$ ______
13. $T_R = 600°$, $T_F =$ ______
14. $T_C = 77°$, $T_K =$ ______
15. $T_C = -50°$, $T_K =$ ______
16. $T_K = 200°$, $T_C =$ ______
17. $T_K = 6000°$, $T_C =$ ______
18. The melting point of pure iron is 1505°C. What Fahrenheit temperature is this?
19. Steel heated to 1650°F is cherry red. What Celsius temperature is this?
20. A welding white heat is approximately 1400°C. What is this temperature expressed in Fahrenheit degrees?

16-3 *HEAT*

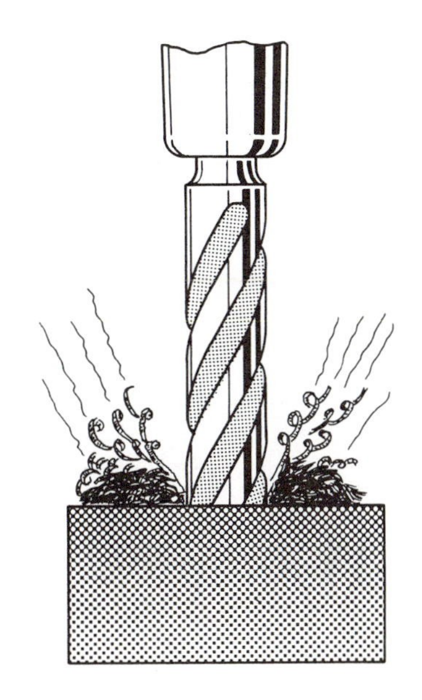

When a machinist drills a hole in a metal block, it becomes very hot. As the drill does mechanical work on the metal, the temperature of the metal increases. How can we explain this? We need to look at the difference between the metal at low temperatures and at high temperatures. At high temperatures the atoms in the metal vibrate more rapidly than at low temperatures. Their velocity is higher at high temperatures, and thus their kinetic energy ($KE = \frac{1}{2} mv^2$) is greater.

To raise the temperature of a material, we must speed up the atoms, that is, we must add energy to them. *Heat is the name given to this energy which is being added to or taken from a material.*

Drilling a hole in a metal block causes a temperature increase. As the drill turns, it collides with atoms of the metal, causing them to speed up. This mechanical work done on the metal has caused an increase in the energy (speed) of the atoms. For this reason, any friction between two surfaces results in a temperature rise of the materials.

Since heat is a form of energy, we could measure it in ft lb or joules, which are energy units. However, it was not always known that heat was a form of energy, and special units for heat were developed and are still in use.

These units are the BTU (British Thermal Unit) in the English system and the calorie in the metric system. The BTU is the amount of heat (energy) necessary to raise the temperature of 1 lb of water 1°F. The calorie is the amount of heat (energy) necessary to raise the temperature of 1 gram of water 1°C.

We said in the last section that temperatures below absolute zero cannot be reached. Now we can see the reason for this. To lower the temperature of a substance, we need to remove some of the energy of motion of the molecules (heat). When we have removed all the heat possible (when the molecules are moving as slowly as possible), we have reached the lowest possible temperature called absolute zero. Lower temperatures cannot be reached because all the heat has been removed. However, there is no upper limit on temperature because we can always add more heat (energy) to a substance to increase its temperature.

16-4 *HEAT TRANSFER*

The movement of heat from a hot engine to the air is necessary to keep the engine from overheating. The heat produced by a furnace is transferred to the various rooms in a house. The movement of heat is a major technical problem.

The transfer of heat from one object to another is always from the warmer object to the colder one or from the warmer part of an object to a colder part.

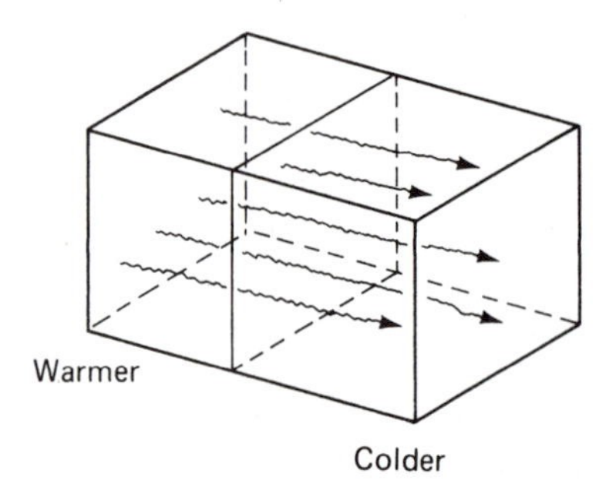

There are three methods of heat transfer: conduction, convection, and radiation. The usual method of heat transfer in solids is conduction. When one end of a metal rod is heated, the molecules in that end move faster than before. These molecules collide with other molecules and cause them to move faster also. In this way the heat is transferred from one end of the metal to the other. Another example of conduction is the transferring of the excess heat produced in the combustion chamber of an engine through the engine block into the water coolant.

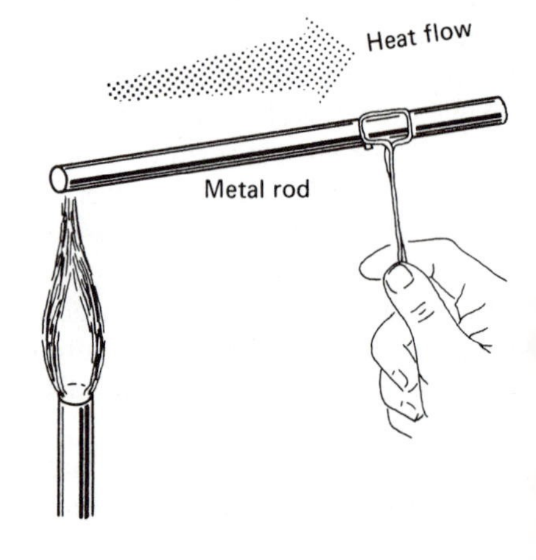

The conduction of heat through some materials is better than through others. A poor conductor of heat is called an insulator. A list of several good conductors and poor conductors is given here.

Good Heat Conductors	Poor Heat Conductors
Copper Aluminum Steel	Asbestos Glass Wood Air

Another method of heat transfer is called convection. This is the movement of warm gases or liquids from one place to another. The wind carries heat along with it. The water coolant in an engine carries hot water from the engine block to the radiator by a convection process. The wind is a natural convection process. The engine coolant is a forced convection process because it depends on a pump.

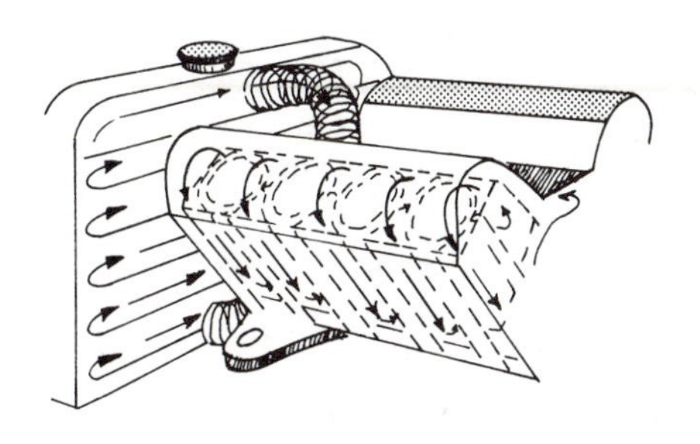

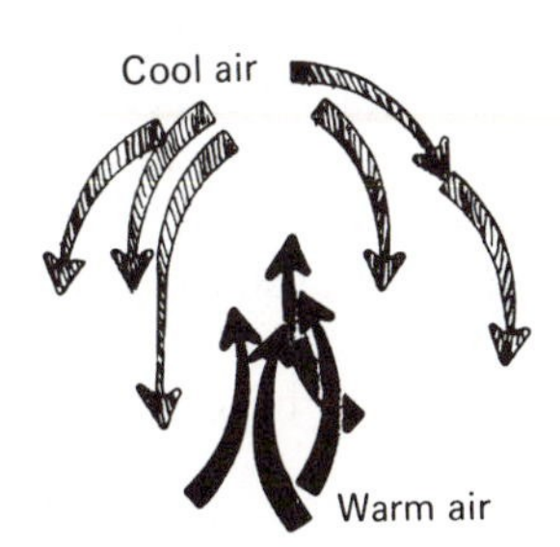

The natural convection process occurs because warm air is lighter (less dense) than cool air. The warm air rises and the cool air falls.

The third method of heat transfer is radiation. Put your hand several inches from a hot iron. The heat you feel is not transferred by conduction because air is a poor conductor. It is not transferred by convection because the hot air rises. This kind of heat transfer is called radiation. This radiant heat is similar to light and passes through air, glass, and the vacuum of space. The energy that comes to us from the sun is in the form of radiant energy.

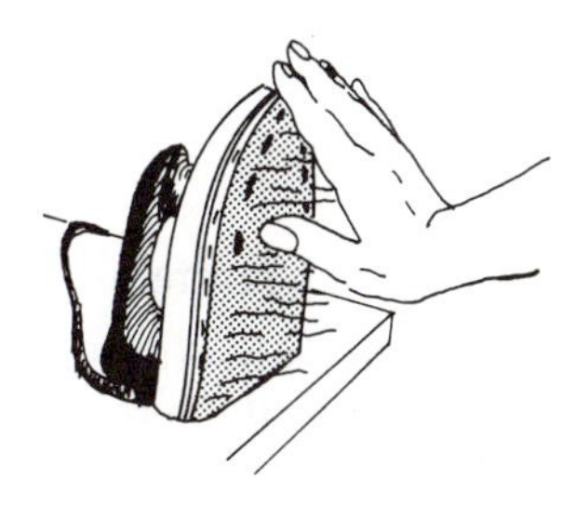

Dark objects absorb radiant heat and light objects reflect radiant heat. This is why we feel cooler on a hot day in light colored clothing than in dark clothing.

16-5 *SPECIFIC HEAT*

If we place a piece of steel and a pan of water in the direct sunlight, we would notice that the water becomes only slightly warmer while the iron gets quite hot. Why should one get so much hotter than the other? If equal masses of steel and water were placed over the same flame for one minute, the temperature of the steel would increase almost ten times more than that of the water. The water has a greater capacity to absorb heat. The specific heat of a substance is a measure of its capacity to absorb or give off heat per degree change in temperature. This property of water to absorb or give up large amounts of heat makes it an effective substance to transfer heat in industrial processes.

The specific heat of a substance is the amount of heat necessary to change the temperature of 1 lb 1°F (1 gram 1°C in the metric system). By formula,

$$c = \frac{Q}{w\,\Delta T} \qquad \text{(English system)}$$

or

$$c = \frac{Q}{m\,\Delta T} \qquad \text{(metric system)}$$

where: c = specific heat
Q = heat
w = weight
m = mass
ΔT = change in temperature

These equations can be rearranged to give the amount of heat added or taken away from a material to produce a certain temperature change.

$$Q = cw\,\Delta T \qquad \text{(English)}$$
$$Q = cm\,\Delta T \qquad \text{(metric)}$$

A short table of specific heats is given below.

Substance	Specific Heat*
Air	0.24
Aluminum	0.22
Brass	0.091
Copper	0.093
Ice	0.51
Steam	0.48
Steel	0.115
Water	1.00

*BTU/lb °F (English system) or cal/g °C (metric system).

EXAMPLE: How much heat must be added to 10.0 lb of steel to raise its temperature 150°F?

DATA:
$w = 10.0$ lb
$\Delta T = 150°$F
$c = 0.115$ BTU/lb°F (from table)
$Q = ?$

BASIC EQUATION: $Q = cw\,\Delta T$

WORKING EQUATION: Same

SUBSTITUTION:
$$Q = 0.115 \frac{\text{BTU}}{\cancel{\text{lb}}\cancel{°\text{F}}} \times 10.0\,\cancel{\text{lb}} \times 150\cancel{°\text{F}}$$
$$Q = 173 \text{ BTU}$$

PROBLEMS

Find Q for the following materials, weights, and temperature changes (1–7).

1. Steel, $w = 3.00$ lb, $\Delta T = 500°$F, $Q =$ _______ BTU
2. Copper, $w = 155$ g, $\Delta T = 170°$C, $Q =$ _______ calories
3. Water, $w = 19.0$ lb, $\Delta T = 200°$F, $Q =$ _______ BTU
4. Water, $m = 250$ g, $\Delta T = 17°$C, $Q =$ _______ calories
5. Ice, $w = 5.00$ kg, $\Delta T = 20°$C, $Q =$ _______ calories
6. Steam, $w = 5.00$ lb, $\Delta T = 40°$F, $Q =$ _______ BTU
7. Aluminum, $m = 79.0$ g, $\Delta T = 16°$C, $Q =$ _______ calories
8. How much heat must be added to 750 g of steel to raise its temperature from 75°C to 300°C?
9. How much heat must be added to 1200 lb of copper to raise its temperature from 100°F to 450°F?
10. How much heat is given off by 500 lb of aluminum when it cools from 650°F to 75°F?
11. How much heat is absorbed by an electric freezer in lowering the temperature of 1850 g of water from 80°C to 10°C?
12. A 520-kg steam boiler is made of steel and contains 300 kg of water at 40°C. Assuming that 75% of the heat is delivered to the boiler and water, how many calories are required to raise the temperature of both the boiler and water to 100°C?

SKETCH
DATA
$a = 1, b = 2,$
$c = ?$
BASIC EQUATION
$a = bc$
WORKING EQUATION
$c = \frac{a}{b}$
SUBSTITUTION
$c = \frac{1}{2}$

16-6 METHOD OF MIXTURES

When two substances at different temperatures are "mixed" together, heat flows from the warmer body to the cooler body until they reach the same temperature. Part of the heat lost by the hotter body is transferred to the colder body and part is lost to the surrounding objects or the air. In many cases most all the heat is transferred to the colder body. We will assume that all the heat lost by the warmer body is transferred to the cooler body. That is, the heat lost by the warmer body equals the heat gained by the cooler body. The amount of heat lost or gained by a body is

$$Q = cw\,\Delta T \quad \text{or} \quad Q = cm\,\Delta T$$

By formula,

$$Q_{\text{lost}} = Q_{\text{gained}}$$

$$c_l w_l(T_l - T_f) = c_g w_g(T_f - T_g)$$

Where the subscript l refers to the warmer body which *loses* heat, the subscript g refers to the cooler body which *gains* heat, and T_f is the final temperature of the mixture.

EXAMPLE 1: A 10.0-lb piece of hot copper is dropped into 30.0 lb of water at 50°F. If the final temperature of the mixture is 65°F, what was the initial temperature of the copper?

DATA:

$$w_l = 10.0 \text{ lb} \qquad w_g = 30.0 \text{ lb}$$

$$c_l = 0.093 \frac{\text{BTU}}{\text{lb°F}} \qquad c_g = 1 \frac{\text{BTU}}{\text{lb°F}}$$

$$T_l = ? \qquad T_g = 50°\text{F}$$

$$T_f = 65°\text{F}$$

BASIC EQUATION: $c_l w_l(T_l - T_f) = c_g w_g(T_f - T_g)$

WORKING EQUATION: In this type of problem, it is advisable to substitute directly into the basic equation.

SUBSTITUTION:

$$0.093 \frac{\text{BTU}}{\cancel{\text{lb}}°\text{F}} \times 10.0\,\cancel{\text{lb}}(T_l - 65°\text{F}) = 1 \frac{\text{BTU}}{\cancel{\text{lb}}\cancel{°\text{F}}} \times 30.0\,\cancel{\text{lb}}(65\cancel{°\text{F}} - 50\cancel{°\text{F}})$$

$$0.930T_l \frac{\text{BTU}}{°\text{F}} - 60.5 \text{ BTU} = 450 \text{ BTU}$$

$$0.930T_l \frac{\text{BTU}}{°\text{F}} = 510.5 \text{ BTU}$$

$$T_l = \frac{510.5\,\cancel{\text{BTU}}}{0.930\,\cancel{\text{BTU}}/°\text{F}}$$

$$T_l = 549°\text{F}$$

EXAMPLE 2: If 200 g of steel at 220°C is added to 500 g of water at 10°C, what is the final temperature of the mixture?

DATA:

$$c_g = 1.00\,\frac{\text{cal}}{\text{g}^\circ\text{C}} \qquad c_l = 0.115\,\frac{\text{cal}}{\text{g}^\circ\text{C}}$$
$$m_g = 500\text{ g} \qquad m_l = 200\text{ g}$$
$$T_g = 10^\circ\text{C} \qquad T_l = 220^\circ\text{C}$$
$$T_f = ?$$

BASIC EQUATION: $c_l m_l(T_l - T_f) = c_g m_g(T_f - T_g)$

WORKING EQUATION: In this type of problem, it is advisable to substitute directly into the basic equation.

SUBSTITUTION:

$$0.115\,\frac{\text{cal}}{\cancel{\text{g}}^\circ\text{C}} \times 200\,\cancel{\text{g}}(220^\circ\text{C} - T_f) = 1.00\,\frac{\text{cal}}{\cancel{\text{g}}^\circ\text{C}} \times 500\,\cancel{\text{g}}(T_f - 10^\circ\text{C})$$
$$5060\,\frac{\text{cal}\cancel{^\circ\text{C}}}{\cancel{^\circ\text{C}}} - 23\,\frac{\text{cal}}{^\circ\text{C}}T_f = 500\,\frac{\text{cal}}{^\circ\text{C}}T_f - 5000\,\frac{\text{cal}\cancel{^\circ\text{C}}}{\cancel{^\circ\text{C}}}$$
$$10{,}060\text{ cal} = 523\,\frac{\text{cal}}{^\circ\text{C}}T_f$$
$$\frac{10{,}060\,\cancel{\text{cal}}}{523\,\cancel{\text{cal}}/^\circ\text{C}} = T_f$$
$$19.2^\circ\text{C} = T_f$$

PROBLEMS

1. Two and one-half pounds of steel is dropped into 11.0 lb of water at 75°F. The final temperature is 84°F. What was the initial temperature of the steel?
2. Five pounds of water at 200°F is mixed with 7.00 lb of water at 65°F. What is the final temperature of the mixture?
3. A 250-g piece of tin at 99°C is dropped in 100 g of water at 10°C. If the final temperature of the mixture is 20°C, what is the specific heat of the tin?
4. How many grams of water at 20°C are necessary to change 800 g of water at 90°C to 50°C?
5. One hundred fifty-nine pounds of aluminum at 500°F are dropped into 400 lb of water at 60°F. What is the final temperature?
6. Forty-two pounds of steel at 670°F is dropped into 100 lb of water at 75°F. What is the final temperature of the mixture?
7. If 1250 g of copper at 20°C is mixed with 500 g of water at 95°C, what is the final temperature of the mixture?
8. If 500 g of brass at 200°C and 300 g of steel at 150°C are added to 900 g of water in a 150 g aluminum pan and both are at 20°C, what is the final temperature of this mixture assuming no loss of heat to the surroundings?

SKETCH
DATA
$a = 1$, $b = 2$,
$c = ?$

BASIC EQUATION
$a = bc$

WORKING EQUATION
$c = \frac{a}{b}$

SUBSTITUTION
$c = \frac{1}{2}$

17 THERMAL EXPANSION OF SOLIDS AND LIQUIDS

17-1 *EXPANSION OF SOLIDS*

Most solids expand when heated and contract when cooled. They expand or contract in all three dimensions—length, width, and thickness.

When a solid is heated, the expansion is due to the increased length of the vibrations of the atoms and molecules. This results in the solid expanding in all directions. This increase in volume results in a decrease in weight density, which is discussed in Chapter 14. Engineers, technicians, and designers must know the effects of thermal expansion.

You have no doubt heard of highway pavements buckling on a hot summer day.

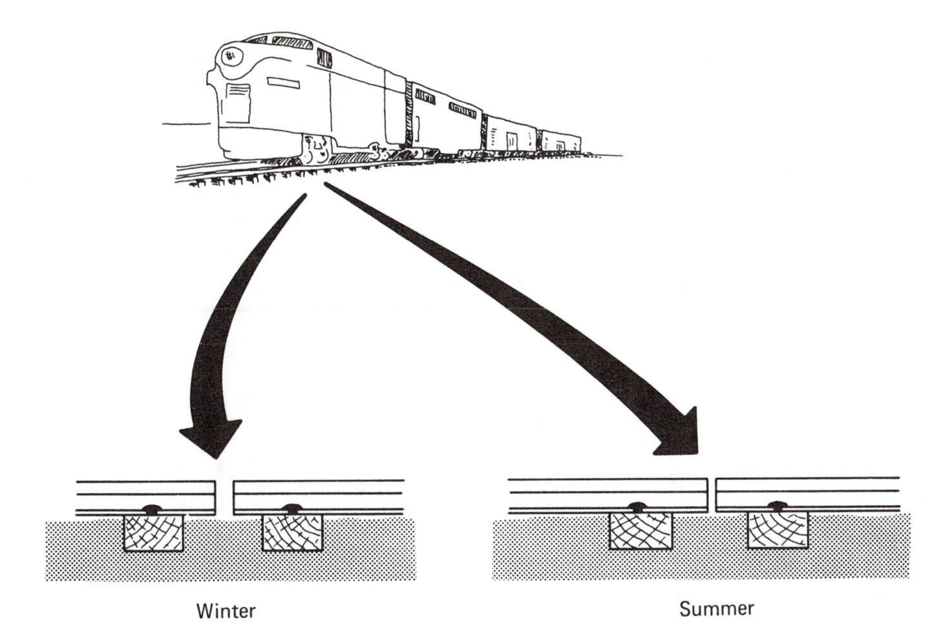

The clicking noise of a train's wheels passing over the rails can be heard more in winter than in summer. The space between rails is larger in winter than in summer. If the rails were placed snugly end to end in the winter, they would buckle in the summer.

Similarly, bridges, pipelines, and buildings must be designed and built to allow for this expansion and contraction.

There are some advantages to solids expanding. A bimetal bar is made by fusing two different metals together side by side as illustrated below. When heated, the brass expands more than the steel, which makes the bar curve as below. The thermostat operates on this principle.

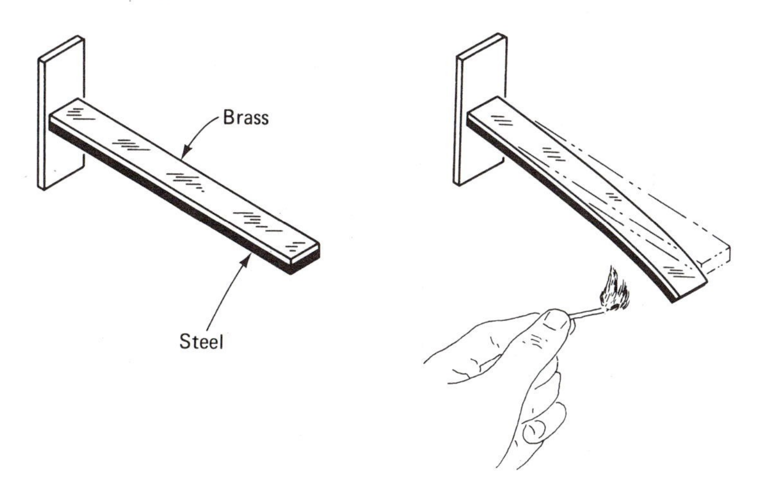

As shown, the basic parts of a thermostat are a bimetal strip on the right and a regular metal strip on the left. The bimetal strip of brass and steel bends with the temperature. The regular metal strip is moved by hand to set the temperature desired.

This particular bimetal bar is made and placed so that it bends to the left when cooled. As a result, when it comes in contact with the strip on the left, it completes a circuit which turns on the furnace. When the room warms to the desired temperature, the bimetal bar moves back to the right, which opens the contacts and shuts off the heat.

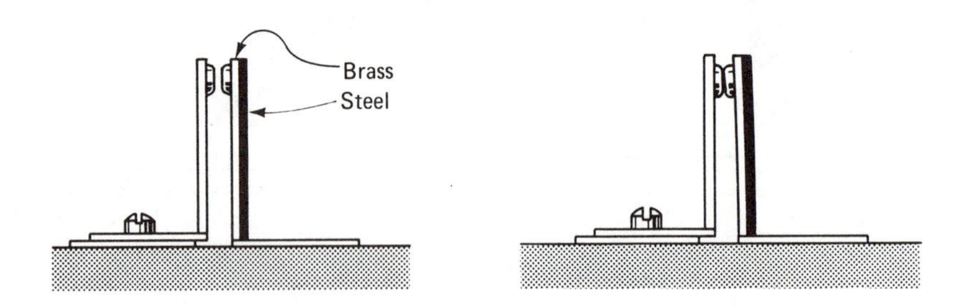

17-2 *LINEAR EXPANSION*

The amount that a solid expands depends on the following:

(a) The material—different materials expand at different rates. Steel expands at a rate less than that of brass.

(b) The length of the solid—the longer the solid, the larger the expansion. A 12-in. steel rod will expand twice as much as a 6-in. steel rod.

(c) The amount of change in temperature—the greater the change in temperature, the greater the expansion.

The above can be written by formula:

$$\Delta l = \alpha l \, \Delta T$$

where: Δl = change in length
α = a constant called the coefficient of linear expansion*
l = the original length or any other linear measurement
ΔT = change in temperature

*Defined as the change in the unit length of a solid when its temperature is changed one degree.

The following table lists the coefficients of linear expansion for some common solids.

Material	α (English)	α (Metric)
Aluminum	$1.3 \times 10^{-5}/F°$	$2.3 \times 10^{-5}/C°$
Brass	$1.0 \times 10^{-5}/F°$	$1.9 \times 10^{-5}/C°$
Concrete	$6.0 \times 10^{-6}/F°$	$1.1 \times 10^{-5}/C°$
Copper	$9.5 \times 10^{-6}/F°$	$1.7 \times 10^{-5}/C°$
Glass	$5.1 \times 10^{-6}/F°$	$9.0 \times 10^{-6}/C°$
Pyrex	$1.7 \times 10^{-6}/F°$	$4.0 \times 10^{-6}/C°$
Steel	$6.5 \times 10^{-6}/F°$	$1.3 \times 10^{-5}/C°$
Zinc	$1.5 \times 10^{-5}/F°$	$2.6 \times 10^{-5}/C°$

EXAMPLE 1: A steel railroad rail is 40.0 ft long at 0°F. How much will it expand when heated to 100°F?

DATA:

$l = 40.0$ ft
$\Delta T = 100°F - 0°F = 100°F$
$\alpha = 6.5 \times 10^{-6}/F°$
$\Delta l = ?$

BASIC EQUATION: $\Delta l = \alpha l \, \Delta T$

WORKING EQUATION: Same

SUBSTITUTION:

$\Delta l = (6.5 \times 10^{-6}/F°)(40.0 \text{ ft})(100°F)$
$\Delta l = 0.026$ ft or 0.312 in.

Pipes which undergo large temperature changes are usually installed with a loop as shown to allow for expansion and contraction.

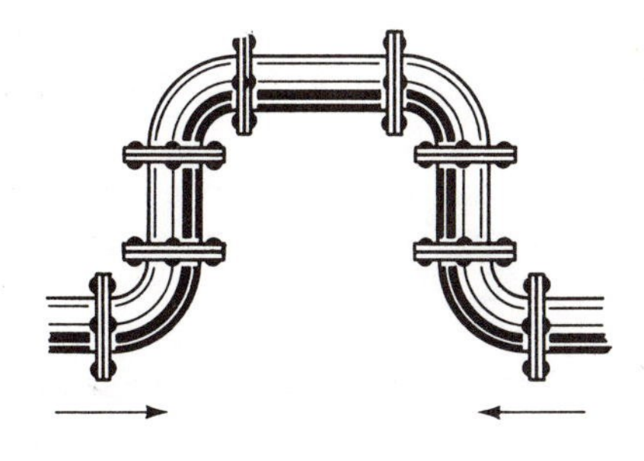

EXAMPLE 2: What allowance for expansion must be made for a steel pipe 120 m long which handles coolants and must undergo temperature changes of 200°C?

DATA:

$$\alpha = 1.3 \times 10^{-5}/\text{C}^\circ$$
$$l = 120 \text{ m}$$
$$\Delta T = 200^\circ\text{C}$$
$$\Delta l = ?$$

BASIC EQUATION: $\Delta l = \alpha l \, \Delta T$

WORKING EQUATION: Same

SUBSTITUTION:

$$\Delta l = (1.3 \times 10^{-5}/\text{C}^\circ)(120 \text{ m})(200^\circ\text{C})$$
$$\Delta l = 0.312 \text{ m} \quad \text{or} \quad 31.2 \text{ cm}$$

PROBLEMS

SKETCH
DATA
$a = 1, b = 2,$
$c = ?$
BASIC EQUATION
$a = bc$
WORKING EQUATION
$c = \frac{a}{b}$
SUBSTITUTION
$c = \frac{1}{2}$

1. Find the increase in length of a copper rod 200 ft long at 40°F when it is heated to 200°F.
2. Find the increase in length of a zinc rod 50.0 m long at 15°C when it is heated to 130°C.
3. Compute the increase in length of 300.00 m of copper wire when its temperature changes from 14°C to 34°C.
4. A steel pipe 25.0 ft long is installed at 80°F. Find the decrease in length when coolants at −90°F pass through the pipe.
5. A steel tape measures 200.00 m at 10°C. What is its length at 50°C?
6. A brass rod 20.0 in. long expands 0.008 in. when it is heated. Find the temperature change.
7. The road bed on a bridge 500 ft long is made of concrete. What allowance is needed for temperatures of −40°F in winter and 140°F in summer?
8. An aluminum plug has a diameter of 10.003 cm at 40°C. At what temperature will it fit exactly into a hole of constant diameter 10.000 cm?
9. The diameter of a steel drill at 50°F is 0.750 in. Find its diameter at 350°F.
10. A brass ball has a diameter 12.00 cm and is 0.011 mm too large to pass through a hole in a copper plate when the ball and plate are at a temperature of 20°C. What is the temperature of the ball when it will just pass through the plate, assuming the temperature of the plate does not change? What is the temperature of the plate when the ball will just pass through, assuming the temperature of the ball does not change?

17-3 *AREA AND VOLUME EXPANSION OF SOLIDS*

Solids expand in width and thickness as well as in length when heated. To allow for this expansion the following formulas are used:

$$\text{Area expansion:}\quad \Delta A = 2\alpha A\,\Delta T$$
$$\text{Volume expansion:}\quad \Delta V = 3\alpha V\,\Delta T$$

EXAMPLE: The top of a circular copper disc has an area of 64.2 in² at 20°F. What is the change in area when the temperature is increased to 150°F?

DATA:

$$\alpha = 9.5 \times 10^{-6}/\text{F}^\circ$$
$$A = 64.2\ \text{in}^2$$
$$\Delta T = 150^\circ\text{F} - 20^\circ\text{F} = 130^\circ\text{F}$$
$$\Delta A = ?$$

BASIC EQUATION: $\Delta A = 2\alpha A\,\Delta T$

WORKING EQUATION: Same

SUBSTITUTION:

$$\Delta A = 2(9.5 \times 10^{-6}/\text{F}^\circ)(64.2\ \text{in}^2)(130^\circ\text{F})$$
$$\Delta A = 0.159\ \text{in}^2$$

PROBLEMS

1. A brass cylinder has a cross-sectional area of 74.8 in² at 10°F. Find its change in area when heated to 230°F.
2. The volume of the cylinder in problem 1 is 237 in³ at 240°F. Find its change in volume when cooled to −75°F.
3. An aluminum pipe has a cross-sectional area of 88.4 cm² at 10°C. What is its cross-sectional area when heated to 150°C?
4. A steel pipe has a cross-sectional area of 27.2 in² at 30°F. What is its cross-sectional area when heated to 180°F?
5. A glass plug has a volume of 60.00 cm³ at 12°C. What is its volume at 76°C?
6. A section of concrete dam is a rectangular solid 20.0 ft by 50.0 ft by 80.0 ft at 110°F. What allowance for change in volume is necessary for a temperature of −20°F?

SKETCH
DATA
$a = 1,\ b = 2,$
$c = ?$

BASIC EQUATION
$a = bc$

WORKING EQUATION
$c = \frac{a}{b}$

SUBSTITUTION
$c = \frac{1}{2}$

Liquids, likewise, generally expand when heated and contract when cooled. The thermometer is made using this principle.

When a thermometer is placed under your tongue, the heat from your mouth causes the mercury in the bottom of the thermometer to expand. Mercury is then forced to rise up the thin calibrated tube. The formula for volume expansion of liquids is:

$$\Delta V = \beta V \Delta T$$

where β is the coefficient of volume expansion for liquids.

The following table lists the coefficients of volume expansion for some common liquids.

Liquid	β (English)	β (Metric)
Acetone	$8.28 \times 10^{-4}/\text{F}^\circ$	$1.49 \times 10^{-3}/\text{C}^\circ$
Alcohol, ethyl	$6.62 \times 10^{-4}/\text{F}^\circ$	$1.12 \times 10^{-3}/\text{C}^\circ$
Carbon tetrachloride	$6.89 \times 10^{-4}/\text{F}^\circ$	$1.24 \times 10^{-3}/\text{C}^\circ$
Mercury	$1.0 \times 10^{-4}/\text{F}^\circ$	$1.8 \times 10^{-4}/\text{C}^\circ$
Petroleum	$5.33 \times 10^{-4}/\text{F}^\circ$	$9.6 \times 10^{-4}/\text{C}^\circ$
Turpentine	$5.39 \times 10^{-4}/\text{F}^\circ$	$9.7 \times 10^{-4}/\text{C}^\circ$
Water	$1.17 \times 10^{-4}/\text{F}^\circ$	$2.1 \times 10^{-4}/\text{C}^\circ$

EXAMPLE 1: If petroleum at 0°C occupies 250 liters, what is its volume at 50°C?

DATA:

$\beta = 9.6 \times 10^{-4}/\text{C}^\circ$
$V = 250$ liters
$\Delta T = 50°\text{C}$
$\Delta V = ?$

BASIC EQUATION: $\Delta V = \beta V \Delta T$

WORKING EQUATION: Same

SUBSTITUTION:

$\Delta V = (9.6 \times 10^{-4}/\cancel{\text{C}^\circ})(250 \text{ liters})(50\cancel{^\circ\text{C}})$
$\Delta V = 12$ liters
Volume at 50°C = 250 liters + 12 liters = 262 liters

EXAMPLE 2: What is the increase in volume of 18.2 in³ of water when heated from 40°F to 180°F?

DATA:
$\beta = 1.17 \times 10^{-4}/\text{F}^\circ$
$V = 18.2 \text{ in}^3$
$\Delta T = 180°\text{F} - 40°\text{F} = 140°\text{F}$
$\Delta V = ?$

BASIC EQUATION: $\Delta V = \beta V \Delta T$

WORKING EQUATION: Same

SUBSTITUTION:
$\Delta V = (1.17 \times 10^{-4}/\text{F}^\circ)(18.2 \text{ in}^3)(140 \text{F}^\circ)$
$\Delta V = 0.298 \text{ in}^3$

PROBLEMS

1. A quantity of carbon tetrachloride occupies 625 liters at 12°C. What is its volume at 48°C?
2. Some mercury occupies 157 in³ at −30°F. What is the change in volume when heated to 90°F?
3. Some petroleum occupies 287 ft³ at 0°F. What volume does it occupy at 80°F?
4. What is the increase in volume of 35 liters of acetone when heated from 28°C to 38°C?
5. Some water at 180°F occupies 3780 ft³. What is its volume at 122°F?
6. A 1200-liter tank of petroleum is completely filled at 10°C. How much spills over if the temperature rises to 45°C?
7. Calculate the increase in volume of 215 cm³ of mercury when its temperature increases from 10°C to 25°C.
8. What is the drop in volume of alcohol in a railroad tank car which contains 2000 ft³ if the temperature drops from 75°F to 54°F?

SKETCH
DATA
$a = 1, b = 2,$
$c = ?$
BASIC EQUATION
$a = bc$
WORKING EQUATION
$c = \frac{a}{b}$
SUBSTITUTION
$c = \frac{1}{2}$

17-5 *EXPANSION OF WATER*

Water is unusual in its expansion characteristics. Most of us have seen the mound in the middle of each ice cube in ice cube trays. This evidence shows the expansion of water during its change of state from liquid to solid form. Nearly all liquids are most dense at their lowest temperature before a change of state to become solids. As the temperature drops, the molecular motion slows and the substance becomes denser.

Water does not follow the general rule stated above. Because of its unusual structural characteristics, water is most dense at 4°C or 39.2°F instead of 0°C or 32°F. A graph of its change in density with an increase in temperature appears at the right.

When ice melts at 0°C or 32°F, the water formed *contracts* as the temperature is raised to 4°C or 39.2°F. Then it begins to *expand* as do most other liquids.

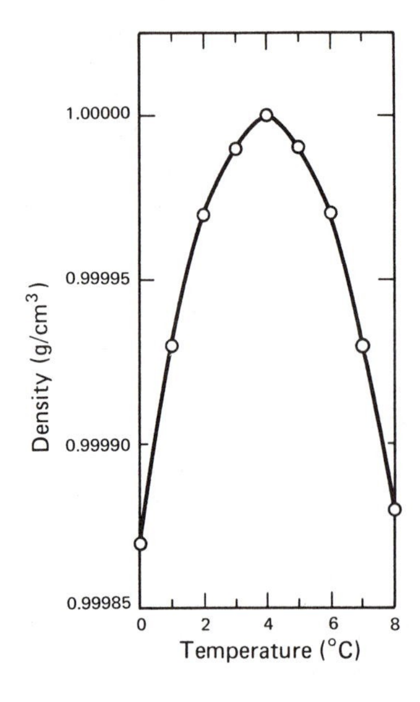

GAS LAWS 18

18-1 *INTRODUCTION*

Suppose before making a long trip in the summer you go to your car and notice that the tires are low. So you stop at a gas station around the corner and have the attendant add air to 28.0-lb/in^2 gage pressure. Then later in the afternoon you need to stop for gas. Since your tires were low that morning, you decide to have them checked again. Now you notice that they look a bit larger and the gage pressure is 40.0 lb/in^2. What happened? (*Note*: Tire companies do *not* recommend bleeding air out when tires are hot due to highway driving.)

When a gas is heated, the increased kinetic energy causes the volume to increase, the pressure to increase, or both to increase.

18-2 *CHARLES' LAW*

RULE

Charles' law: If the pressure on a gas is constant, the volume is directly proportional to its Kelvin or Rankine temperature.*

---

*Directly proportional means that as temperature increases, volume increases, and as temperature decreases, volume decreases.

By formula,

$$\frac{V}{T} = \frac{V'}{T'}$$

where: V = the original volume
T = the original temperature
V' = the final volume
T' = the final temperature

EXAMPLE 1: A gas occupies 450 cm^3 at 30°C. At what temperature will the gas occupy 480 cm^3?

DATA:
$V = 450\ cm^3$
$T = 30° + 273° = 303°K$ (Note: We must use Kelvin temperature.)
$V' = 480\ cm^3$
$T' = ?$

BASIC EQUATION: $\frac{V}{T} = \frac{V'}{T'}$

WORKING EQUATION: $T' = \frac{TV'}{V}$

SUBSTITUTION:
$$T' = \frac{(303°K)(480\ \cancel{cm^3})}{450\ \cancel{cm^3}}$$
$$= 323°K \quad \text{or} \quad 50°C$$

EXAMPLE 2: At 40°F, a gas occupies 15.0 ft^3. What will be the volume of this gas at 90°F?

DATA:
$V = 15.0\ ft^3$
$T = 40° + 460° = 500°R$ (*Note*: We must use Rankine temperature.)
$T' = 90° + 460° = 550°R$
$V' = ?$

BASIC EQUATION: $\frac{V}{T} = \frac{V'}{T'}$

WORKING EQUATION: $V' = \frac{VT'}{T}$

SUBSTITUTION: $V' = \frac{(15.0\text{ ft}^3)(550^{\circ})}{500^{\circ}}$

$V' = 16.5\text{ ft}^3$

PROBLEMS

1. Change 15°C to °K.
2. Change −13°C to °K.
3. Change 317°K to °C.
4. Change 235°K to °C.
5. Change 72°F to °R.
6. Change −50°F to °R.
7. Change 550°R to °F.
8. Change 375°R to °F.

Use $\frac{V}{T} = \frac{V'}{T'}$ to find the following:

9. $T = 300°\text{K}$, $V' = 200\text{ cm}^3$, $T' = 270°\text{K}$, find V.
10. $T = 600°\text{R}$, $V = 60.3\text{ in}^3$, $T' = 450°\text{R}$, find V'.
11. $V = 200\text{ ft}^3$, $T' = 95°\text{F}$, $V' = 250\text{ ft}^3$, find T.
12. $V = 19.7$ liters, $T = 51°\text{C}$, $V' = 25.2$ liters, find T'.
13. Some gas occupies a volume of 325 cm³ at 41°C. What is its volume at 94°C?
14. Some oxygen occupies 275 in³ at 30°F. What is its volume at 95°F?
15. Some methane occupies 1500 liters at 45°C. What is its volume at 20°C?
16. Some helium occupies 1200 ft³ at 70°F. At what temperature will its volume be 600 ft³?

SKETCH
DATA
$a = 1, b = 2,$
$c = ?$

BASIC
EQUATION
$a = bc$

WORKING
EQUATION
$c = \frac{a}{b}$

SUBSTITUTION
$c = \frac{1}{2}$

18-3 *BOYLE'S LAW*

RULE

Boyle's Law: If the temperature on a gas is constant, the volume is inversely proportional to the pressure.*

By formula,

$$\frac{V}{V'} = \frac{P'}{P}$$

*Inversely proportional means that as volume increases, pressure decreases, and as volume decreases, pressure increases.

where: $V =$ the original volume
$V' =$ the final volume
$P =$ the original pressure
$P' =$ the final pressure

Note: The pressure must be expressed in terms of *absolute pressure.*

EXAMPLE: Some oxygen occupies 500 in³ at a pressure of 40.0 lb/in² (psi). What is its volume at a pressure of 100 psi?

DATA: $V = 500 \text{ in}^3$
$P = 40.0 \text{ psi}$
$P' = 100 \text{ psi}$
$V' = ?$

BASIC EQUATION: $\frac{V}{V'} = \frac{P'}{P}$

WORKING EQUATION: $V' = \frac{VP}{P'}$

SUBSTITUTION: $V' = \frac{(500 \text{ in}^3)(40.0 \cancel{\text{psi}})}{100 \cancel{\text{psi}}}$
$V' = 200 \text{ in}^3$

18-4 *DENSITY AND PRESSURE*

If the pressure of a given amount of gas is increased, its density increases as the gas molecules are forced closer together. Also, if the pressure is decreased, the density decreases. That is, the density of a gas is directly proportional to its pressure. In equation form,

$$\frac{D}{D'} = \frac{P}{P'}$$

where: $D =$ the original density
$D' =$ the final density
$P =$ the original pressure (absolute)
$P' =$ the final pressure (absolute)

EXAMPLE: A given gas has a density of 1600 g/cm³ at a pressure of 75.0 cm of mercury. What is the density when the pressure is decreased to 71.0 cm?

DATA: $D = 1600 \text{ g/cm}^3$
$P = 75.0 \text{ cm}$

$P' = 71.0 \text{ cm}$
$D' = ?$

BASIC EQUATION: $\dfrac{D}{D'} = \dfrac{P}{P'}$

WORKING EQUATION: $D' = \dfrac{DP'}{P}$

SUBSTITUTION: $D' = \dfrac{(1600 \text{ g/cm}^3)(71.0 \text{ cm})}{75.0 \text{ cm}}$
$= 1500 \text{ g/cm}^3$

PROBLEMS

Use $V/V' = P'/P$ or $D/D' = P/P'$ to find the following:
(All pressures are absolute unless otherwise stated.)

1. $V' = 300 \text{ cm}^3$, $P = 76.0$ cm, $P' = 60.0$ cm; find V.
2. $V = 450$ liters, $V' = 700$ liters, $P = 750$ mm; find P'.
3. $V = 76.0 \text{ in}^3$, $V' = 139 \text{ in}^3$, $P' = 41.0$ psi; find P.
4. $V = 439 \text{ in}^3$, $P' = 38.7$ psi, $P = 47.1$ psi; find V'.
5. $D = 1.80 \text{ g/cm}^3$, $P = 80.0$ cm, $P' = 70.0$ cm; find D'.
6. $D = 1.65 \text{ g/cm}^3$, $P = 735$ mm, $D' = 1.85 \text{ g/cm}^3$; find P'.
7. $P = 51.0$ psi, $P' = 65.3$ psi, $D' = 0.231 \text{ lb/ft}^3$; find D.
8. Some air at 22.5 psi occupies 1400 in³. What is its volume at 18.0 psi?
9. Some gas at a pressure of 76.0 cm of mercury occupies 1850 cm³. What is its pressure if its volume is changed to 2250 cm³?
10. Some gas at 750 mm occupies 65.0 liters. What is its volume at a pressure of 650 mm of mercury?
11. Some gas has a density of 2750 g/cm³ at a pressure of 810 mm of mercury. What is its density if the pressure is decreased to 720 mm?
12. A gas has a density of 1.75 g/liter at a pressure of 760 mm of mercury. What is its pressure when the density is changed to 1.45 g/liter?
13. Some methane at 50.0 psi occupies 750 in³. What is its pressure if its volume is 500 in³?
14. Some helium at 15.0 psi gage pressure occupies 20.0 ft³. What is its volume at 20.0 psi gage pressure?
15. Some nitrogen at 80.0 psi gage pressure occupies 13.0 ft³. What is its volume at 50.0 psi gage pressure?
16. A gas at 300 psi occupies 40.0 ft³. What is its pressure if its volume is doubled? tripled? halved?

SKETCH
DATA
$a = 1, b = 2,$
$c = ?$
BASIC EQUATION
$a = bc$
WORKING EQUATION
$c = \dfrac{a}{b}$
SUBSTITUTION
$c = \dfrac{1}{2}$

18-5 *CHARLES' AND BOYLE'S LAWS COMBINED*

Most of the time it is very difficult to keep the pressure constant or the temperature constant. In this case we combine Charles' law and Boyle's law as follows:

$$\frac{VP}{T} = \frac{V'P'}{T'}$$

EXAMPLE: Assume we have 500 in^3 of acetylene at 40°F at 2000 psi. What is the pressure if its volume is changed to 800 in^3 at 100°F?

DATA:

$V = 500 \text{ in}^3$
$P = 2000 \text{ psi}$
$T = 40° + 460° = 500°\text{R}$
$V' = 800 \text{ in}^3$
$P' = ?$
$T' = 100° + 460° = 560°\text{R}$

BASIC EQUATION:

$$\frac{VP}{T} = \frac{V'P'}{T'}$$

WORKING EQUATION:

$$P' = \frac{VPT'}{TV'}$$

SUBSTITUTION:

$$P' = \frac{(500 \text{ in}^3)(2000 \text{ psi})(560°)}{(500°)(800 \text{ in}^3)}$$

$$P' = 1400 \text{ psi}$$

SKETCH
DATA
$a = 1, b = 2,$
$c = ?$
BASIC EQUATION
$a = bc$
WORKING EQUATION
$c = \frac{a}{b}$
SUBSTITUTION
$c = \frac{1}{2}$

PROBLEMS

Use $VP/T = V'P'/T'$ to find the following:
(All pressures are absolute unless otherwise stated.)

1. $P = 800$ psi, $T_R = 570°$, $V' = 1500 \text{ in}^3$, $P' = 600$ psi, $T'_R = 500°$; find V.
2. $V = 500 \text{ in}^3$, $T_R = 500°$, $V' = 800 \text{ in}^3$, $P' = 800$ psi, $T'_R = 450°$; find P.
3. $V = 900 \text{ cm}^3$, $P = 755$ mm, $T = 300°\text{K}$, $P' = 785$ mm, $T' = 265°\text{K}$; find V'.
4. $V = 18.0 \text{ m}^3$, $P = 77.5$ cm, $V' = 15.0 \text{ m}^3$, $P' = 81.3$ cm, $T' = 235°\text{K}$; find T.

5. $V = 532\ \text{ft}^3$, $P = 21.2$ psi gage pressure, $T = 487°\text{R}$, $V' = 379\ \text{ft}^3$, $P' = 123$ psi gage pressure; find T'.
6. We have 600 in^3 of oxygen at 1500 psi at 65°F. What is the volume at 1200 psi at 90°F?
7. We have 800 m^3 of natural gas at 1200 mm at 30°C. What is the temperature if the volume is changed to 1200 m^3 at 1500 mm?
8. We have 1400 liters of nitrogen at 51.0 cm at 53°C. What is the temperature if the volume changes to 800 liters at 81.5 cm?
9. An acetylene welding tank has a pressure of 2000 psi at 40°F. If the temperature rises to 90°F, what is the new pressure?
10. What is the new pressure in problem 9 if the temperature falls to −30°F?

CHANGE OF STATE 19

19-1 *FUSION*

Many industries are concerned with a change of state in the materials they use. In foundries the principle activity is to change the state of solid metals to liquid, pour the liquid metal into molds, and allow it to become solid again.

This first change of state from solid to liquid is called *melting* or *fusion*. The change from the liquid to solid state is called *freezing* or *solidification*. Most solids have a crystal structure and a definite melting point at any given pressure. Fusion and solidification of these substances occur at the same temperature. For example, water at 32° Farenheit (0° Celsius) changes to ice and ice changes to water at the same temperature. There is no temperature change during change of state. Ice at 32°F changes to water at 32°F. Only a few substances, like butter and glass, have no particular melting temperature but change state gradually.

Although there is no temperature change during a change of state, *there is a transfer of heat*. A melting solid *absorbs* heat and a solidifying liquid *gives off* heat. When one gram of ice at 0°C melts, it absorbs 80 calories of heat. Similarly, when one gram of water freezes at 0°C, ice at 0°C is produced and 80 calories of heat are released.

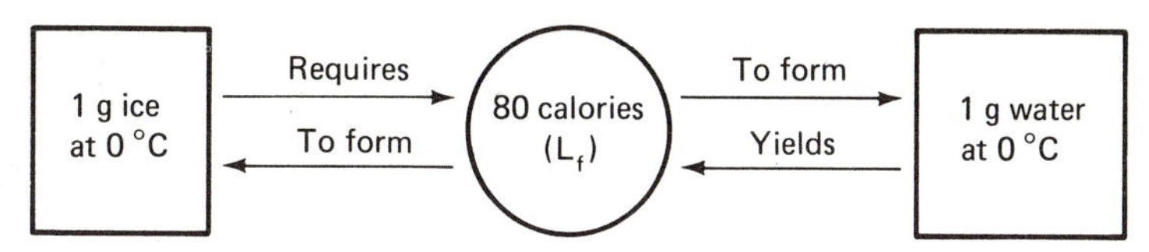

When one pound of ice at 32°F melts, it absorbs 144 BTU of heat. Similarly, when one pound of water freezes at 32°F, ice at 32°F is produced and 144 BTU of heat are released.

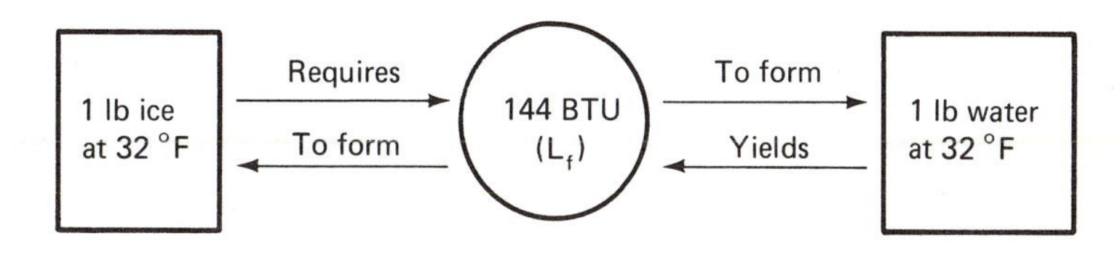

The amount of heat required to melt 1 g or 1 lb of a liquid is called its heat of fusion, designated L_f.

$$L_f = \frac{Q}{w} \text{ (English system)} \qquad L_f = \frac{Q}{m} \text{ (Metric system)}$$

where: L_f = heat of fusion
Q = quantity of heat (BTU or cal)
w = weight of substance (English system)
m = mass of substance (metric system)

EXAMPLE 1: If 864 BTU are required to melt 6.00 lb of ice at 32°F into water at 32°F, what is the heat of fusion of water?

DATA: $Q = 864$ BTU
$w = 6.00$ lb
$L_f = ?$

BASIC EQUATION: $L_f = \frac{Q}{w}$

WORKING EQUATION: Same

SUBSTITUTION: $L_f = \frac{864 \text{ BTU}}{6.00 \text{ lb}}$

$L_f = 144 \frac{\text{BTU}}{\text{lb}}$

$$\text{Heat of fusion (water)} = 144 \frac{\text{BTU}}{\text{lb}} \text{ or } 80 \frac{\text{cal}}{\text{g}}$$

19-2 *VAPORIZATION*

Steam heating systems in homes and factories are important applications of the principles of change of state from liquid to the gaseous or vapor state. This change of state is called *vaporization.* Boiling water shows this change of state. The reverse process (change from gas to liquid) is called *condensation.* As steam condenses in radiators, large amounts of heat are released.

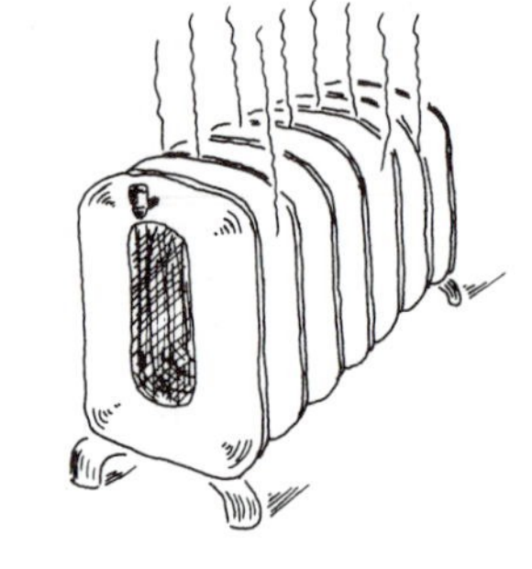

While a liquid is boiling, the temperature of the liquid does not change. There is, however, a transfer of heat. A liquid being vaporized (boiling) *absorbs* heat. As a vapor condenses, heat is given off.

The amount of heat required to vaporize 1 g or 1 lb of a liquid is called its *heat of vaporization*, designated L_v. Water has a heat of vaporization of 540 cal/g or 970 BTU/lb. So, when one gram of water at 100°C changes to steam at 100°C, it absorbs 540 calories; when one gram of steam at 100°C condenses to water at 100°C, 540 calories of heat are given off.

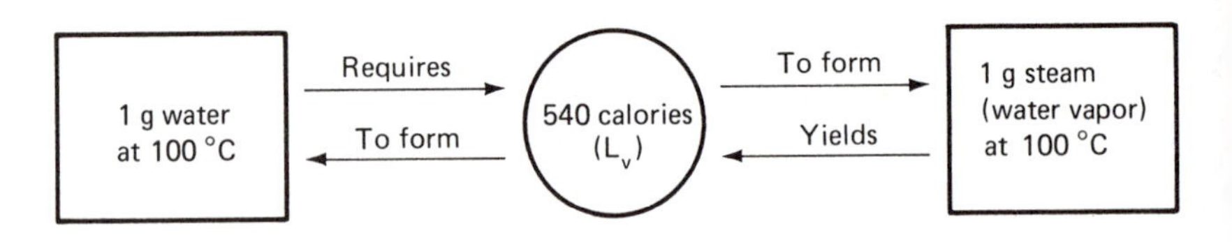

When 1 lb of water at 212°F changes to steam at 212°F, 970 BTU of heat are absorbed; when 1 lb of steam at 212°F condenses to water at 212°F, 970 BTU are given off.

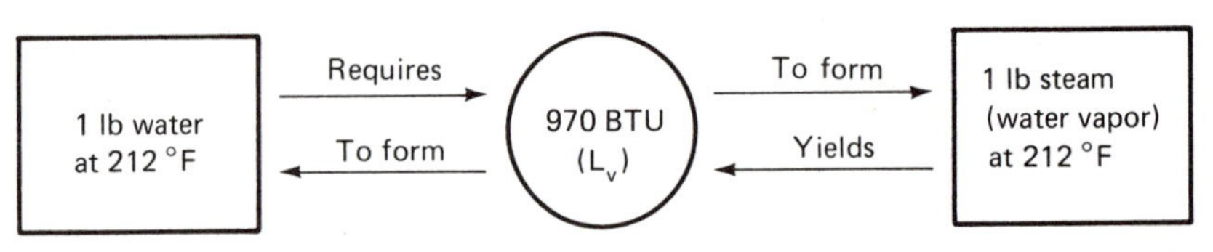

$$L_v = \frac{Q}{w} \quad \text{(English system)}$$

$$L_v = \frac{Q}{m} \quad \text{(Metric system)}$$

where: L_v = heat of vaporization
Q = quantity of heat

$w =$ weight of substance (English system)
$m =$ mass of substance (metric system)

EXAMPLE 1: If 135,000 calories are required to vaporize 250 grams of water at 100°C, what is the heat of vaporization of water?

DATA:

$$Q = 135{,}000 \text{ cal}$$
$$m = 250 \text{ g}$$
$$L_v = ?$$

BASIC EQUATION: $L_v = \dfrac{Q}{m}$

WORKING EQUATION: Same

SUBSTITUTION:

$$L_v = \frac{135{,}000 \text{ cal}}{250 \text{ g}} = 540 \frac{\text{cal}}{\text{g}}$$

> Heat of vaporization (water) $= 540 \dfrac{\text{cal}}{\text{g}}$ or $970 \dfrac{\text{BTU}}{\text{lb}}$

We may now calculate the amount of heat released when a given quantity of steam is cooled through the change of state. We will use the heat of vaporization and the method of mixtures studied in Chapter 16. From Chapter 16, heat lost during cooling is given by $Q = cw\,\Delta T$. The total amount of heat released by changing steam to water equals the amount of heat lost by the steam during cooling. ($c_{\text{steam}}w_{\text{steam}}\,\Delta T_{\text{steam}}$), plus the quantity of heat lost during condensation ($L_v w_{\text{steam}}$), plus the heat lost by the water during cooling ($c_{\text{water}}w_{\text{water}}\,\Delta T_{\text{water}}$). Therefore:

$$Q = (c_{\text{steam}})(w_{\text{steam}})(\Delta T_{\text{steam}}) + L_v(w_{\text{steam}}) + (c_{\text{water}})(w_{\text{water}})(\Delta T_{\text{water}})$$

EXAMPLE 2: How many BTUs are released when 4.00 lb of steam at 222°F is cooled to water at 82°F?

DATA:

$$w = 4.00 \text{ lb}$$
$$T_i \text{ of steam} = 222°\text{F}$$
$$T_f \text{ of water} = 82°\text{F}$$
$$Q = ?$$

BASIC EQUATION: $Q = (c_{\text{steam}})(w_{\text{steam}})(\Delta T_{\text{steam}}) + L_v(w_{\text{steam}}) + (c_{\text{water}})(w_{\text{water}})(\Delta T_{\text{water}})$

WORKING EQUATION: Same

SUBSTITUTION:
$$Q = \left(0.48\,\frac{\text{BTU}}{\text{lb}\,^\circ\text{F}}\right)(4.00\,\text{lb})(10\,^\circ\text{F}) + \left(970\,\frac{\text{BTU}}{\text{lb}}\right)(4.00\,\text{lb}) + \left(1\,\frac{\text{BTU}}{\text{lb}\,^\circ\text{F}}\right)(4.00\,\text{lb})(130\,^\circ\text{F})$$
$$Q = 4419.2\ \text{BTU}$$

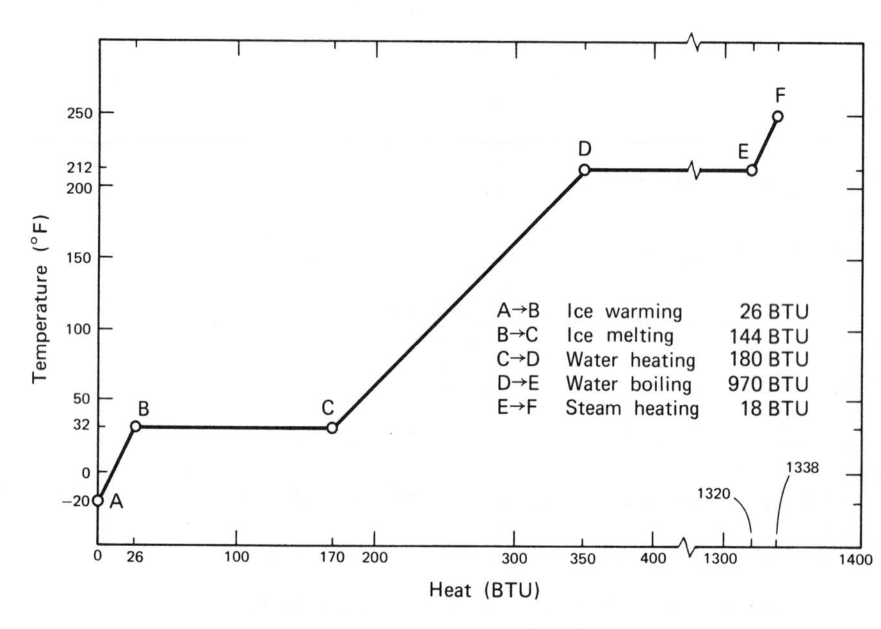

Heat gained by one pound of ice at −20 °F as it is converted to steam at 250 °F.

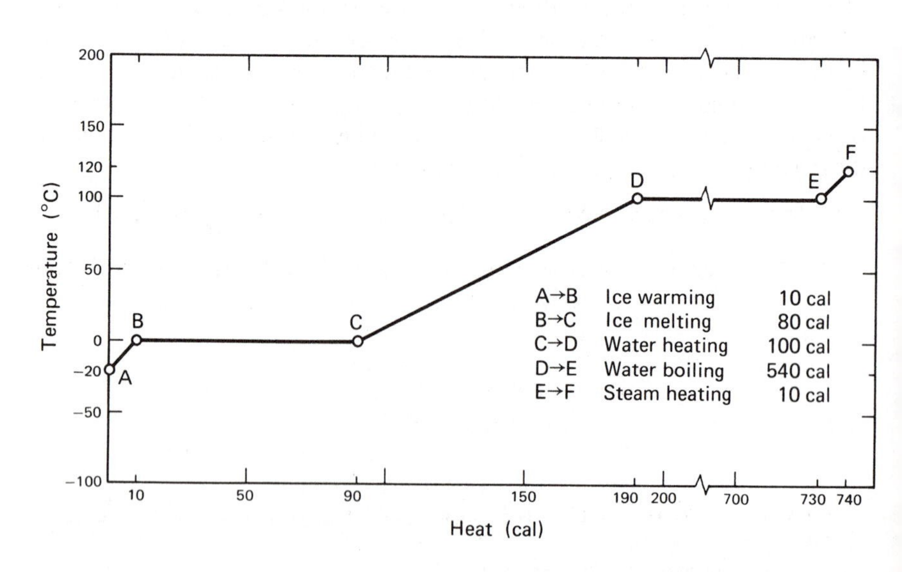

Heat gained by one gram of ice at −20 °C as it is converted to steam at 120 °C.

Note that during changes of state there are no temperature changes.

The table below shows some common substances and their heat characteristics.

HEAT CONSTANTS					
	Melting Point (°C)	Boiling Point (°C)	Specific Heat (cal/g°C)	Heat of Fusion (cal/g)	Heat of Vapori-zation (cal/g)
Alcohol, ethyl	−117	78.5	0.581	24.9	204
Aluminum	660		0.22	76.8	
Brass	840		0.092		
Copper	1080		0.093	49.0	
Glass			0.199		
Ice	0		0.51	80	
Iron	1540		0.107	7.89	
Lead	327		0.031	5.86	
Mercury	−38.9	357	0.033	2.82	65.0
Silver	961		0.056	26.0	
Steam			0.48		
Water (liquid)	0	100	1.00		540
Zinc	420		0.093	23.0	

19-3 *EFFECTS OF PRESSURE AND IMPURITIES ON CHANGE OF STATE*

Automobile cooling systems present important problems concerning change of state. Most substances contract on solidifying. Water and a few other substances, however, expand. The tremendous force exerted by this expansion is shown by the

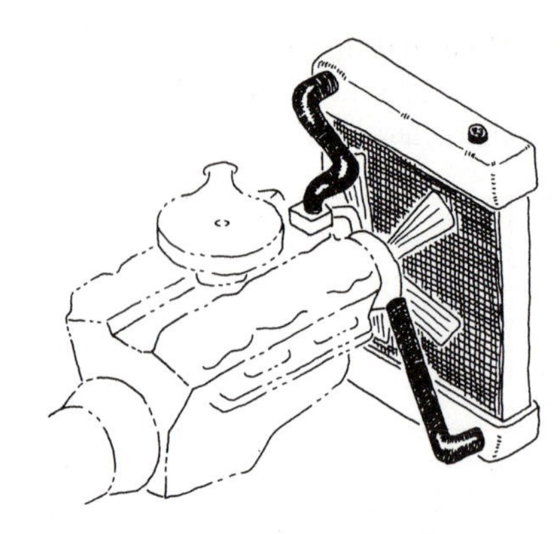

number of cracked automobile blocks and burst radiators suffered by careless motorists every winter.

Impurities in water tend to *lower* the freezing point. Alcohol has a lower freezing point than water and is used in some types of antifreeze. By mixing antifreeze with water in the cooling system, the freezing point of the water may be lowered to avoid freezing in winter. Automobile engines may also be ruined in winter by overheating if the water in the radiator is frozen, preventing the engine from being cooled by circulation in the system.

An increase in the pressure on a liquid *raises* the boiling point. Automobile manufacturers utilize this fact by pressurizing their cooling systems and thereby raising the boiling point of the coolant used.

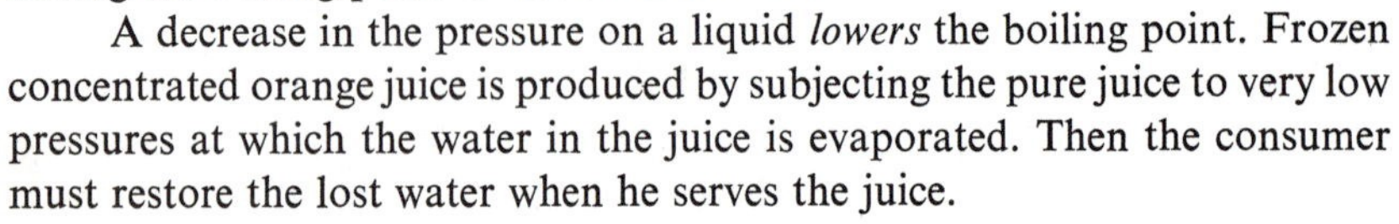

A decrease in the pressure on a liquid *lowers* the boiling point. Frozen concentrated orange juice is produced by subjecting the pure juice to very low pressures at which the water in the juice is evaporated. Then the consumer must restore the lost water when he serves the juice.

PROBLEMS

1. How many calories are required to melt 14.0 g of ice at 0°C?
2. How many pounds of ice at 32°F can be melted by the addition of 635 BTU of heat?
3. How many BTUs are required to vaporize 11.0 lb of water at 212°F?
4. How many grams of steam in a boiler at 100°C can be condensed to water at 100°C by the removal of 1520 calories of heat?
5. How many calories are required to melt 320 g of ice at 0°C?
6. How many calories are given off when 3250 g of steam is condensed to water at 100°C?
7. How many BTUs are required to melt 33.0 lb of ice and to raise the temperature of the melted ice to 72°F?
8. How many BTUs are liberated when 20.0 lb of water at 80°F is cooled to 32°F and then frozen in an ice plant?
9. How many BTUs are required to change 9.00 lb of ice at 10°F to steam at 232°F?
10. How many calories are liberated when 200 g of steam at 120°C are changed to ice at −12°C?

SKETCH
DATA
$a = 1, b = 2,$
$c = ?$

BASIC
EQUATION
$a = bc$

WORKING
EQUATION
$c = \frac{a}{b}$

SUBSTITUTION
$c = \frac{1}{2}$

STATIC ELECTRICITY 20

20-1 *ELECTRIFICATION*

The discovery of electrification was made when an amber rod was rubbed with a wool cloth and the rod attracted small bits of paper. When two objects are rubbed together, they become electrified.

When you slide rubber soled shoes on a wool rug on a dry day, you become electrified. We say you have acquired a static charge. This static charge is usually lost when you touch a metal object or some other ground such as a person.

Trucks which carry flammable liquids prevent a buildup of static charge by dragging a chain on the pavement. Otherwise, a spark caused by a discharge could cause an explosion.

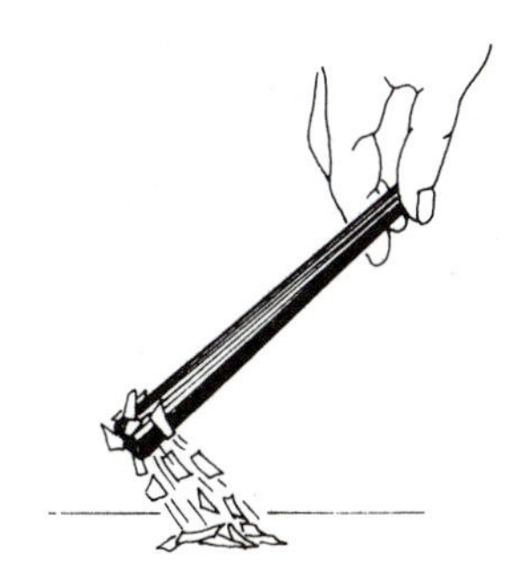

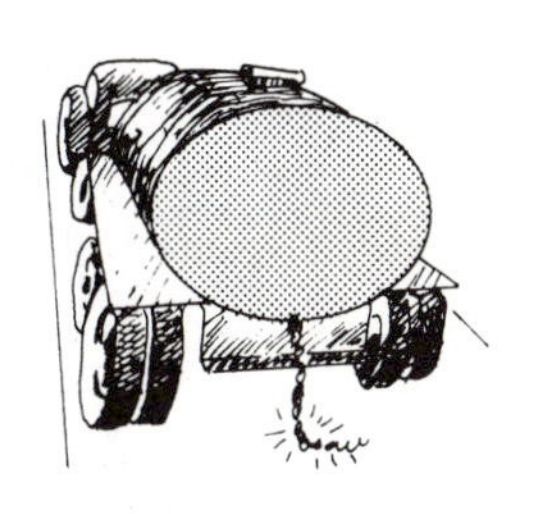

There are two types of electrical charges that can be produced. They can be observed indirectly by using an electroscope. A very simple electroscope is a ball of wood pith on a silk thread (Fig. 1).

We can produce a charge on a hard rubber rod by rubbing it with a wool cloth. Now transfer some of this charge to the pith ball (Fig. 2).

Fig. 1

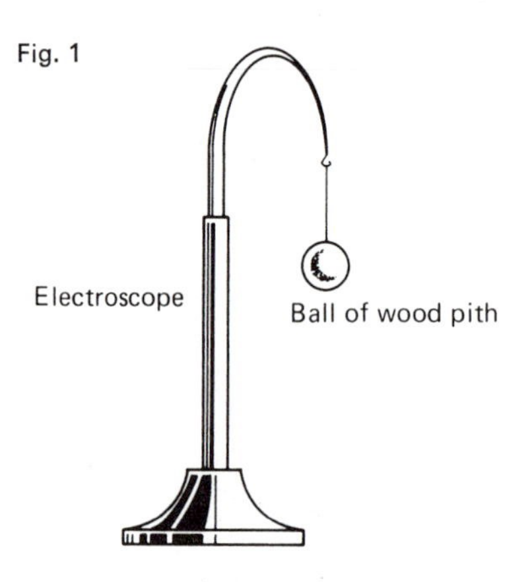

Fig. 2

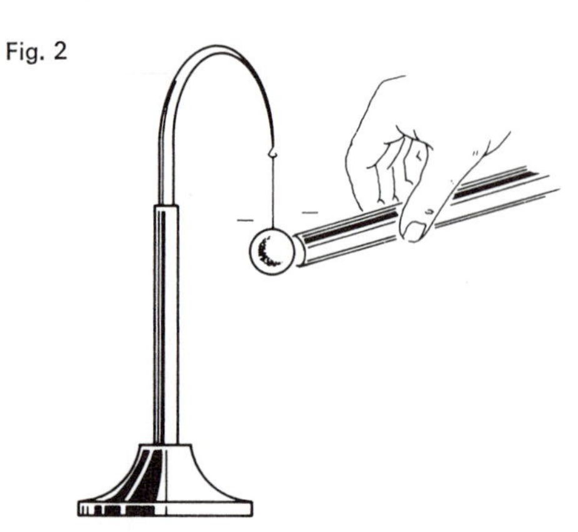

Another pith ball charged in the same way is repelled by the other pith ball (Fig. 3). This charge is called a negative charge (—).

Now rub a glass rod with silk. Transfer this charge to a pith ball (Fig. 4).

This pith ball is attracted to the negatively charged pith ball (Fig. 5).

Fig. 3

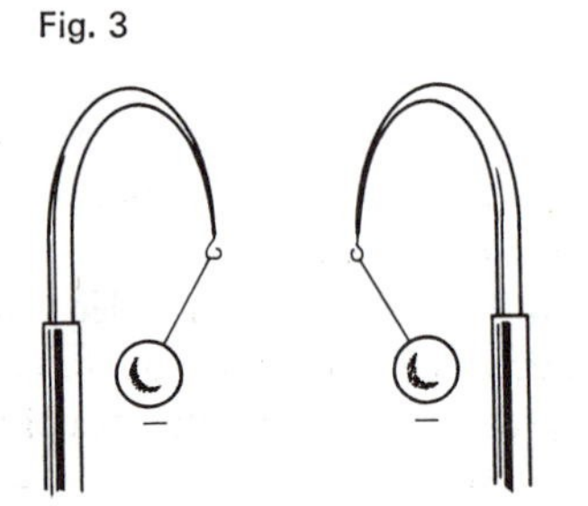

Fig. 4

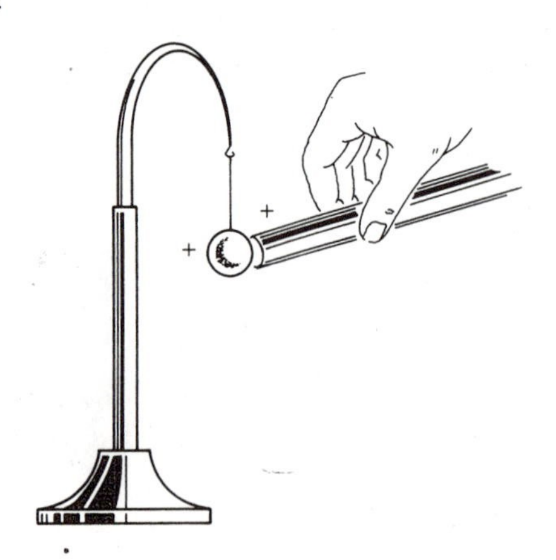

Fig. 5

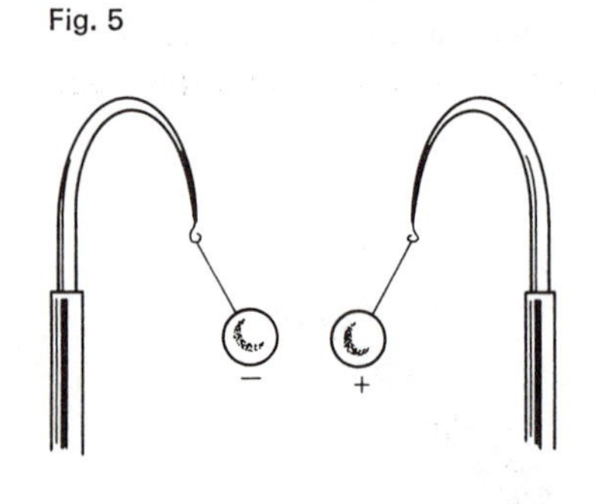

The charge produced by glass and silk is called a positive charge (+). Two pith balls which are positively charged will repel each other.

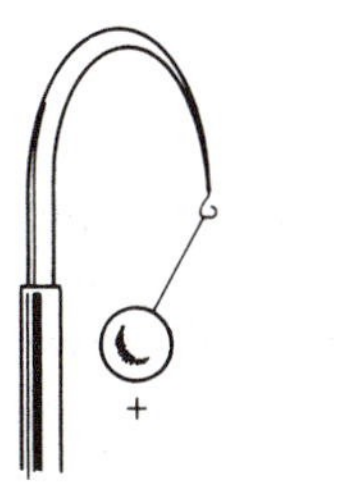

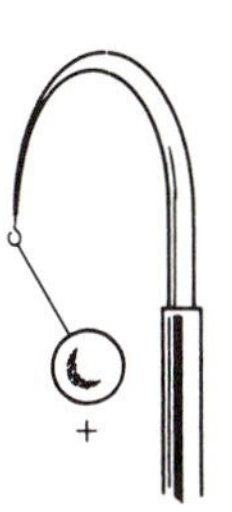

In summary,

Like charges repel each other, and unlike charges attract each other.

To understand electricity we need to know more about the structure of matter. We have seen that all matter is made up of atoms. These atoms are made of electrons, protons, and neutrons. Each proton has one unit of positive charge, and each electron has one unit of negative charge. The neutron has no charge. The protons and neutrons are tightly packed into what is called the nucleus. Electrons may be thought of as small "planets" which orbit the nucleus.

An atom normally has the same number of electrons as protons and thus is uncharged.

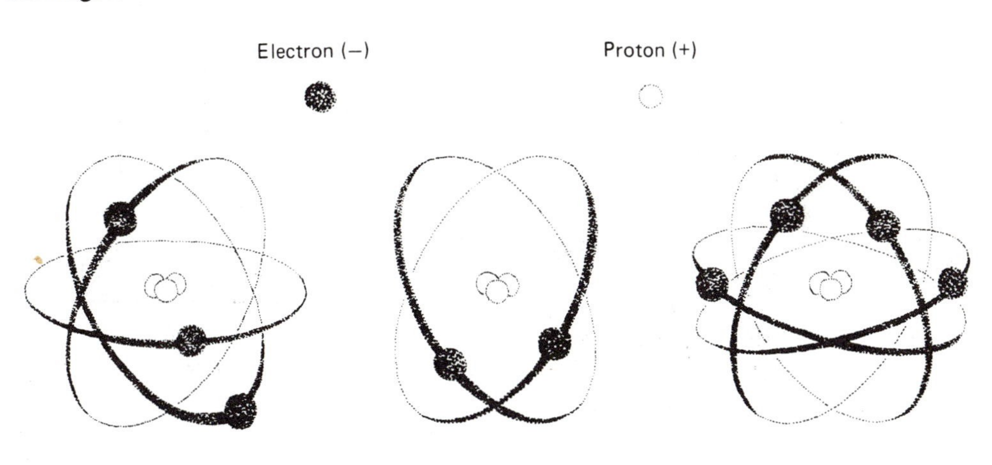

Normal atom (uncharged) | Atom with a positive charge | Atom with a negative charge

If an electron is removed, the atom is left with a positive charge.

If an extra electron is added, the atom has a negative charge.

When two materials are rubbed together, the atoms on the two surfaces move across each other and brush off electrons. The electrons are transferred from one surface to the other. One surface is then left with a positive charge and the other is negative. This is the process we have called electrification.

20-3 *ELECTROSTATIC GENERATORS*

There are two common electrostatic generators in use in physics laboratories. They are the Van de Graff generator and the Wimshurst machine. The Van de Graff consists of an electron source, a rubber belt driven by a motor, and a metal sphere supported by an insulating stand. The electron source charges the rubber belt as it passes by and carries the electrons up to the sphere and deposits them there. This builds up a high voltage (several hundred thousand volts) between the sphere and the ground.

The Wimshurst machine produces charge by a more complicated method which will not be discussed here. The voltage built up by a Wimshurst machine is not as large as that produced by a Van de Graff. However, the current delivered by a Wimshurst machine can be very large.

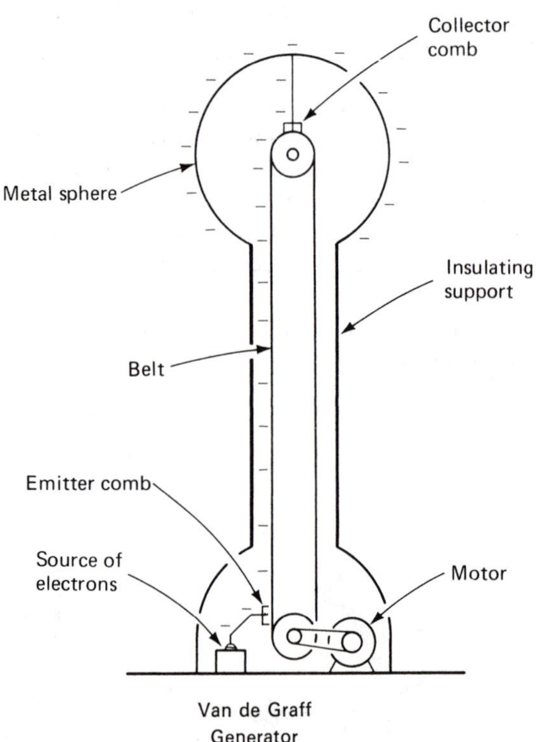

Van de Graff
Generator

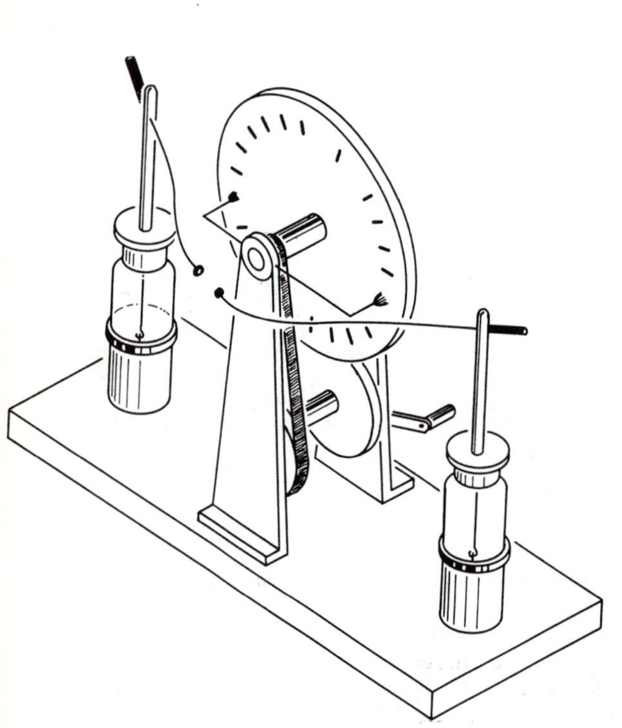

Winshurst
Machine

DIRECT CURRENT ELECTRICITY 21

21-1 *MOTION OF ELECTRONS*

Electrons moving in a wire make up a current in the wire. When the electron current flows in only one direction, it is called *direct current* (dc). Current which changes direction is called *alternating current* (ac). Alternating current will be considered in later chapters.

An electric current is a convenient means of transmitting energy. Technicians face many situations every day which require energy to do a particular task. To drill a hole in a metal block, energy must be supplied and transformed into mechanical energy to turn the drill bit. The problem the technician faces is how to supply energy to the machine being used in a form which the machine can turn into useful work. Electricity is often the most satisfactory means of transmitting energy. We will first study the use of electricity in transferring energy by looking at an example.

At the right below is a circuit like that of a simple flashlight. An electric circuit is a conducting loop in which electrons carrying electric energy may be transferred from a suitable source to do useful work and back to the source. Energy is stored in the battery. When the switch is closed, energy is transmitted to the light and the light glows.

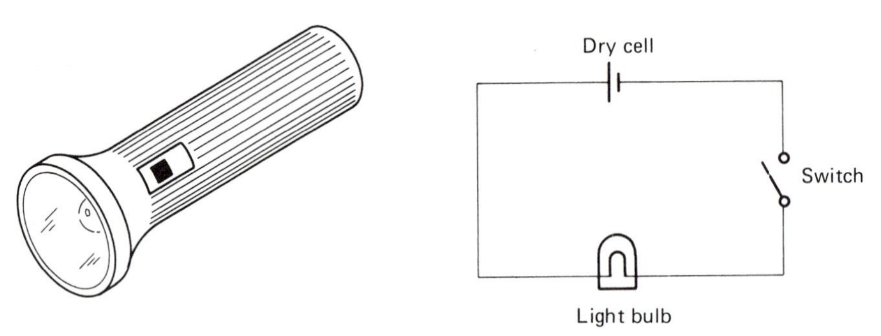

Compared with static electricity, current electricity is the flow of energized electrons through an electron carrier called a *conductor*. The electrons move from the energy *source* (the battery, here) to the *load* (the place where the transmitted energy is turned into useful work). There they lose energy picked up in the source.

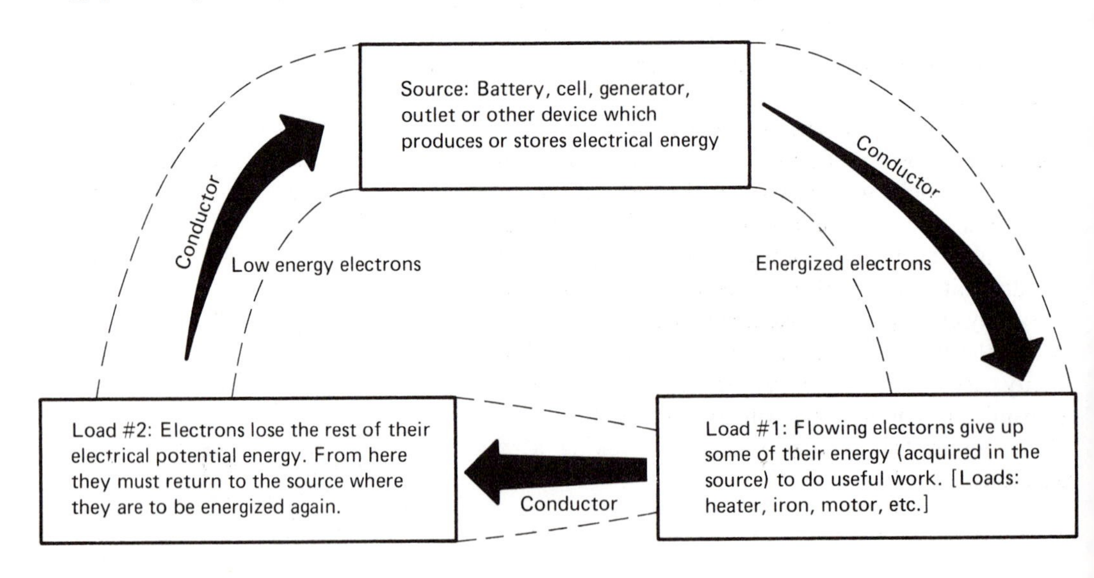

Let's consider each part of the circuit and determine its function.

21-3 *THE SOURCE OF ENERGY*

The dry cell is a device which converts chemical energy to electrical energy. How the cell does this will be studied in Chapter 23. Here, we will simply state that, by chemical action, electrons are given energy in the cell. When energy is given to electrons in this manner, their potential electrical energy is raised.

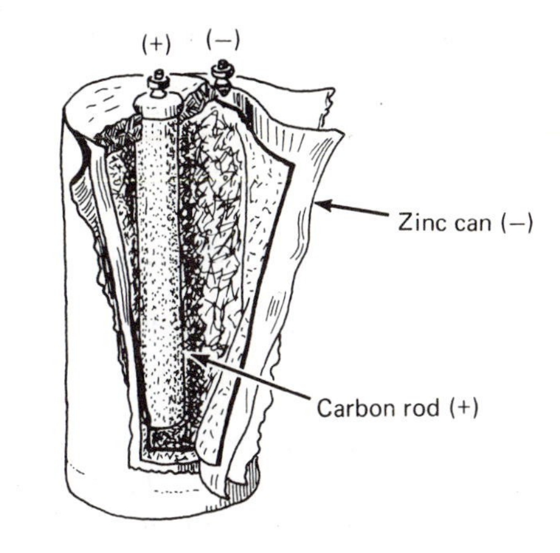

What does "potential electrical energy" mean? Remember, when a stone is lifted off the ground it is given gravitational potential energy. (It has the ability to do work in falling back to the ground under the pull of gravity—stored energy.) There is a difference in potential due to its position or condition. Work has been done in lifting it against gravity to the higher position.

In a source of electrical energy something similar happens. In the source (the battery), work is done on electrons which gives them potential energy. This potential difference between the energized electrons (at the negative (−) pole) and the low potential energy electrons (at the positive (+) pole) causes the electrons to flow from one point (−) to the other (+) when connected. The energized electrons collect at the negative pole, repel each other, and flow through the circuit to the positive pole. They lose their potential electrical energy to the load.

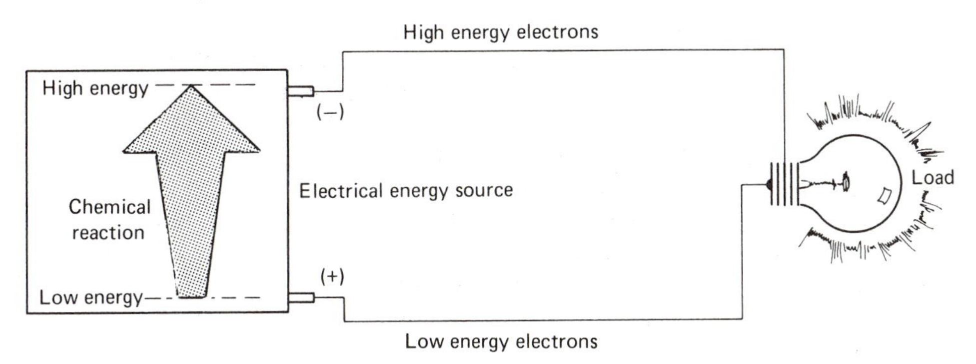

21-4 *THE CONDUCTOR*

A conductor carries or transfers the electrical charge to the load (here the lightbulb). *Conductors* are substances (such as copper) which have large numbers of free electrons

(electrons which are free to move throughout the conductor). As high-energy electrons from the dry cell pass through the conductor, they collide with other electrons in the conductor. These electrons then carry the energy farther along the wire until they collide again and transfer energy on through the wire.

Silver, copper, and aluminum are metals which allow electrons to pass freely through and thus are good conductors. Other metals offer more opposition to electrons and are poorer conductors.

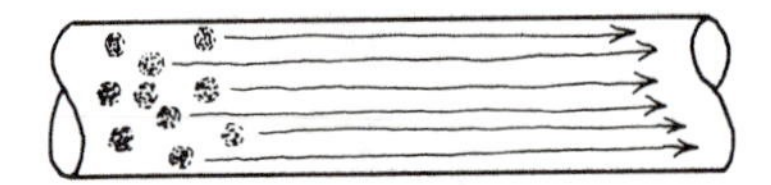

Good conductor

Poor conductor

Substances which do not allow electrons to pass readily are called *insulators*. Common insulators are rubber, wool, silk, glass, wood, distilled water, and dry air.

A small number of materials, called *semiconductors*, fall between conductors and insulators. Their importance is due to the fact that these materials under certain conditions allow current to flow only in one direction and not in the other. Silicon and germanium are examples of semiconductors and are used in transistors. More about semiconductors is discussed in Chapter 29.

21-5 *THE LOAD*

In the load electrons lose their energy. The load converts the electrical energy to other useful forms. In a lightbulb electrical energy is changed to light and heat. An electric motor changes electrical energy to mechanical energy. The load may be a complex motor or only a simple resistor with heat the only new form of energy. The electrons do not collect and remain in the load, but continue back to the low-energy side of the battery (+). There they may be energized again for another trip through the circuit.

21-6 *CURRENT*

The flow of electrons through a conductor is called *current*. We could count the electrons passing a point during a certain time to get the rate of flow. This is impractical because the number of electrons is so large (about 10^{18}/s).

To have a workable unit of electric charge, we will define the charge on 6.25×10^{18} electrons as one *coulomb*. The *ampere* (A) is the rate of flow of one coulomb of charge passing a point in one second. Now we can define a unit for the rate of flow of charge.

$$1 \text{ ampere (A)} = \frac{1 \text{ coulomb (C)}}{1 \text{ second (s)}}$$

As mentioned above, the charge carriers in metals are electrons. In some other conductors, such as electrolytes (conducting liquid solutions), the charge carriers may be positive or negative or both. An agreement must be made to determine which charge carriers should be assumed in our following discussions. Note that positive charges would flow in the opposite direction (toward the negative terminal) from that followed by negative charges (toward the positive terminal) when a battery is connected to a circuit. A positive current moving in one direction is equivalent for almost all measurements to a current of negative charges flowing in the opposite direction.

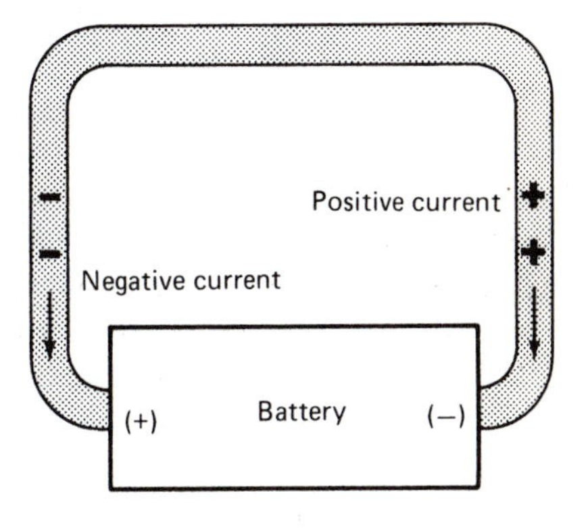

We will assume in all the following discussions that the charge carriers are positive, and we will draw our current arrows in the direction that a positive charge would flow. This is the practice of the majority of engineers and technicians. If the student should encounter the negative current convention at a later date, he should remember that a negative current flows in the opposite direction from that of a positive current. Regardless of the method used, the analysis of a situation by either method will give the correct result. Some of the rules discussed later, such as the right-hand rule for finding the direction of the magnetic field, will be different if the negative current convention is used.

21-7 *VOLTAGE*

We have seen that current flows in a circuit because of the difference in potential of the different points in the circuit. In *sources*, the raising of the potential energy of electrons which results in a potential difference is called *emf* (E). In *circuits*, the lowering of the potential energy of electrons in a load is called *voltage drop*.

The *volt* (V) is the unit of both emf and voltage drop. We define the volt as the potential difference between two points if one joule (J) of work is produced or used in moving one coulomb of charge from one point to another.

$$1 \text{ volt (V)} = \frac{1 \text{ joule (J)}}{1 \text{ coulomb (C)}}$$

21-8 *RESISTANCE*

Not all substances and not even all metals are good conductors of electricity. Those with few free electrons tend to have greater opposition to the flow of charge. This opposition is called *resistance*. The unit of resistance is the ohm (Ω). It is not a fundamental unit and will be discussed in a later section.

The resistance of a wire is determined by several factors. Among these are:

1. *Temperature:* An increase in temperature results in an increase in resistance in a wire.
2. *Length:* Resistance varies directly with length. If we double the length of a given wire, the resistance is doubled.
3. *Cross-sectional area:* Resistance varies inversely with cross-sectional area. If we double the cross section of a wire, the resistance is *halved*. This is similar to water flowing through two pipes. It flows more easily through the larger pipe. (Note that doubling the radius of a wire *more than* doubles the cross-sectional area: $A = \pi r^2$).

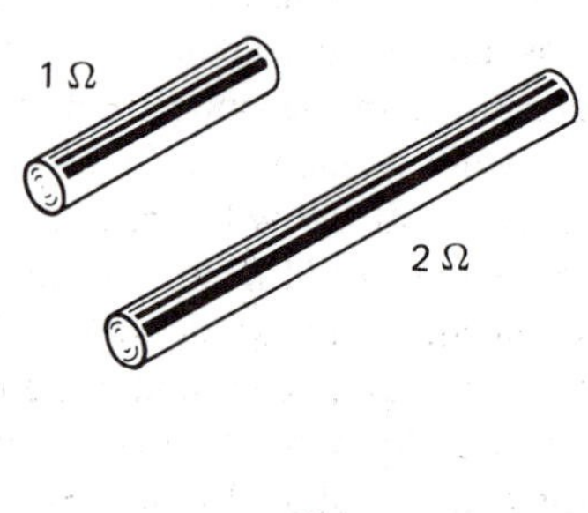

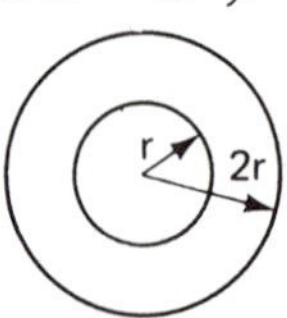

4. *Material.* Resistance depends on the nature of the material. For example, copper is a better conductor than steel. The conducting characteristic of various materials is described by resistivity. Resistivity (ρ) is the resistance of a specified amount of the material. These factors are related by the equation:

$$R = \frac{\rho l}{A}$$

where: R = resistance (Ω)
ρ = resistivity (Ω cm)
l = length (cm)
A = cross-sectional area (cm^2)

EXAMPLE: What is the resistance of a copper wire 20.0 m long with cross-sectional area of 6.56×10^{-3} cm^2 at 20°C? The resistivity of copper at 20°C is $1.72 \times 10^{-6}\ \Omega$ cm.

DATA:
$l = 20.0$ m
$A = 6.56 \times 10^{-3}$ cm^2
$\rho = 1.72 \times 10^{-6}\ \Omega$ cm
$R = ?$

BASIC EQUATION: $R = \frac{\rho l}{A}$

WORKING EQUATION: Same

SUBSTITUTION: $R = \frac{(1.72 \times 10^{-6}\ \Omega\ \cancel{\text{cm}})(2.00 \times 10^{3}\ \cancel{\text{cm}})}{6.56 \times 10^{-3}\ \cancel{\text{cm}^2}}$

$R = 0.524\ \Omega$

SKETCH
DATA
$a = 1, b = 2,$
$c = ?$
BASIC EQUATION
$a = bc$
WORKING EQUATION
$c = \frac{a}{b}$
SUBSTITUTION
$c = \frac{1}{2}$

PROBLEMS

1. What is the resistance of 78.0 m of No. 20 gage aluminum wire at 20°C? ($\rho = 2.83 \times 10^{-6}\ \Omega$ cm, $A = 2.07 \times 10^{-2}$ cm^2)
2. What is the resistance of 375 ft of No. 24 copper wire if it has a resistance of 0.0262 Ω/ft?
3. What is the resistance per foot of No. 22 copper wire if 580 ft has a resistance of 9.57 Ω?
4. At 77°F, 100 ft of No. 18 gage copper wire has a resistance of 0.651 Ω. What is the resistance of 500 ft of this wire?

5. What is the resistance of 125 m of No. 20 gage copper wire at 20°C? ($\rho = 1.72 \times 10^{-6}\ \Omega$ cm, $A = 2.07 \times 10^{-2}$ cm^2)
6. What is the resistance of 100 m of No. 20 gage copper wire at 20°C? ($\rho = 1.72 \times 10^{-6}\ \Omega$ cm, $A = 2.07 \times 10^{-2}$ cm^2)
7. What is the resistance of 50.0 m of No. 20 gage aluminum wire at 20°C? ($\rho = 2.83 \times 10^{-6}\ \Omega$ cm, $A = 2.07 \times 10^{-2}$ cm^2)

22 OHM'S LAW AND DC CIRCUITS

22-1 *OHM'S LAW*

When a current flows through a conductor, there is a definite relationship between current, voltage drop, and resistance. A German physicist, Georg Simon Ohm, studied this relationship and formulated

RULE

Ohm's Law:

$$I = \frac{V}{R}$$

where: I = *current*
V = *voltage drop*
R = *resistance*

Ohm's law can also be written:

$$I = \frac{E}{R}$$

where: E = *the emf of the source of electrical energy.*

EXAMPLE 1: A soldering iron operating on a 115-V outlet has a resistance of 15.0 Ω. What current does it draw?

DATA:

$E = 115\text{ V}$
$R = 15.0\ \Omega$
$I = ?$

BASIC EQUATION: $I = \frac{E}{R}$

WORKING EQUATION: Same

SUBSTITUTION:

$$I = \frac{115\text{ V}}{15.0\ \Omega}$$

$$I = 7.67\frac{\text{V}}{\Omega}$$

$$I = 7.67\text{ A} \qquad \left(\text{A} = \frac{\text{V}}{\Omega}\right)$$

Ohm's law applies to all dc circuits and those ac circuits containing only resistance. It may be applied to the whole circuit or to any part of it.

Ohm's law should aid us now in understanding resistance. As we mentioned earlier, the ohm (Ω) is a derived unit. From Ohm's law,

$$I = \frac{V}{R}$$

Solving for R,

$$R = \frac{V}{I}$$

Substituting units,

$$\Omega = \frac{V}{A}$$

EXAMPLE 2: A flashlight bulb is connected to a 1.50-V dry cell. If it draws 0.250 A, what is its resistance?

SKETCH:

DATA:

$E = 1.50\text{ V}$
$I = 0.250\text{ A}$
$R = ?$

BASIC EQUATION: $I = \frac{E}{R}$

WORKING EQUATION: $R = \dfrac{E}{I}$

SUBSTITUTION: $R = \dfrac{1.50 \text{ V}}{0.250 \text{ A}}$
$= 6.00 \dfrac{\text{V}}{\text{A}}$
$= 6.00\ \Omega$

PROBLEMS

1. A heating element operates on a 115-V line. If it has a resistance of 12.0 Ω, what current does it draw?
2. A given coffee pot operates on 12.0 V. If it draws 2.50 A, what is its resistance?
3. An electric heater draws a maximum of 14.0 A. If its resistance is 15.7 Ω, to what voltage line should it be connected?
4. A heating coil operates on a 220-V line. If it draws 11.0 A, what is its resistance?

SKETCH
DATA
$a = 1, b = 2,$
$c = ?$

BASIC EQUATION
$a = bc$

WORKING EQUATION
$c = \dfrac{a}{b}$

SUBSTITUTION
$c = \dfrac{1}{2}$

22-2 *CIRCUIT DIAGRAMS*

In order to communicate about problems in electricity, technicians have developed a "language" of their own. It is a picture language using symbols and diagrams. Some of the most often used symbols appear below and on the following page.

ELECTRICAL SYMBOLS

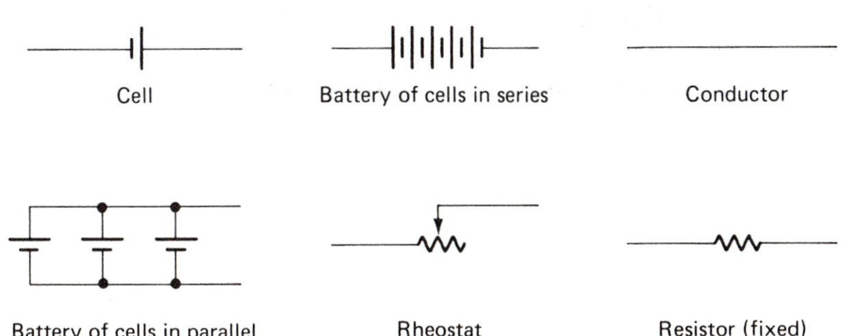

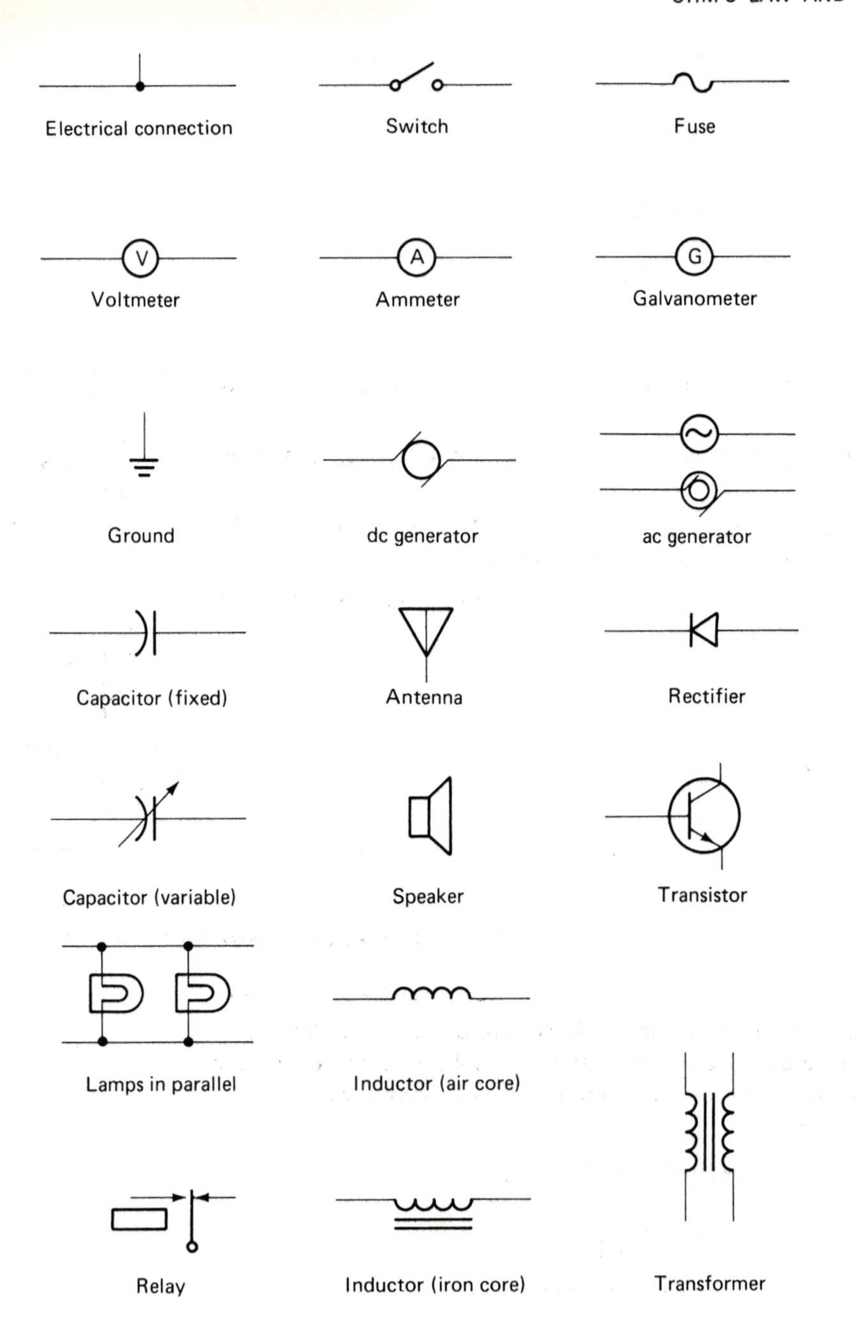

Electrical connection

Switch

Fuse

Voltmeter

Ammeter

Galvanometer

Ground

dc generator

ac generator

Capacitor (fixed)

Antenna

Rectifier

Capacitor (variable)

Speaker

Transistor

Lamps in parallel

Inductor (air core)

Relay

Inductor (iron core)

Transformer

The circuit diagram is the most common and useful way to show a circuit. Note how each component (part) of the picture is represented by its symbol in the circuit diagram in its relative position.

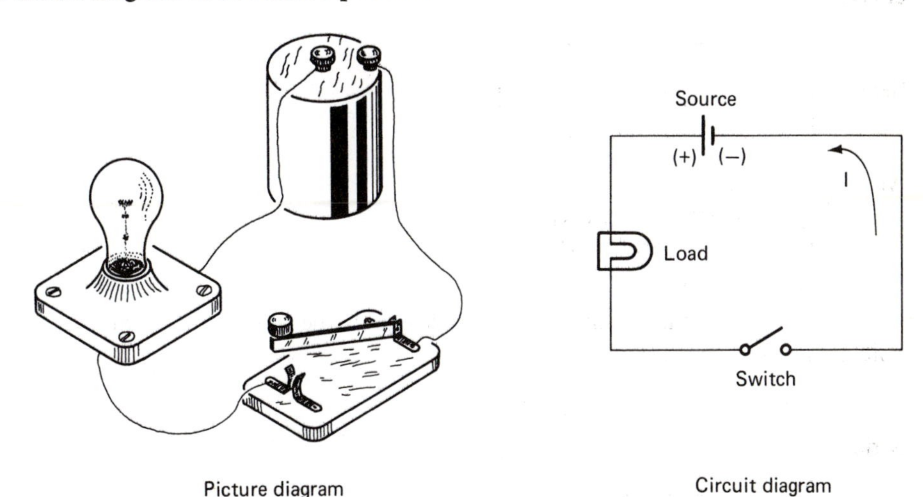

Picture diagram

Circuit diagram

The light bulb may be represented as a resistance. Then the circuit diagram above would appear:

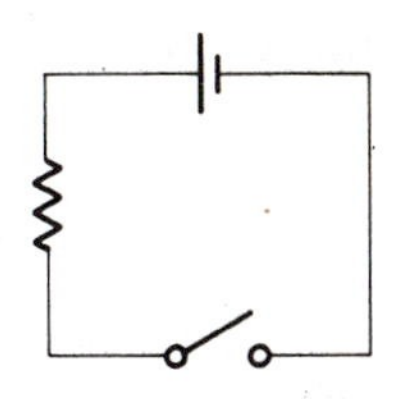

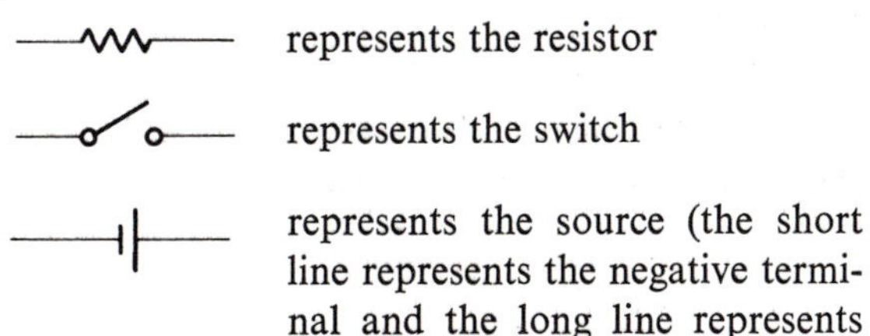

22-3 *SERIES CIRCUITS*

There are two basic types of circuits: series and parallel. A fuse in a house is wired in series with its outlets. The outlets, themselves, are wired in parallel. A study of series and parallel circuits are basic to a study of electricity.

An electrical circuit with only one path for the current to flow is called a *series* circuit.

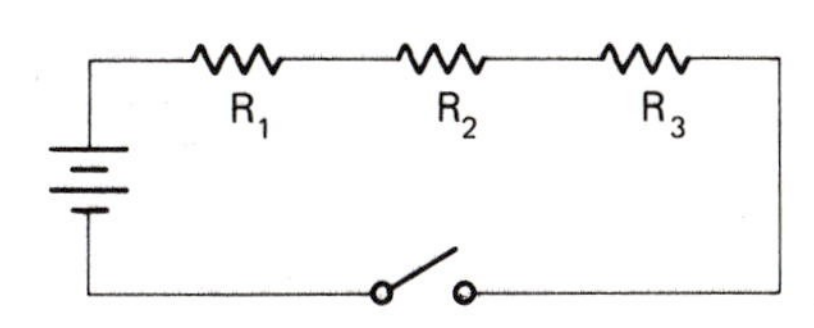

The current in a series circuit is the same throughout. That is, the total current is the same as the current flowing through each resistance in the circuit.

Series $$I = I_1 = I_2 = I_3 = \cdots$$

where: I = total current
I_1 = current through R_1
I_2 = current through R_2
I_3 = current through R_3

In a series circuit the emf of the source equals the sum of the separate voltage drops in the circuit.

Series $$E = V_1 + V_2 + V_3 + \cdots$$

where: E = emf of the source
V_1 = voltage drop across R_1
V_2 = voltage drop across R_2
V_3 = voltage drop across R_3

The resistance of the conducting wires is very small and will be neglected here. The total resistance of a series circuit is equal to the sum of all the resistances in the circuit.

Series $$R = R_1 + R_2 + R_3 + \cdots$$

where: R = total resistance of the circuit
R_1 = resistance of first load
R_2 = resistance of second load
R_3 = resistance of third load

EXAMPLE 1: What is the total resistance of the circuit below?

SKETCH:

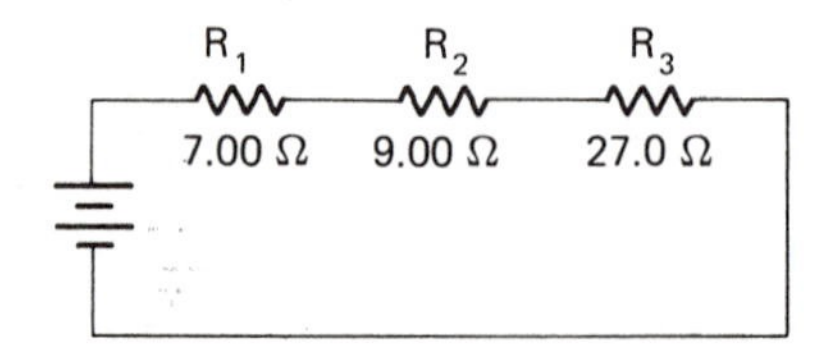

DATA: $R_1 = 7.00\ \Omega$
$R_2 = 9.00\ \Omega$
$R_3 = 27.0\ \Omega$
$R = ?$

BASIC EQUATION: $R = R_1 + R_2 + R_3$

WORKING EQUATION: Same

SUBSTITUTION: $R = 7.00\ \Omega + 9.00\ \Omega + 27.0\ \Omega$
$R = 43.0\ \Omega$

EXAMPLE 2: What is the current in the circuit below?

SKETCH:

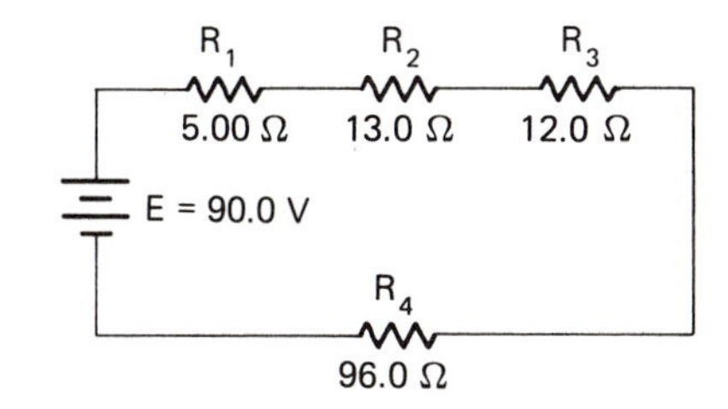

DATA: $R_1 = 5.00\ \Omega$
$R_2 = 13.0\ \Omega$
$R_3 = 12.0\ \Omega$
$R_4 = 96.0\ \Omega$
$E = 90.0\ \text{V}$
$I = ?$

BASIC EQUATIONS: $R = R_1 + R_2 + R_3 + R_4$ and $I = \dfrac{E}{R}$

WORKING EQUATIONS: Same

SUBSTITUTIONS: $R = 5.00\ \Omega + 13.0\ \Omega + 12.0\ \Omega + 96.0\ \Omega$
$R = 126\ \Omega$

$$I = \frac{90.0\ \text{V}}{126\ \Omega}$$
$$I = 0.714\ \text{A}$$

EXAMPLE 3: What is the value of R_3 in the circuit below?

SKETCH:

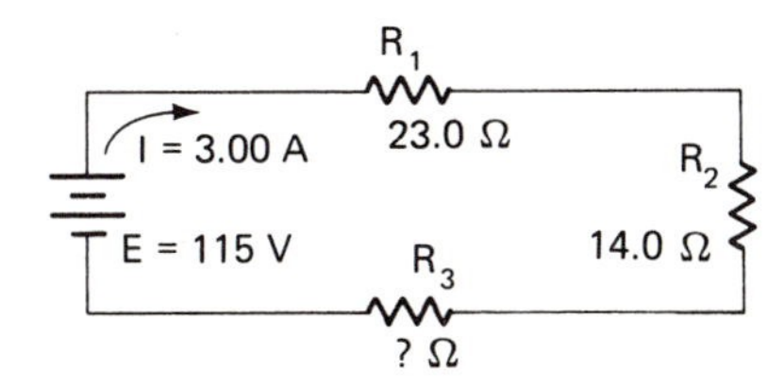

DATA: $I = 3.00\text{ A}$
$E = 115\text{ V}$
$R_1 = 23.0\ \Omega$
$R_2 = 14.0\ \Omega$
$R_3 = ?$

BASIC EQUATIONS: $I = \frac{E}{R}$ and $R = R_1 + R_2 + R_3$

WORKING EQUATIONS: $R = \frac{E}{I}$ and $R_3 = R - R_1 - R_2$

SUBSTITUTIONS: $R = \frac{115\text{ V}}{3.00\text{ A}}$
$R = 38.3\ \Omega$

$R_3 = 38.3\ \Omega - 23.0\ \Omega - 14.0\ \Omega$
$R_3 = 1.3\ \Omega$

EXAMPLE 4: What is the voltage drop across R_3 in Example 3?

DATA: $I = 3.00\text{ A}$
$I_3 = 3.00\text{ A}$
$R_3 = 1.3\ \Omega$
$V_3 = ?$

BASIC EQUATION: $I_3 = \frac{V_3}{R_3}$

WORKING EQUATION: $V_3 = I_3 R_3$

SUBSTITUTION: $V_3 = (3.00\text{ A})(1.3\ \Omega)$
$V_3 = 3.90\text{ V}$

CHARACTERISTICS OF SERIES CIRCUITS

	Series
Current	$I = I_1 = I_2 = I_3 = \cdots$
Resistance	$R = R_1 + R_2 + R_3 + \cdots$
Voltage	$E = V_1 + V_2 + V_3 + \cdots$

PROBLEMS

1. Three resistors of 2.00 Ω, 5.00 Ω and 7.00 Ω are connected in series with a 24.0-V battery. What is the total resistance of the circuit?
2. What is the current in problem 1?
3. What is the total resistance in the circuit below?

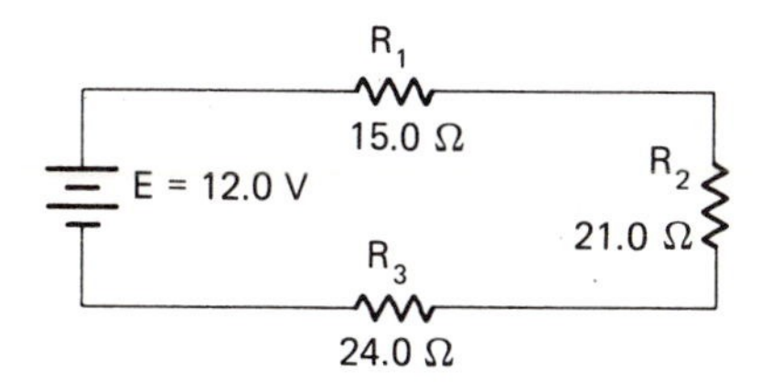

4. Find the current through R_2 in problem 3.
5. Find the current in the circuit below.

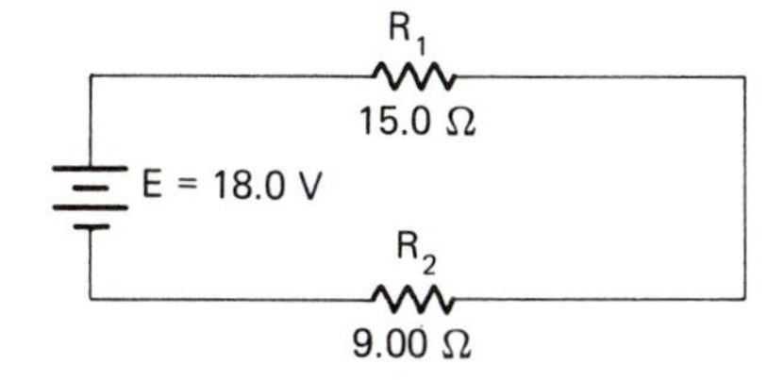

6. What is the voltage drop across R_1 in problem 5?
7. What emf is needed for the circuit below?

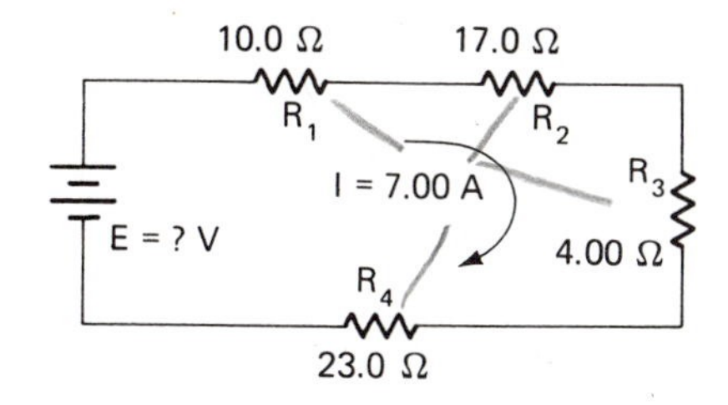

8. What is the voltage drop across R_3 in problem 7?
9. Find the total resistance in the circuit below.
10. Find R_3 in the circuit in problem 9.

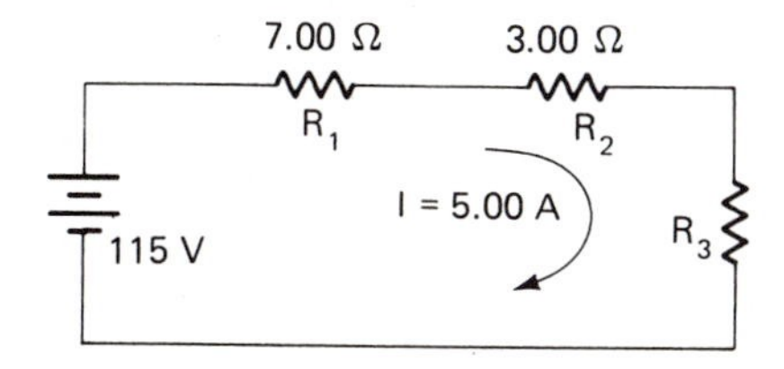

22-4 *PARALLEL CIRCUITS*

An electrical circuit with more than one path for the current to flow is called a *parallel* circuit.

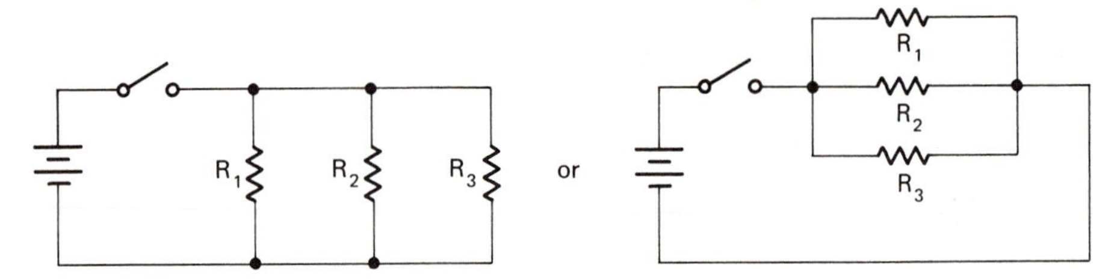

The current in a parallel circuit is divided among the branches of the circuit. How it is divided depends on the resistance of each branch. Since the current divides, the current from the source is equal to the sum of the currents through each of the branches.

Parallel $$I = I_1 + I_2 + I_3 + \cdots$$

where: I = total current in the circuit
I_1 = current through R_1
I_2 = current through R_2
I_3 = current through R_3

The emf of the source is the same as the voltage drop across each resistance in the circuit. Therefore, several different loads requiring the same voltage may be connected in parallel.

Parallel $$E = V_1 = V_2 = V_3 = \cdots$$

where: E = emf of the source
V_1 = voltage drop across R_1
V_2 = voltage drop across R_2
V_3 = voltage drop across R_3

The equivalent resistance of a parallel circuit is less than the resistance of any single branch of the circuit. To find the equivalent resistance, use the formula:

Parallel $$\frac{1}{R} = \frac{1}{R_1} + \frac{1}{R_2} + \frac{1}{R_3} + \cdots$$

where: R = equivalent resistance
R_1 = resistance of R_1
R_2 = resistance of R_2
R_3 = resistance of R_3

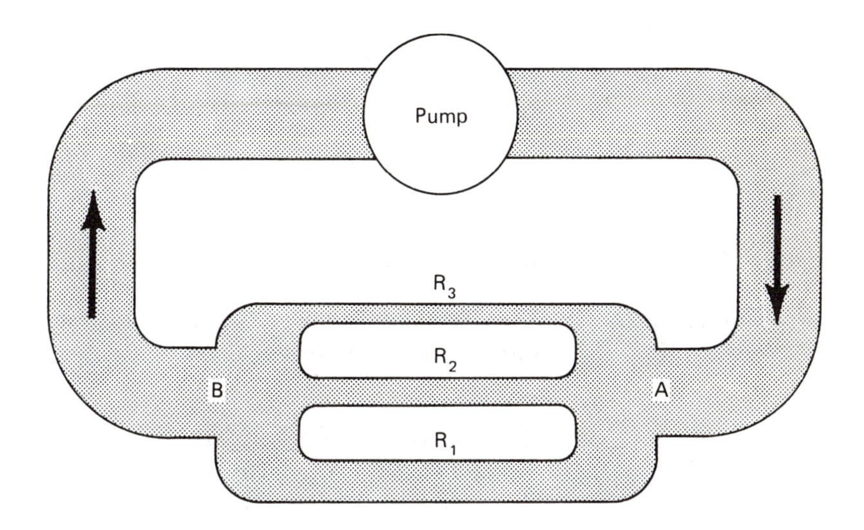

For comparison to parallel circuits, consider the water system shown above.

1. The total amount of water flowing through $R_1 + R_2 + R_3$ is equal to the amount flowing through A or B.
2. The water flowing past point A divides into the three branches R_1, R_2, and R_3.
3. The larger pipes have *less* opposition to water flow than do the smaller pipes.
4. Because R_1 has a larger cross-sectional area than R_2 or R_3, it has less opposition to the flow of water and, therefore, carries more water than R_2 or R_3.

Similarly in a parallel circuit:

1. The total amount of current flowing through $R_1 + R_2 + R_3$ is equal to the amount flowing through A or B.
2. The current flowing past point A divides into the three branches R_1, R_2, and R_3.
3. The smaller resistors have *less* opposition to current flow and therefore carry larger currents.

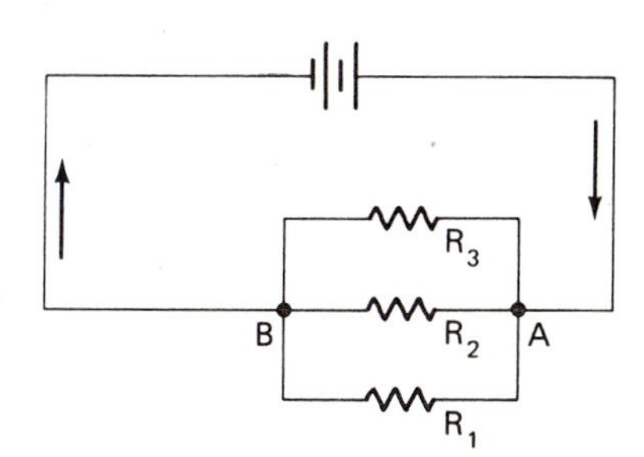

EXAMPLE 1: What is the equivalent resistance of the circuit below?

SKETCH:

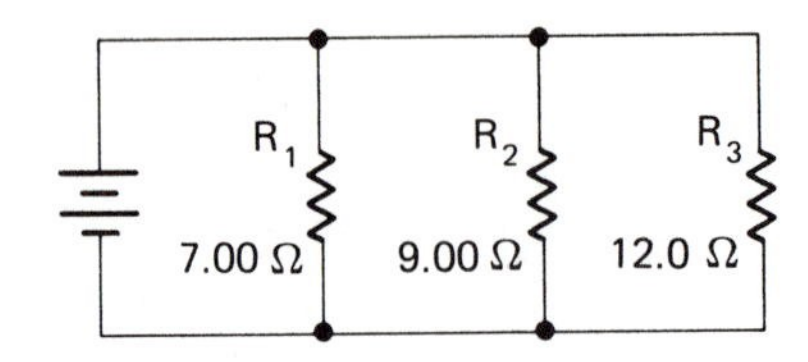

DATA:
$R_1 = 7.00\ \Omega$
$R_2 = 9.00\ \Omega$
$R_3 = 12.0\ \Omega$
$R = ?$

BASIC EQUATION: $$\frac{1}{R} = \frac{1}{R_1} + \frac{1}{R_2} + \frac{1}{R_3}$$

WORKING EQUATION: When using this formula, it is advisable to substitute directly and then solve for the unknown.

SUBSTITUTION:
$$\frac{1}{R} = \frac{1}{7.00\ \Omega} + \frac{1}{9.00\ \Omega} + \frac{1}{12.0\ \Omega}$$
$$\frac{1}{R} = \frac{0.143}{\Omega} + \frac{0.111}{\Omega} + \frac{0.0833}{\Omega}$$
$$\frac{1}{R} = \frac{0.3373}{\Omega}$$
$$R = \frac{1\ \Omega}{0.3373}$$
$$R = 2.97\ \Omega$$

EXAMPLE 2: What is the total current in the circuit below.

SKETCH:

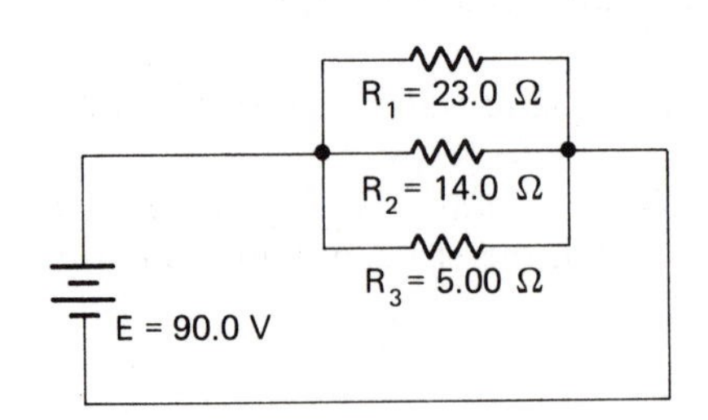

DATA:
$R_1 = 23.0\ \Omega$
$R_2 = 14.0\ \Omega$
$R_3 = 5.00\ \Omega$
$E = 90.0\ \text{V}$
$I = ?$

First, find the equivalent resistance, R. Second, find the total current, I. To find R:

BASIC EQUATION: $$\frac{1}{R} = \frac{1}{R_1} + \frac{1}{R_2} + \frac{1}{R_3}$$

SUBSTITUTION:
$$\frac{1}{R} = \frac{1}{23.0\,\Omega} + \frac{1}{14.0\,\Omega} + \frac{1}{5.00\,\Omega}$$
$$\frac{1}{R} = \frac{0.0435}{\Omega} + \frac{0.0714}{\Omega} + \frac{0.200}{\Omega}$$
$$\frac{1}{R} = \frac{0.3149}{\Omega}$$
$$R = \frac{1\,\Omega}{0.3149}$$
$$R = 3.18\,\Omega$$

To find I:

BASIC EQUATION: $$I = \frac{E}{R}$$

WORKING EQUATION: Same

SUBSTITUTION:
$$I = \frac{90.0\text{ V}}{3.18\,\Omega}$$
$$I = 28.3\text{ A}$$

EXAMPLE 3: What is the equivalent resistance and the value of R_3 in the circuit?

SKETCH:

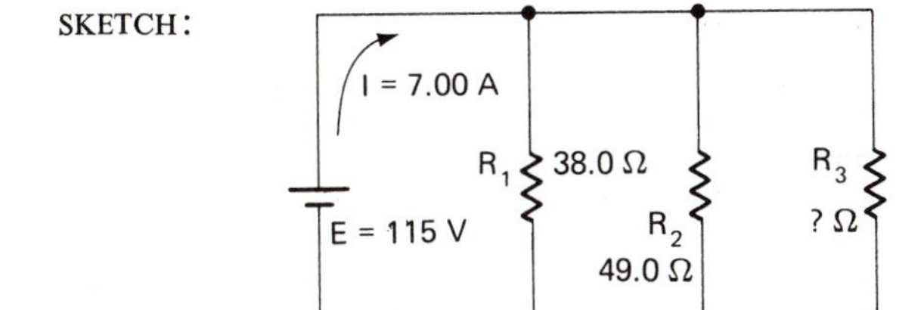

DATA:
$$E = 115\text{ V}$$
$$I = 7.00\text{ A}$$
$$R_1 = 38.0\,\Omega$$
$$R_2 = 49.0\,\Omega$$

To find R:

BASIC EQUATION: $$I = \frac{E}{R}$$

WORKING EQUATION: $R = \frac{E}{I}$

SUBSTITUTION: $R = \frac{115\text{ V}}{7.00\text{ A}}$

$R = 16.4\ \Omega$

To find R_3:

BASIC EQUATION: $\frac{1}{R} = \frac{1}{R_1} + \frac{1}{R_2} + \frac{1}{R_3}$

SUBSTITUTION: $\frac{1}{16.4\ \Omega} = \frac{1}{38.0\ \Omega} + \frac{1}{49.0\ \Omega} + \frac{1}{R_3}$

$\frac{1}{R_3} = \frac{1}{16.4\ \Omega} - \frac{1}{38.0\ \Omega} - \frac{1}{49.0\ \Omega}$

$\frac{1}{R_3} = \frac{0.0610}{\Omega} - \frac{0.0263}{\Omega} - \frac{0.0204}{\Omega}$

$\frac{1}{R_3} = \frac{0.0143}{\Omega}$

$R_3 = \frac{1\ \Omega}{0.0143}$

$R_3 = 69.9\ \Omega$

CHARACTERISTICS OF PARALLEL CIRCUITS

	Parallel
Current	$I = I_1 + I_2 + I_3 + \cdots$
Resistance	$\frac{1}{R} = \frac{1}{R_1} + \frac{1}{R_2} + \frac{1}{R_3} + \cdots$
Voltage	$E = V_1 = V_2 = V_3 = \cdots$

PROBLEMS

1. Find the equivalent resistance in the circuit at the right.

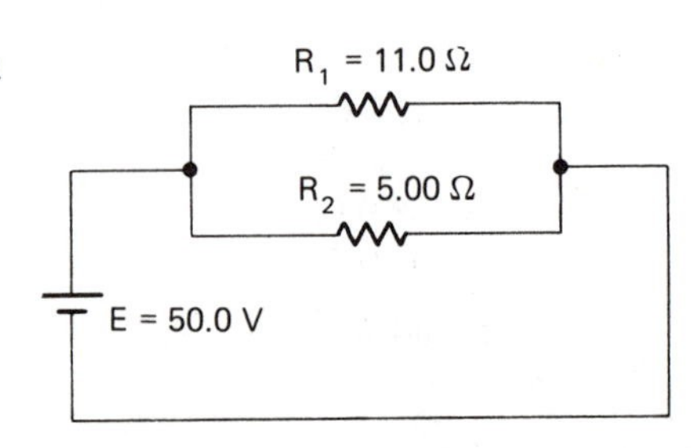

2. What is the total current in the circuit in problem 1?
3. Find I_2 (current through R_2) in the following circuit.

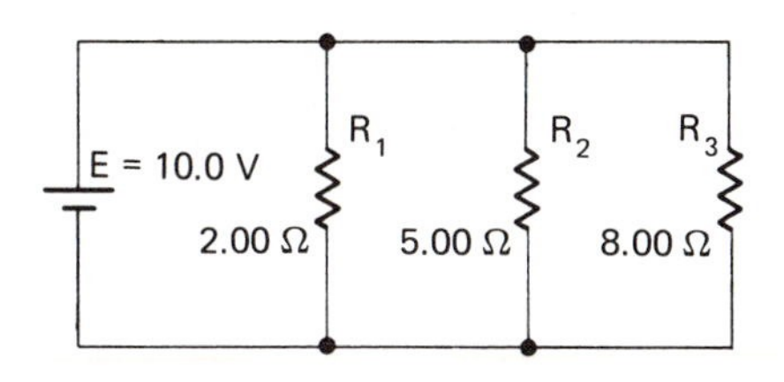

4. Find the total current in the circuit in problem 3.
5. Find the equivalent resistance in the circuit in problem 3.
6. Find the resistance of R_3 in the following circuit.

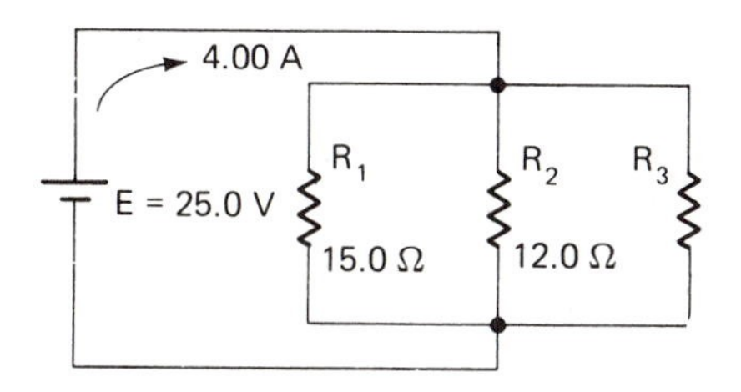

7. What is the current through R_1 in problem 6?
8. What is the equivalent resistance in the following circuit?

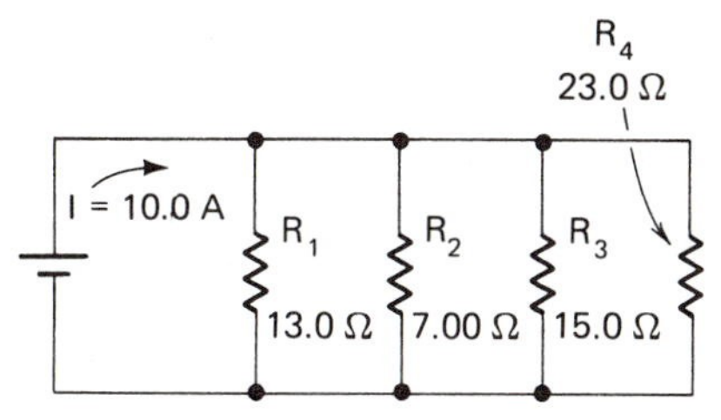

9. What emf is required for the circuit in problem 8?
10. What is the current through R_2 in problem 8?

22-5 *SERIES-PARALLEL CIRCUITS*

Most circuits cannot be solved directly because of the number and arrangement of the resistances. We usually apply the rules for series and parallel circuits to find an equivalent circuit which reduces to a circuit with one resistance.

Example 1: Circuit A is equivalent to circuit B, where $R_4 = R_1 + R_2$. Then, circuit B is equivalent to circuit C, where $\frac{1}{R_5} = \frac{1}{R_3} + \frac{1}{R_4}$

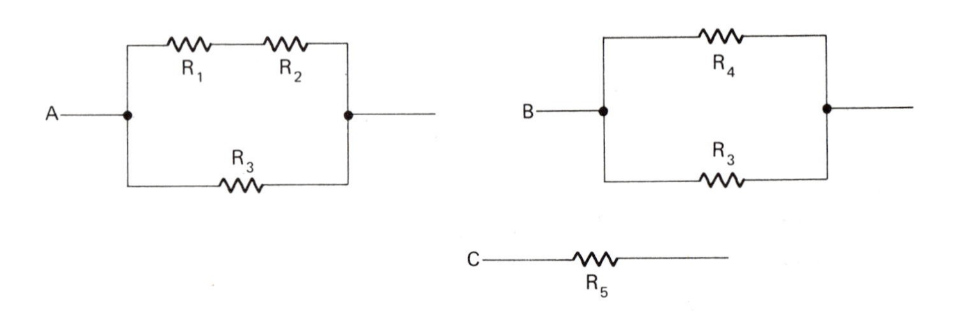

Example 2: Circuit A is equivalent to circuit B, where $\frac{1}{R_4} = \frac{1}{R_2} + \frac{1}{R_3}$

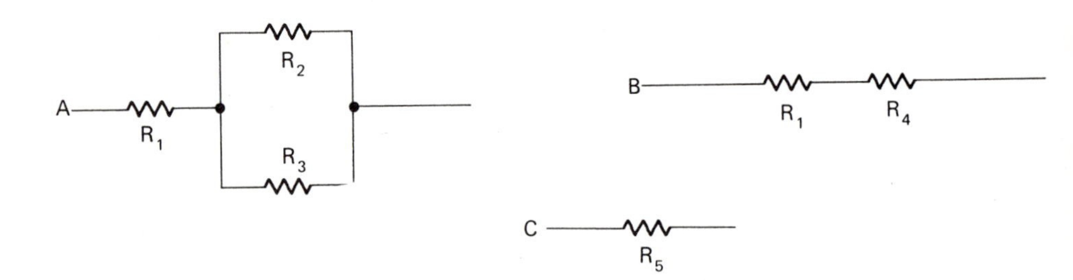

Then, circuit B is equivalent to circuit C, where $R_5 = R_1 + R_4$.

Example 3: Find the total current in the circuit below.

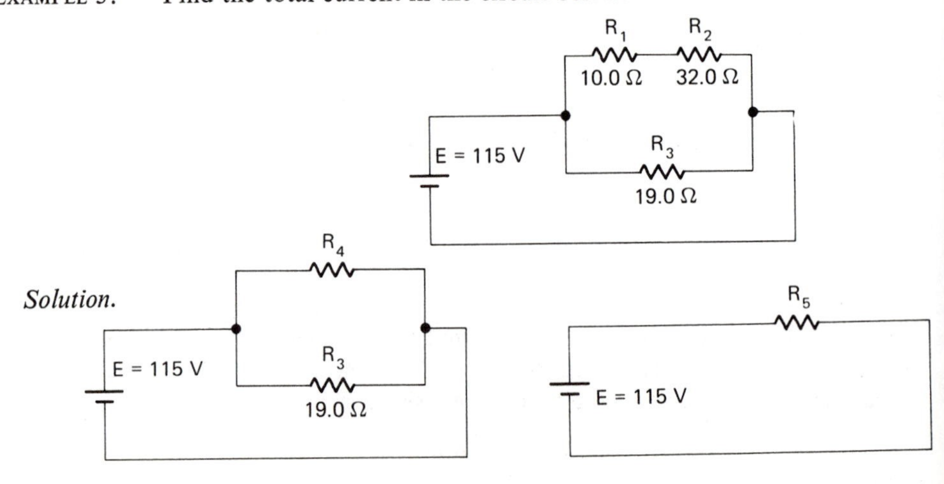

DATA: $E = 115 \text{ V}$
$R_1 = 10.0\ \Omega$
$R_2 = 32.0\ \Omega$
$R_3 = 19.0\ \Omega$
$R_4 = R_1 + R_2 = 10.0\ \Omega + 32.0\ \Omega = 42.0\ \Omega$
$I = ?$

First, find the equivalent resistance, R_5. Second, find the total current, I. To find R_5:

BASIC EQUATION: $$\frac{1}{R_5} = \frac{1}{R_3} + \frac{1}{R_4}$$

$$\frac{1}{R_5} = \frac{1}{19.0\ \Omega} + \frac{1}{42.0\ \Omega}$$

$$\frac{1}{R_5} = \frac{0.0526}{1\ \Omega} + \frac{0.0238}{1\ \Omega}$$

$$\frac{1}{R_5} = \frac{0.0764}{1\ \Omega}$$

$$R_5 = \frac{1\ \Omega}{0.0764}$$

$$R_5 = 13.1\ \Omega$$

To find I:

BASIC EQUATION: $$I = \frac{E}{R_5}$$

WORKING EQUATION: Same

SUBSTITUTION: $$I = \frac{115 \text{ V}}{13.1\ \Omega}$$

$$I = 8.78 \text{ A}$$

EXAMPLE 4: Find the equivalent resistance in the circuit at the right.

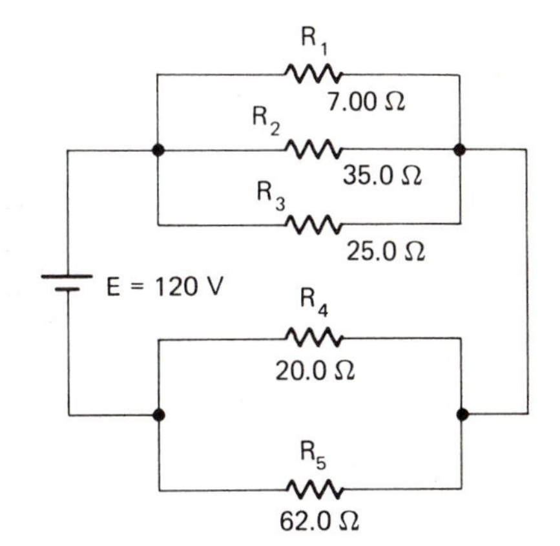

Solution.

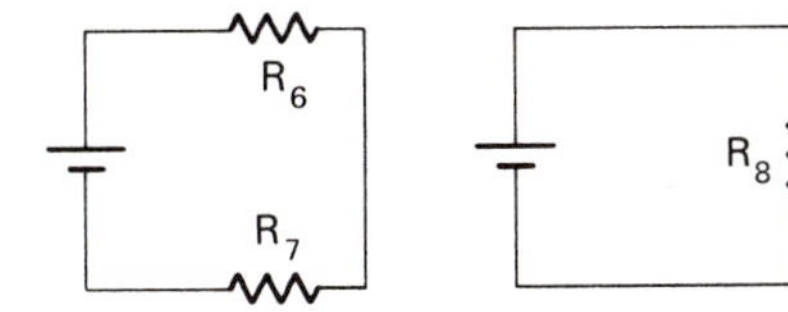

DATA: $R_1 = 7.00\ \Omega$
$R_2 = 35.0\ \Omega$
$R_3 = 25.0\ \Omega$
$R_4 = 20.0\ \Omega$
$R_5 = 62.0\ \Omega$
$E = 120\ \text{V}$
$R_8 = ?$

First, find R_6. Second, find R_7. Third, find the equivalent resistance, R_8. To find R_6:

BASIC EQUATION: $$\frac{1}{R_6} = \frac{1}{R_1} + \frac{1}{R_2} + \frac{1}{R_3}$$

SUBSTITUTION: $$\frac{1}{R_6} = \frac{1}{7.00\ \Omega} + \frac{1}{35.0\ \Omega} + \frac{1}{25.0\ \Omega}$$

$$\frac{1}{R_6} = \frac{0.143}{1\ \Omega} + \frac{0.0286}{1\ \Omega} + \frac{0.0400}{1\ \Omega}$$

$$\frac{1}{R_6} = \frac{0.2116}{1\ \Omega}$$

$$R_6 = \frac{1\ \Omega}{0.2116}$$

$$R_6 = 4.72\ \Omega$$

To find R_7:

BASIC EQUATION: $$\frac{1}{R_7} = \frac{1}{R_4} + \frac{1}{R_5}$$

$$\frac{1}{R_7} = \frac{1}{20.0\ \Omega} + \frac{1}{62.0\ \Omega}$$

$$\frac{1}{R_7} = \frac{0.0500}{1\ \Omega} + \frac{0.0161}{1\ \Omega}$$

$$\frac{1}{R_7} = \frac{0.0661}{1\ \Omega}$$

$$R_7 = \frac{1\ \Omega}{0.0661}$$

$$R_7 = 15.1\ \Omega$$

To find R_8:

BASIC EQUATION: $R_8 = R_6 + R_7$

WORKING EQUATION: Same

SUBSTITUTION: $R_8 = 4.72\ \Omega + 15.1\ \Omega$
$R_8 = 19.82\ \Omega$ or $19.8\ \Omega$

EXAMPLE 5: Find the total current in Example 4.

DATA:
$E = 120\text{ V}$
$R_8 = 19.8\ \Omega$
$I = ?$

BASIC EQUATION: $I = \dfrac{E}{R_8}$

WORKING EQUATION: Same

SUBSTITUTION:
$I = \dfrac{120\text{ V}}{19.8\ \Omega}$
$I = 6.06\text{ A}$

CHARACTERISTICS OF SERIES AND PARALLEL CIRCUITS

	Series	Parallel
Current	$I = I_1 = I_2 = I_3 = \cdots$	$I = I_1 + I_2 + I_3 + \cdots$
Resistance	$R = R_1 + R_2 + R_3 + \cdots$	$\dfrac{1}{R} = \dfrac{1}{R_1} + \dfrac{1}{R_2} + \dfrac{1}{R_3} + \cdots$
Voltage	$E = V_1 + V_2 + V_3 + \cdots$	$E = V_1 = V_2 = V_3 = \cdots$

PROBLEMS

Circuit A: For use in problems 1–5.

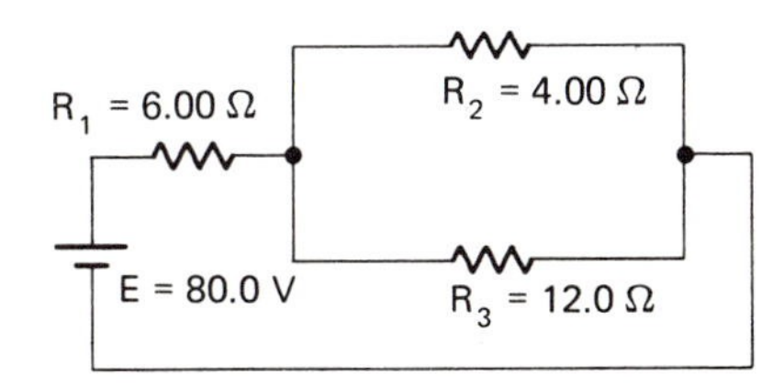

1. What is the equivalent resistance of the resistances connected in parallel?
2. What is the equivalent resistance of the entire circuit?
3. What is the current in R_1?
4. What is the voltage drop across R_1?
5. Find the current through R_3.

Circuit B: For use in problems 6–12.

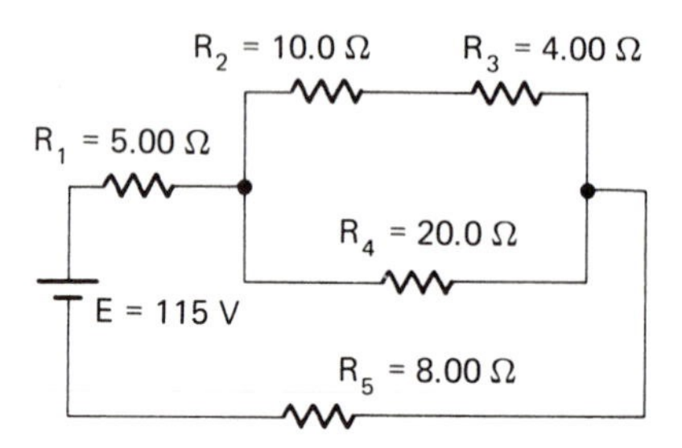

6. What is the equivalent resistance of the resistances connected in parallel?
7. What is the equivalent resistance of the circuit?
8. What is the current in R_1?
9. What is the voltage drop across the parallel part of the circuit?
10. What is the current through R_3?
11. What is the current through R_5?
12. What is the voltage drop across R_3?

Circuit C: For use in problems 13–20.

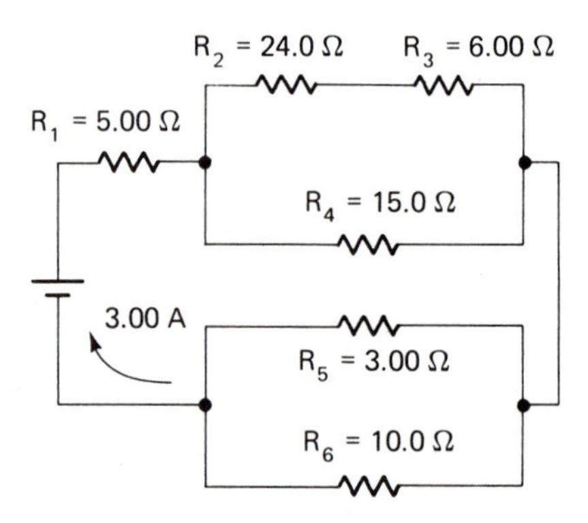

13. What is the equivalent resistance of the parallel arrangement in the upper branch?
14. What is the equivalent resistance of the parallel arrangement in the lower branch?
15. What is the equivalent resistance of the entire circuit?
16. What emf is required for the given current flow in the circuit?
17. What is the voltage drop across the parallel arrangement in the upper branch?

18. What is the voltage drop across R_4?
19. What is the voltage drop across R_6?
20. What is the current through R_6?

Circuit D: For use in problems 21–25.

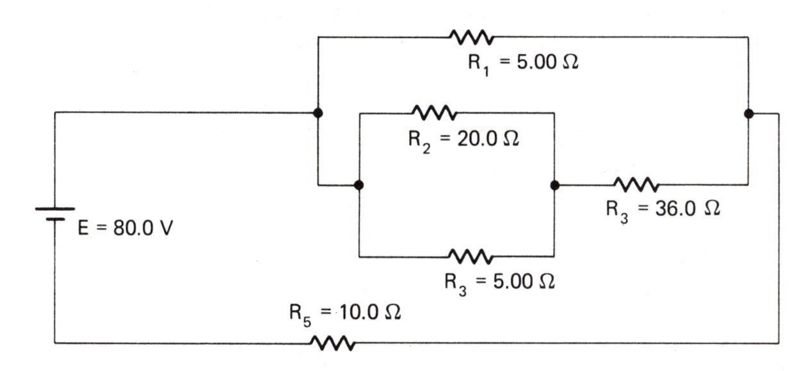

21. What is the equivalent resistance in the circuit?
22. What is the current in R_5?
23. What is the voltage drop across R_5?
24. What is the voltage drop across R_4?
25. What is the current through R_2?

22-6 *ELECTRICAL INSTRUMENTS*

In the laboratory we will use several kinds of electric meters for measurements. Great care must be taken to avoid passing a large current through the meter. Meters are fragile instruments and abuse will ruin them. A large current will burn out the meter.

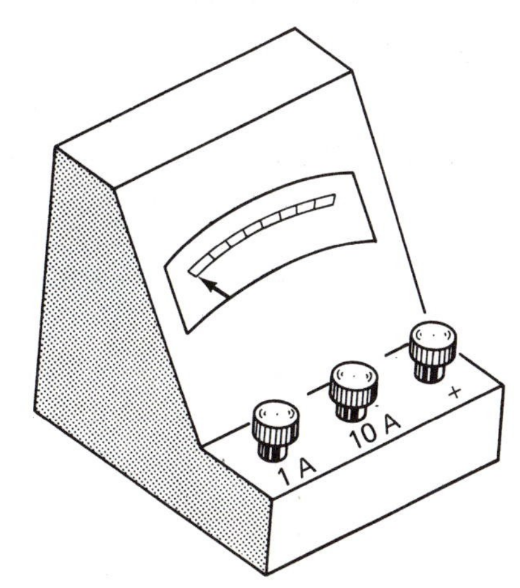

Measurements in ac circuits and dc circuits require different kinds of meters. One type of meter may *not* be used in the other type of circuit. Many instruments have more than one range on which readings are to be made. If you connect the lead wires to the (+) and 10-A posts, each scale division represents 1 A. If you connect the lead wires to the (+) and 1-A posts, each scale division represents 0.1 A.

Because of the construction of the meter, the most accurate readings are obtained from the middle of the scale. You should *start with the highest range* and then adjust downward until you get a mid-scale reading. The reading and use of the different kinds of meters will be studied in the laboratory.

The Voltmeter. This instrument measures the difference in potential between two points in a circuit. It should *always* be connected in *parallel* with the part of the circuit over which we wish to measure the voltage drop. The voltmeter is a high-resistance instrument and draws very little current.

The Ammeter. The ammeter measures the current flowing in a circuit. Therefore, it is connected in *series* in the circuit. Since all the current flows through the meter, it has very low resistance so that its effect on the circuit will be as small as possible.

The Galvanometer. The galvanometer is a very sensitive instrument which is used to detect the presence and direction of *very small* currents.

The Ohmmeter. The ohmmeter is used to measure the resistance of a circuit component. It should only be used when there is *no current* flowing in the circuit.

PROBLEMS

Using the formulas for series and parallel circuits, fill in the blanks in the tables below. In the blanks across from Batt. (Battery) under

V: write the emf of the battery.
I: write the total current in the circuit.
R: write the equivalent or total resistance of the entire circuit.

In the blanks across from R_1 under

V: write the voltage drop across R_1.
I: write the current flowing through R_1.
R: write the resistance of R_1.

In the blanks across from R_2, R_3, . . . , fill in the appropriate numbers under V, I, and R. (Begin by looking for key information given in the table and work from there.)

1.

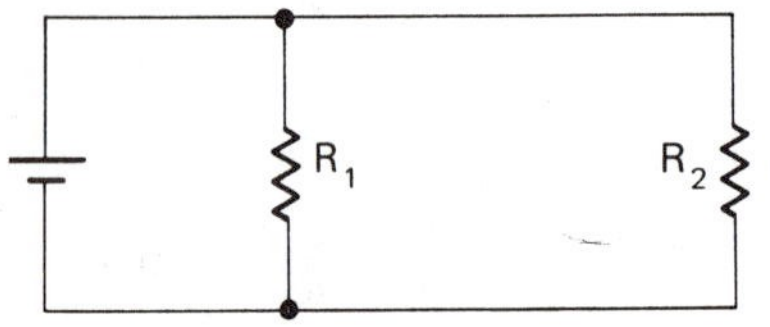

	V	I	R
Batt.	12.0 V	A	Ω
R_1	V	A	2.00 Ω
R_2	V	A	4.00 Ω

2.

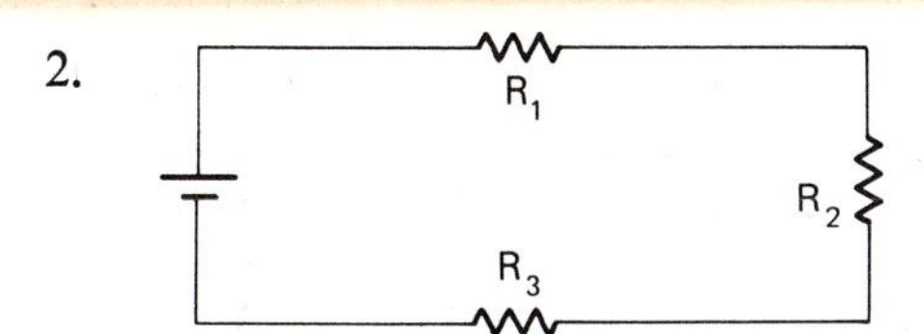

	V	I	R
Batt.	V	A	Ω
R_1	V	2.00 A	4.00 Ω
R_2	V	A	6.00 Ω
R_3	V	A	8.00 Ω

3.

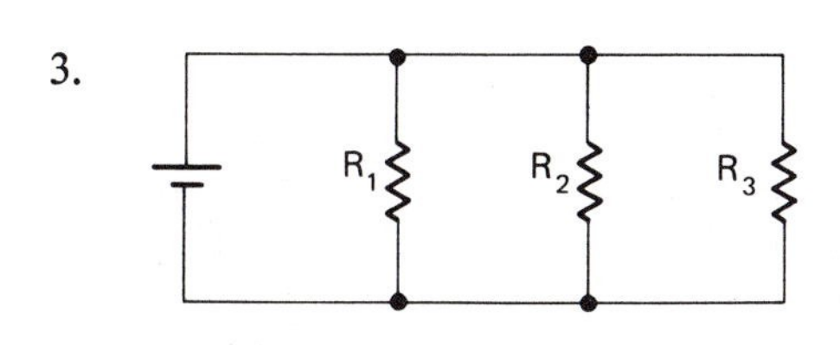

	V	I	R
Batt.	V	A	Ω
R_1	V	2.00 A	Ω
R_2	V	3.00 A	12.0 Ω
R_3	V	1.00 A	Ω

4.

	V	I	R
Batt.	12.0 V	2.00 A	Ω
R_1	V	A	6.00 Ω
R_2	V	A	4.00 Ω
R_3	V	A	15.0 Ω

5.

	V	I	R
Batt.	50.0 V	5.00 A	Ω
R_1	V	2.00 A	Ω
R_2	25.0 V	A	Ω
R_3	10.0 V	A	Ω
R_4	V	3.00 A	Ω

6.

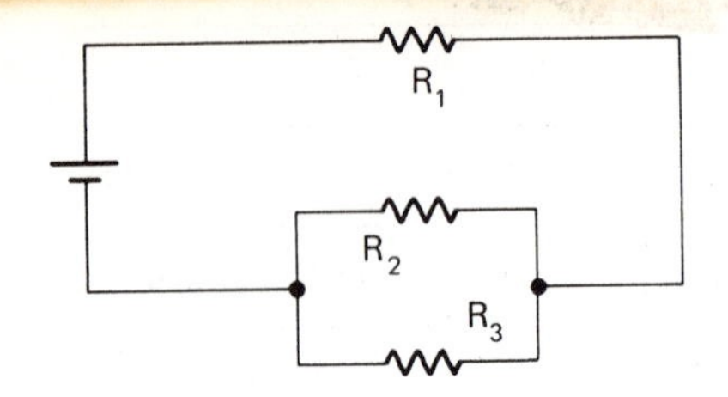

	V	I	R
Batt.	24.0 V	A	Ω
R_1	8.00 V	A	Ω
R_2	V	4.00 A	Ω
R_3	V	2.00 A	Ω

7.

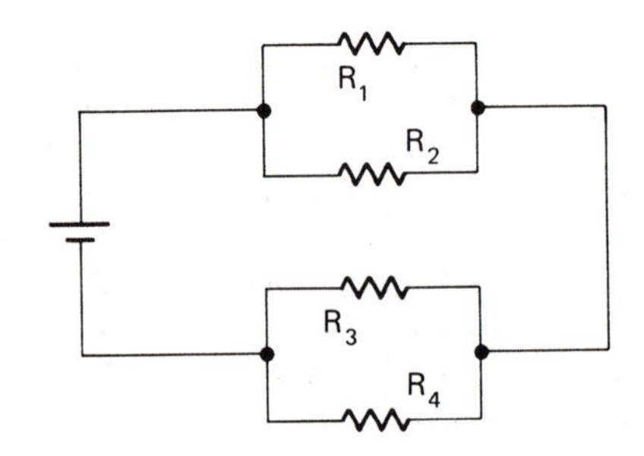

	V	I	R
Batt.	V	A	Ω
R_1	12.0 V	A	2.00 Ω
R_2	V	A	4.00 Ω
R_3	24.0 V	A	4.00 Ω
R_4	V	A	8.00 Ω

8.

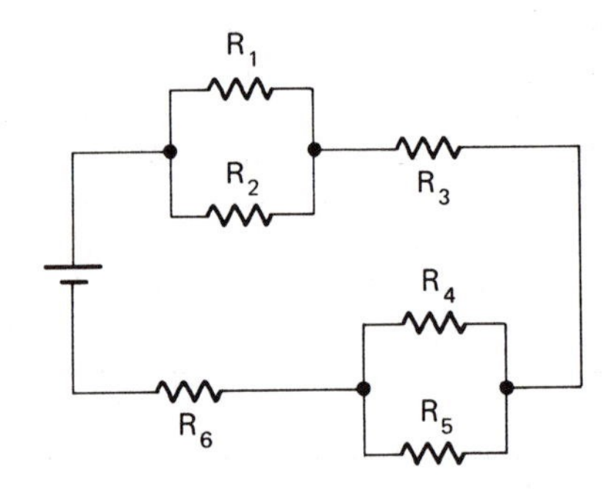

	V	I	R
Batt.	30.0 V	A	Ω
R_1	6.00 V	3.00 A	Ω
R_2	V	2.00 A	Ω
R_3	V	A	3.00 Ω
R_4	V	1.00 A	Ω
R_5	8.00 V	A	Ω
R_6	V	A	Ω

9.

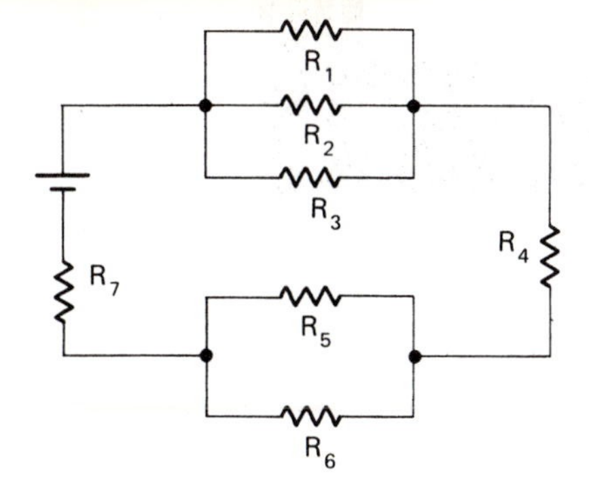

	V	I	R
Batt.	V	12.0 A	Ω
R_1	V	A	Ω
R_2	18.0 V	2.00 A	Ω
R_3	V	A	3.00 Ω
R_4	V	A	4.00 Ω
R_5	V	A	2.00 Ω
R_6	V	8.00 A	Ω
R_7	6.00 V	A	Ω

10.

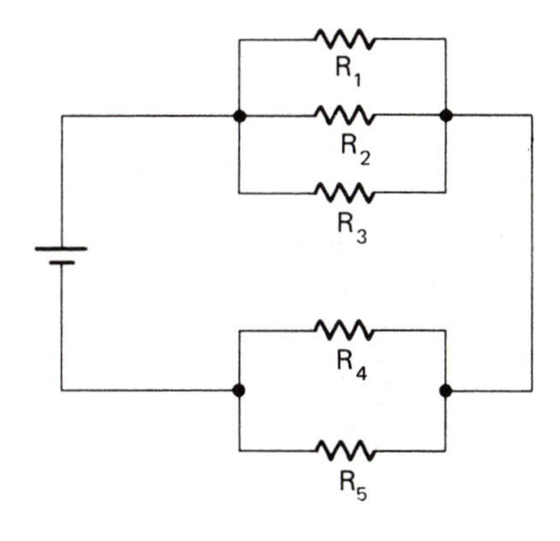

	V	I	R
Batt.	46.0 V	A	Ω
R_1	V	3.00 A	Ω
R_2	V	4.00 A	Ω
R_3	V	A	6.00 Ω
R_4	V	3.00 A	Ω
R_5	V	7.00 A	Ω

11.

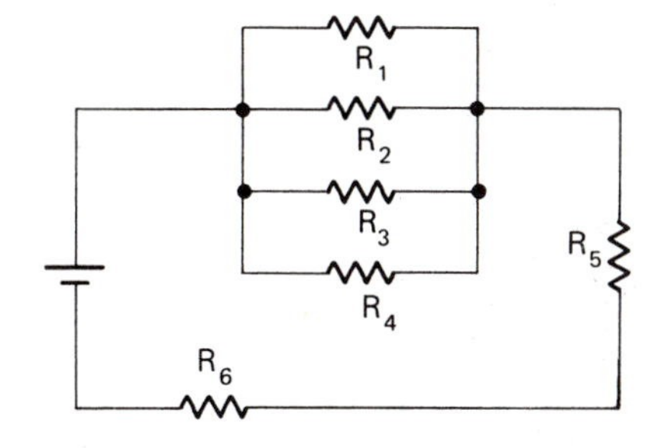

	V	I	R
Batt.	V	A	Ω
R_1	V	A	20.0 Ω
R_2	10.0 V	A	Ω
R_3	V	A	4.00 Ω
R_4	V	1.00 A	Ω
R_5	V	5.00 A	5.00 Ω
R_6	V	A	6.00 Ω

DC SOURCES 23

23-1 *THE LEAD STORAGE CELL*

Lead storage batteries are used in automobiles and many other types of vehicles and machinery. A battery is a group of cells connected together. These lead cells are secondary cells, which means they are rechargeable. The passing of an electric current through the cell to restore the original chemicals is called recharging. Cells, like the dry cell, which cannot be efficiently recharged are called primary cells.

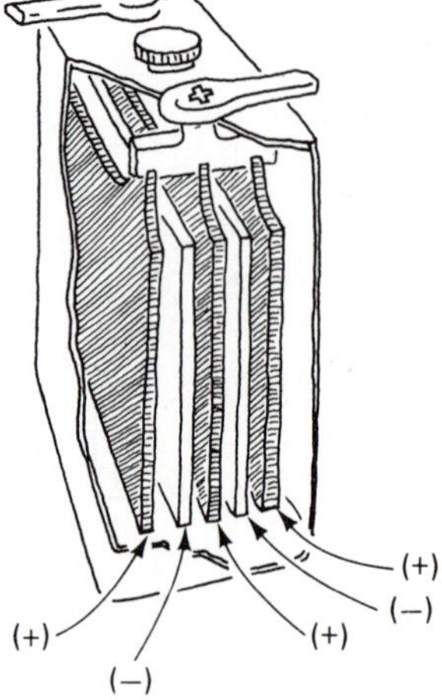

(+)
(−)
(+)
(−)
(+)

Lead storage cells are made up of two kinds of lead plates (lead and lead oxide) submerged in a solution of distilled water and sulfuric acid. This acid solution is called an electrolyte.

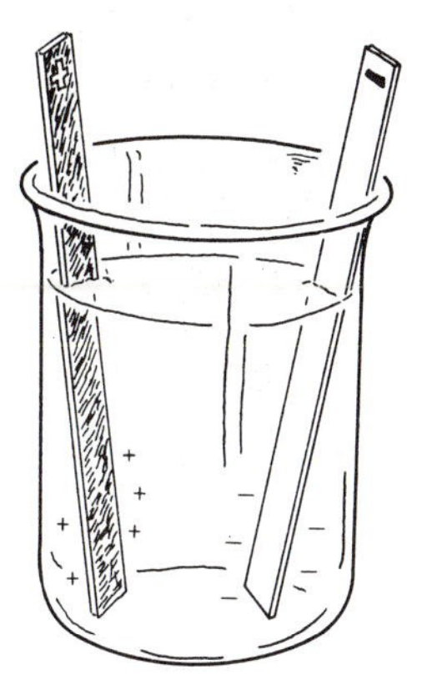

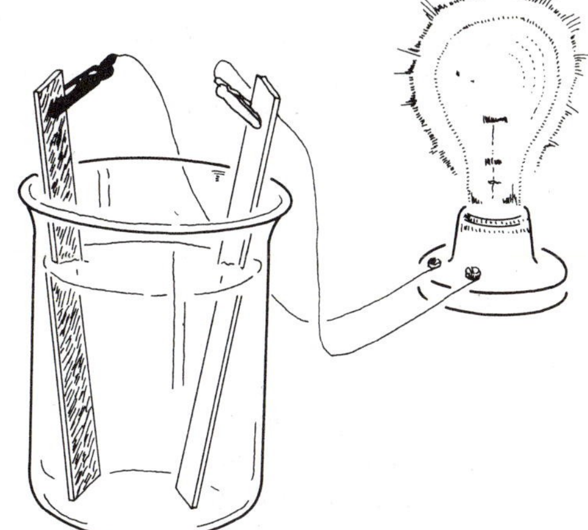

Acid solution

The chemical action between the lead plates and the acid solution produces large numbers of free electrons at the negative (—) pole of the battery. These electrons have a large amount of electrical potential energy which is used in the load in the circuit (for instance, to operate headlights or to turn a starter motor).

As the electrical energy is used in the load, the battery must be recharged. This is done by a generator or an alternator. Such devices provide an electric current to reverse the chemical reaction taking place in the battery. The recharging process extends the life of the battery which would otherwise be very short.

23-2 *THE DRY CELL*

The dry cell is the most widely used primary cell. This is the kind of cell we use in flashlights and portable radios. A dry cell is made of a carbon rod, which is the positive (+) terminal or pole, and a zinc can, which acts as the negative (—) terminal. In between is a paste of chemicals and water which reacts with the terminals to provide energized electrons. These cells are available in a wide range of sizes. Common sizes are usually 1.5 volts to 9 volts.

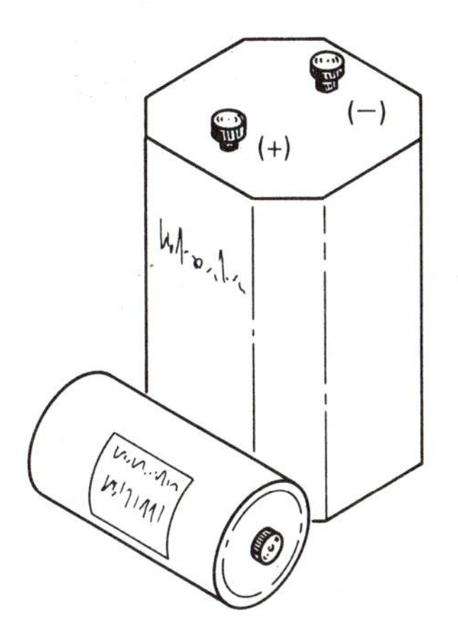

The dry cell, as well as the lead cell, has resistance within the cell itself which opposes the movement of the electrons from the (+) to (−) poles. This is called the *internal resistance* (r) of the cell. Because current flows in the cell, the emf of the cell is reduced by the voltage drop across the internal resistance. The voltage applied to the external circuit is then:

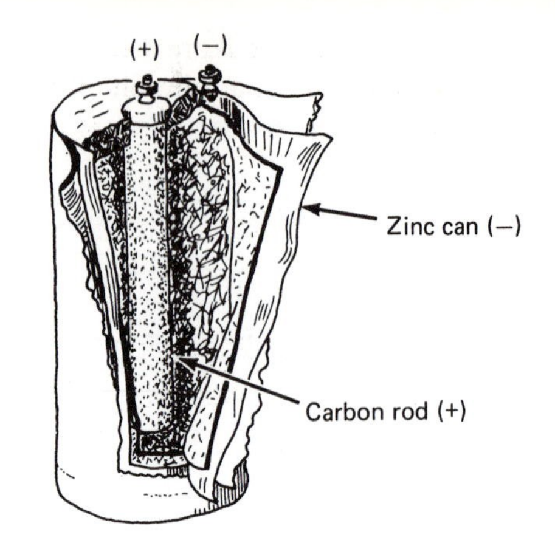

$$V = E - Ir$$

where: $V =$ voltage applied to circuit
$E =$ emf of the cell
$I =$ current through cell
$r =$ internal resistance of cell

Many times, the current or voltage available from a single cell is inadequate to do a particular job. Then we usually connect two or more cells in series or parallel.

23-3 *CELLS IN SERIES*

To connect cells in series, the positive terminal of one is connected to the negative terminal of the next cell. This procedure is continued until the desired number of cells are all connected.

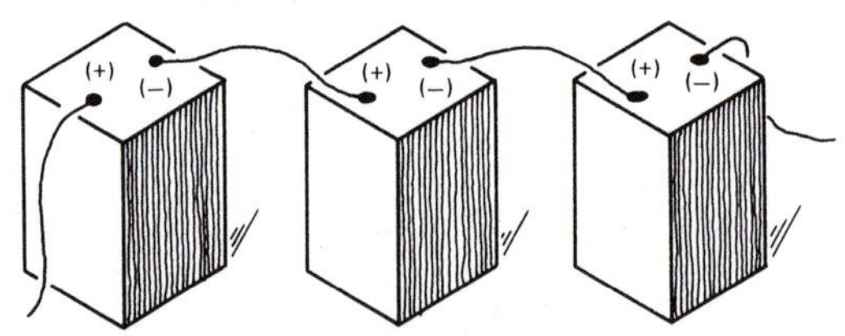

Series connected dry cells

The rules for cells connected in series and parallel are similar to those for simple resistances.

RULE

CELLS IN SERIES

1. The current in the circuit is the same as in any single cell:

$$I = I_1 = I_2 = I_3 = \cdots$$

2. The internal resistance of the battery is equal to the sum of the individual internal resistances of the cells:

$$r = r_1 + r_2 + r_3 + \cdots$$

3. The emf of the battery is equal to the sum of the emfs of the individual cells:

$$E = E_1 + E_2 + E_3 + \cdots$$

EXAMPLE: Two 6.00-V cells with internal resistance of 0.100 Ω each are connected in series to form a battery with a current of 0.750 A in each cell.

(a) What is the emf of the battery?
(b) What is the internal resistance of the battery?
(c) What is the current in the external circuit?

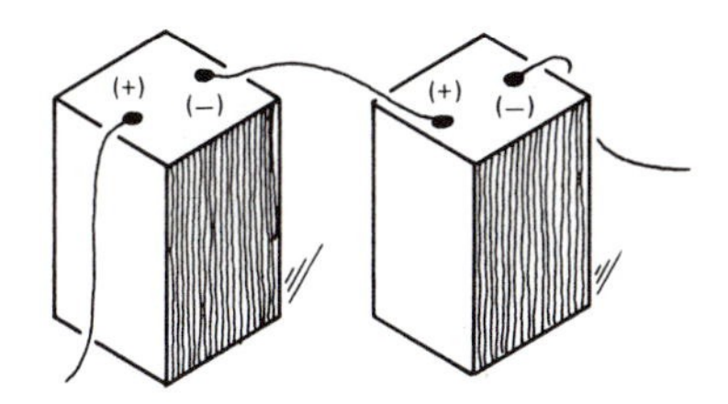

Series

(+) (−)

Circuit diagram

(a) $E = E_1 + E_2 = 6.00\text{ V} + 6.00\text{ V} = 12.0\text{ V}$ (Rule 3)
(b) $r = r_1 + r_2 = 0.100\ \Omega + 0.100\ \Omega = 0.200\ \Omega$ (Rule 2)
(c) 0.750 A (Rule 1)

23-4 *CELLS IN PARALLEL*

To connect cells in parallel, the positive terminals of all the cells are connected together and the negative terminals are all connected together. The leads from the external circuit may be connected to any positive and negative terminals. (The external circuit is all of the circuit *outside* the battery or cell.)

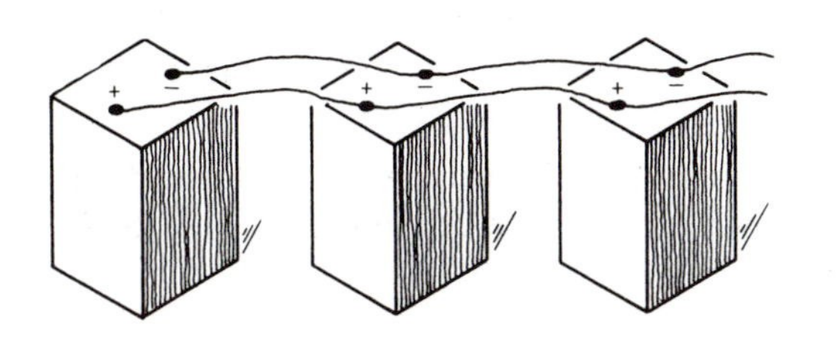

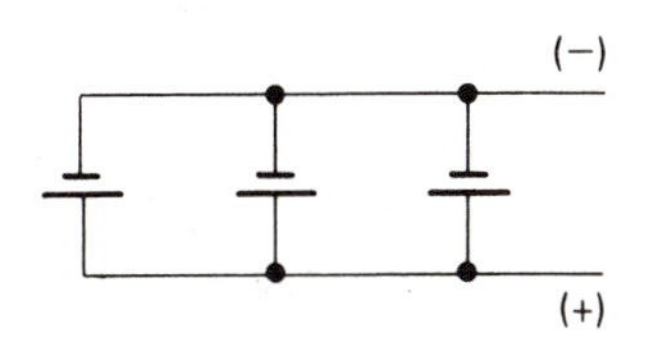

RULE

Cells in Parallel

1. The total current is the sum of the individual currents in each cell:
$$I = I_1 + I_2 + I_3 + \cdots$$
2. The internal resistance is equal to the resistance of one cell divided by the number of cells:
$$r = \frac{r \text{ of one cell}}{\text{no. of cells}}$$
3. The emf of the battery is equal to the emf of any single cell:
$$E = E_1 = E_2 = E_3 = \cdots$$

Example: Four cells, each 1.50 V and internal resistance of 0.0500 Ω, are connected in parallel to form a battery with a current output of 0.250 A in each cell.

(a) What is the emf of the battery?
(b) What is the internal resistance of the battery?
(c) What is the current in the external circuit?

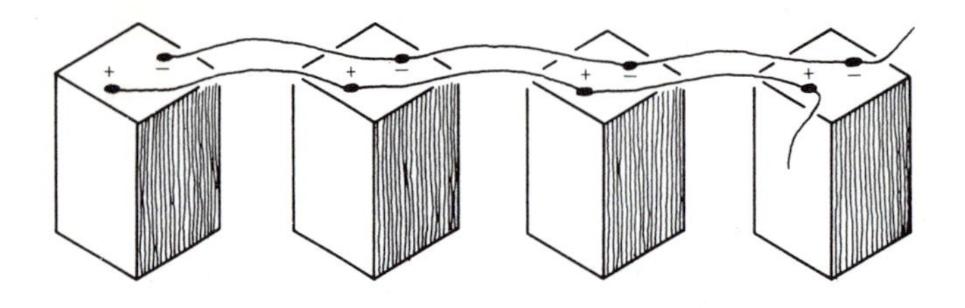

Parallel

(−)
(+)

Circuit diagram

(a) 1.50 V (Rule 3)

(b) $\frac{r \text{ of one cell}}{\text{no. of cells}} = \frac{0.0500\ \Omega}{4} = 0.0125\ \Omega$ (Rule 2)

(c) $I = I_1 + I_2 + I_3 + I_4 = 0.25\text{ A} + 0.25\text{ A} + 0.25\text{ A} + 0.25\text{ A}$
$= 1.00\text{ A}$ (Rule 1)

PROBLEMS

1. A cell has an emf of 1.50 V and an internal resistance of 0.0500 Ω. If there is 0.250 A in the cell, what voltage is applied to the external circuit?
2. The voltage applied to a circuit is 11.8 V when the current through the battery is 0.500 A. If the internal resistance of the battery is 0.300 Ω, what is the emf of the battery?

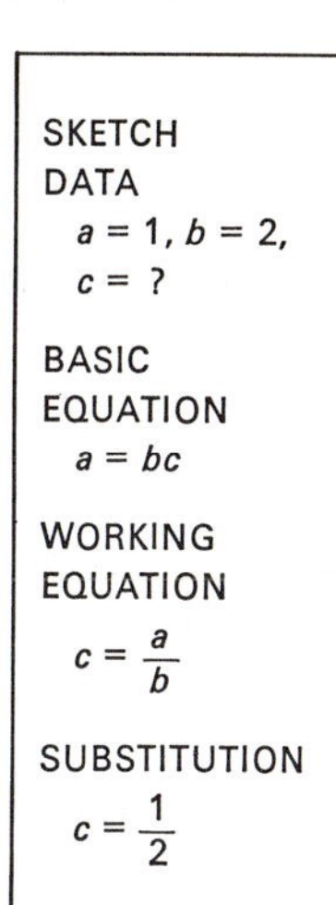

3. Three 1.50-V cells, each with internal resistance of 0.0500 Ω, are connected in series to form a battery with a current of 0.850 A in each cell.
   (a) What is the current in the external circuit?
   (b) What is the emf of the battery?
   (c) What is the internal resistance of the battery?
4. Five 9.00-V cells with internal resistance of 0.100 Ω each are connected in parallel to form a battery with a current output of 0.750 A in a certain circuit.
   (a) What is the current in the external circuit?
   (b) What is the emf of the battery?
   (c) What is the internal resistance of the battery?
5. What is the current in the circuit on the left below?

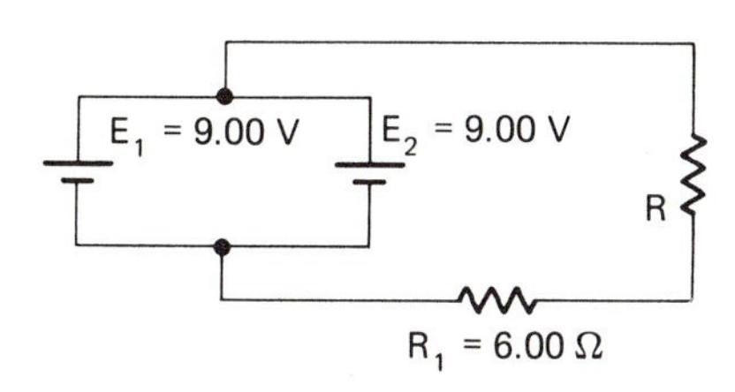

6. Find the current in the circuit at the right.

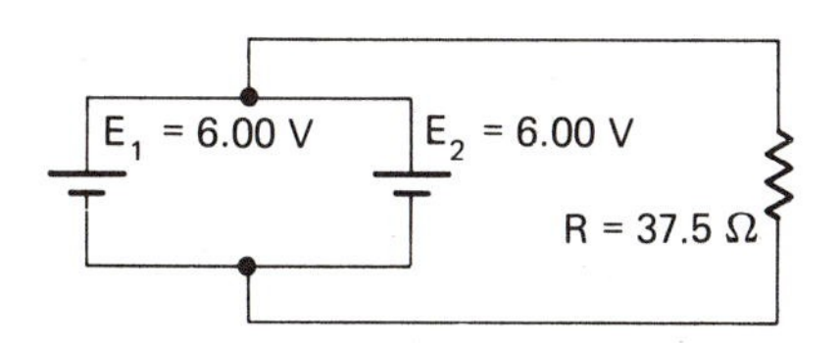

7. If the current in the circuit below is 1.20 A, what is the value of R?

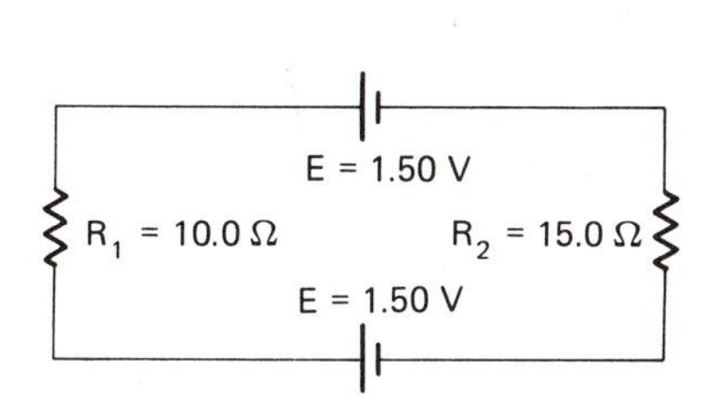

8. Find the current in the circuit at the right.

23-5 *ELECTRICAL POWER*

Tremendous quantities of energy are used by industry. This energy is, of course, not free but is sold by power companies. The rate of consuming energy is called power. The unit of power is the watt. One watt is the power generated by a current of 1 A flowing because of a potential difference of 1 V.

$$P = VI$$

where: P = power (watts)
V = voltage drop
I = current

Recalling Ohm's law: $I = V/R$, we can find two other equations for power:

Given: $P = VI$

Substitute: $V = IR \qquad P = (IR)I$

Result: $$P = I^2R$$

Also given: $P = I^2R$

Substitute: $I = \frac{V}{R} \qquad P = \left(\frac{V}{R}\right)^2 R = \frac{V^2}{R^2} \cdot R$

Result: $$P = \frac{V^2}{R}$$

EXAMPLE 1: A soldering iron draws 7.50 A on a 115 V circuit. What is its wattage rating?

DATA: $I = 7.50$ A
$V = 115$ V
$P = ?$

BASIC EQUATION: $P = VI$

WORKING EQUATION: Same

SUBSTITUTION: $P = (115\text{ V})(7.50\text{ A})$
$= 863$ W

EXAMPLE 2: A hand drill draws 4.00 A and has a resistance of 14.6 Ω. What power does it use?

DATA: $I = 4.00$ A
$R = 14.6\ \Omega$
$P = ?$

BASIC EQUATION: $P = I^2R$

WORKING EQUATION: Same

SUBSTITUTION: $P = (4.00\ \text{A})^2(14.6\ \Omega)$
$P = 234\ \text{W}$

Since the watt is a relatively small unit, the kilowatt (kW = 1000 watts) is commonly used in industry. Although we speak of "paying our power bill," what power companies actually sell is energy. Energy is sold in kilowatt-hours (kW hr). The amount of energy consumed is equal to the power used times the time it is used. Therefore,

$$\text{Energy} = \text{power} \times \text{time}$$

or

$$\text{Energy (in kW hr)} = (VI)t$$
$$\text{No. of kW hr} = VIt$$

This equation is useful in finding the cost of electrical energy. Cost is measured in cents per kilowatt-hour.

The cost of operating an electrical device may be found:

$$\text{Cost} = (\text{kW hr})\left(\frac{\text{cents}}{\text{kW hr}}\right)$$

$$\boxed{\text{Cost} = \text{power} \times \text{hours} \times \frac{\text{cents}}{\text{kW hr}} \times \frac{\text{kW}}{1000\ \text{W}}}$$

conversion factor

EXAMPLE 3: An iron is rated at 550 watts. How much would it cost to operate it for 1.50 hours at \$.05/kW hr?

DATA: $P = 550\ \text{W}$
$t = 1.50\ \text{hr}$
$\text{Rate} = \frac{\$.05}{\text{kW hr}}$
$\text{Cost} = ?$

BASIC EQUATION: $\text{Cost} = Pt\left(\frac{\text{cents}}{\text{kW hr}}\right)\left(\frac{\text{kW}}{1000\ \text{W}}\right)$

WORKING EQUATION: Same

SUBSTITUTION: $\text{Cost} = (550\ \cancel{\text{W}})(1.50\ \cancel{\text{hr}})\left(\frac{\$.05}{\cancel{\text{kW}}\ \cancel{\text{hr}}}\right)\left(\frac{\cancel{\text{kW}}}{1000\ \cancel{\text{W}}}\right)$
$\text{Cost} = \$.04$

PROBLEMS

1. A heater draws 8.70 A on a 110-V line. What is its wattage rating?
2. What power is needed for a sander which draws 3.50 A and has a resistance of 6.70 Ω?
3. How many amperes will a 50.0-W lamp on a 110-V line draw?
4. What is the resistance of the lamp in problem 3?
5. How many amperes will a 750-W lamp draw on a 110-V circuit?
6. What would it cost to operate the lamp in problem 5 for 40.0 hours if the cost of energy is $.04 per kW hr?
7. Six 50.0-W bulbs are operated on a 115-V circuit. They are in use for 25.0 hours in a certain month. If energy costs $.045 per kW hr, what is the cost of operating them for the month?
8. A small furnace expends 2.00 kW of power. If the cost of operation of the furnace is $2.40 for a 24.0-hr period, what is the cost of energy per kW hr?

SKETCH
DATA
$a = 1$, $b = 2$,
$c = ?$

BASIC EQUATION
$a = bc$

WORKING EQUATION
$c = \frac{a}{b}$

SUBSTITUTION
$c = \frac{1}{2}$

MAGNETISM 24

24-1 *INTRODUCTION*

Many devices which use or produce electrical energy depend on the relation of magnetism and electric currents. Motors and meters are designed to use the fact that electric currents in wires behave like magnets. Generators produce electrical current due to the movement of wires near very large magnets.

We will investigate the basic properties of magnets and the relation between currents and magnetism in this chapter. The later chapters on generators, motors, and transformers will use the basic principles of magnetism that are developed here.

24-2 *MAGNETIC MATERIALS*

Certain kinds of metal have been found to have the ability to attract pieces of iron, steel, and some other metals. Metals that have this ability are said to be magnetic. Deposits of iron ore which are naturally magnetic have been found. This ore is called lodestone.

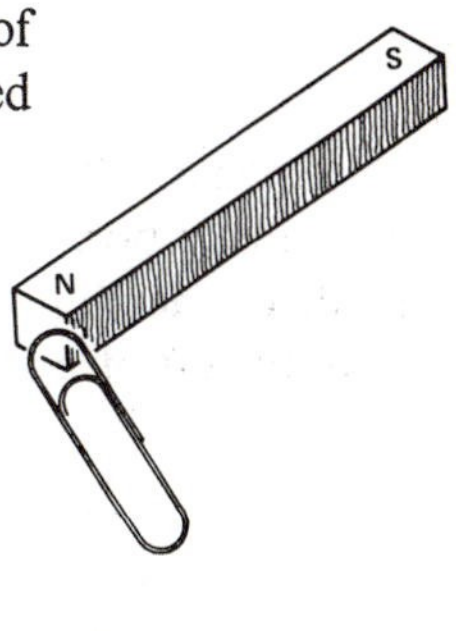

Artificial magnets can be made from iron, steel, and several special alloys such as permalloy and alnico. We will discuss the process of creating artificial magnets later. Materials which can be made into magnets are called magnetic materials. Most materials are nonmagnetic (examples are wood, aluminum, copper, and zinc).

24-3 *FORCES BETWEEN MAGNETS*

Suppose a bar magnet is suspended by a string so that it is free to rotate. It will rotate until one end points north and the other south. The end which points north is called the north-seeking pole, or just north (N) pole. The other end is the south-seeking pole or south (S) pole.

If the north pole of another bar magnet is brought near the north pole of this magnet, the two like poles will repel. The south pole of one magnet will attract the north pole of the other.

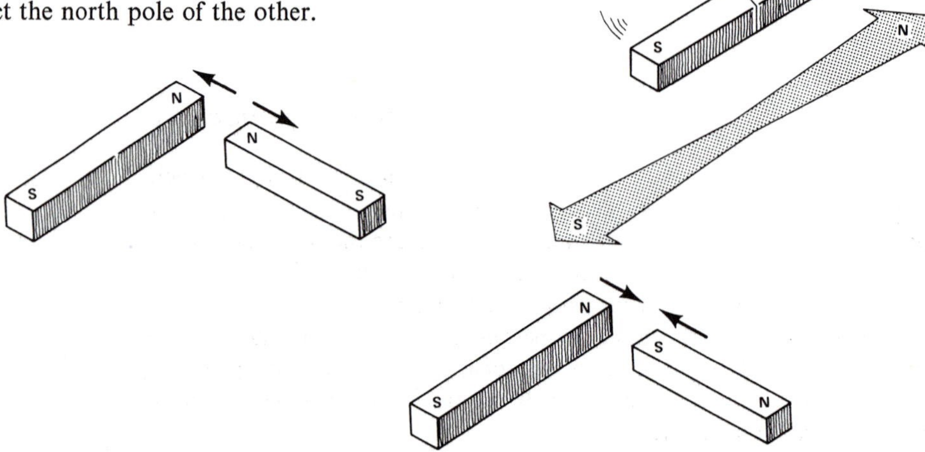

In summary:

Like magnetic poles repel each other, and unlike magnetic poles attract each other.

A magnet lines up along a north-south line because of the attraction of the magnetic poles of the earth. The earth has a south magnetic pole which attracts the north pole of a magnet and a north magnetic pole which attracts the south end of a magnet.

The earth's south magnetic pole is near its north geographic pole and vice versa.

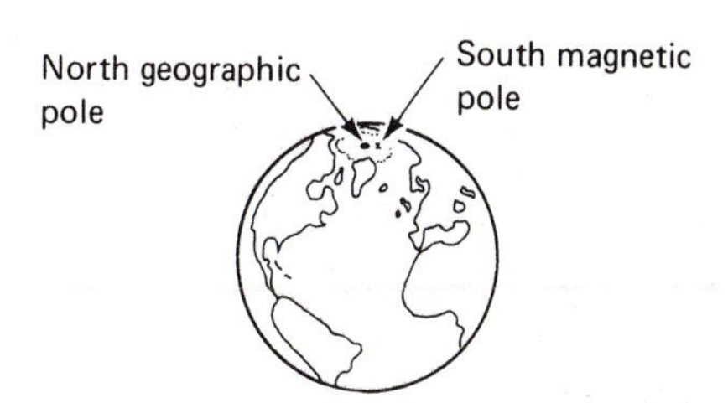

A compass is simply a small magnetic needle which is free to rotate on a bearing. The needle's north pole always points to the south magnetic pole of the earth.

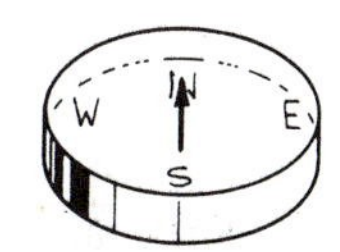

24-4 *MAGNETIC FIELDS OF FORCE*

There is a magnetic field near a magnetic pole. The existence of this field can be detected by using another magnet. We can represent this field of force by drawing lines which indicate the direction of the force exerted on a north pole placed there.

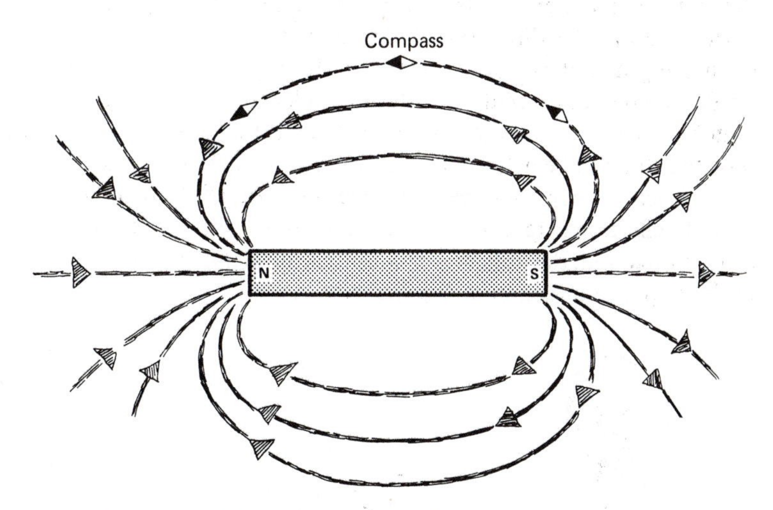

The field of a bar magnet can be mapped by moving a small compass around the magnet as shown. These resulting lines are called flux lines (lines of force).

The flux lines can also be found by sprinkling iron filings on a sheet of paper laid over a magnet. The fields of combinations of magnets can also be found in this way.

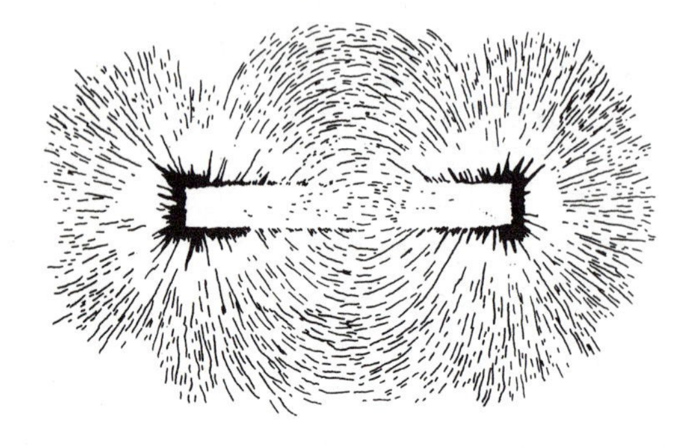

The magnetic field around a magnet is a three-dimensional field as shown at the right.

24-5 *MAGNETIC EFFECTS OF CURRENTS*

When a current passes through a conductor, it sets up a magnetic field. A compass placed near the current shows the direction of this magnetic field. We can show this by connecting a battery to a wire (refer to Figs. 1–6 below for this discussion). A compass needle is placed under the wire (Fig. 1). When the switch is closed, the needle deflects (Fig. 2). If the terminals of the battery are reversed, the needle deflects in the opposite direction (Fig. 3).

When the compass needle is placed on top of the conductor, the direction of deflection is reversed in each case. When the current in a wire flows in a given direction, the flux lines point in one direction below the wire and point in the opposite direction above the wire.

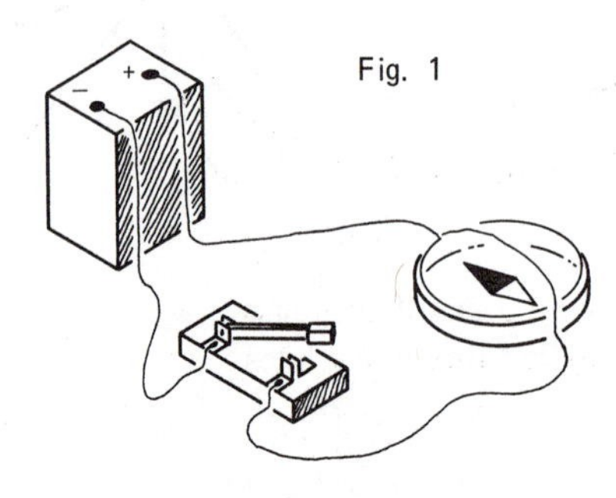

Compass below conductor, SW open

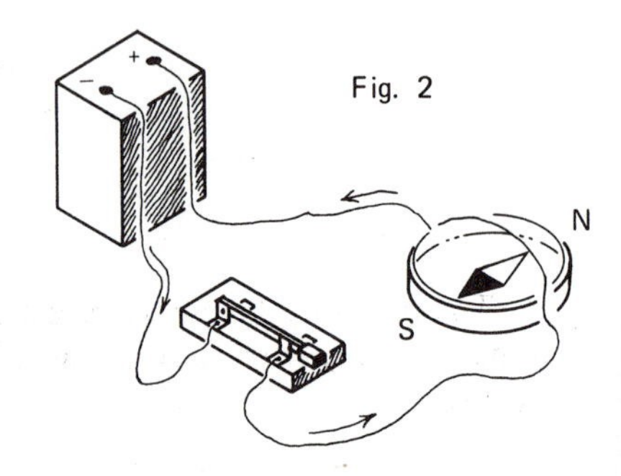

Compass below conductor, SW closed

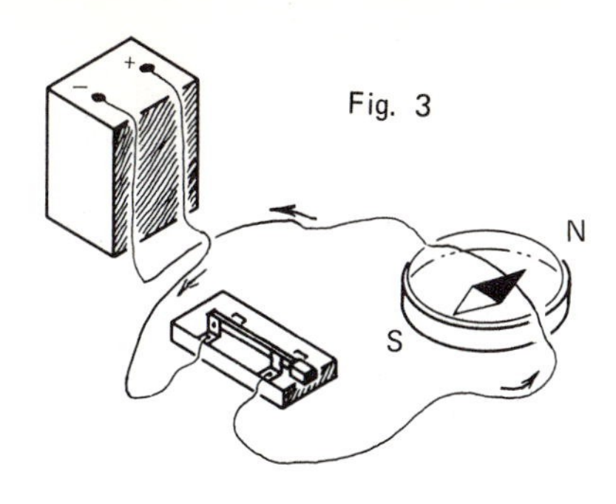

Compass below conductor, SW closed
Battery terminals reversed

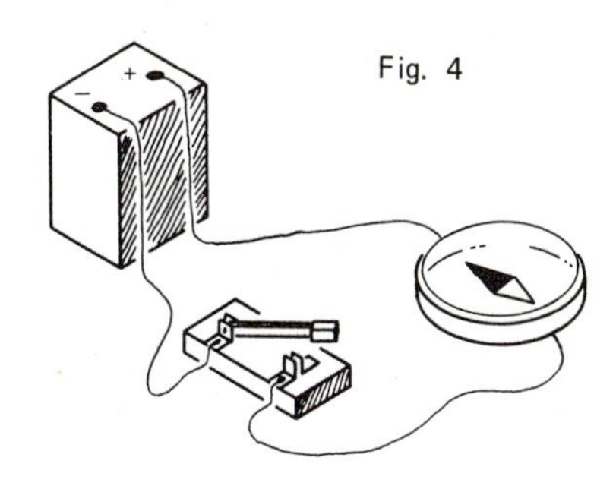

Compass above conductor, SW open

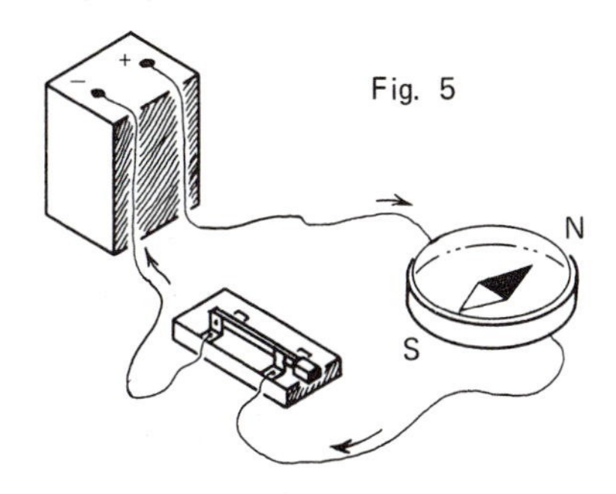

Compass above conductor, SW closed

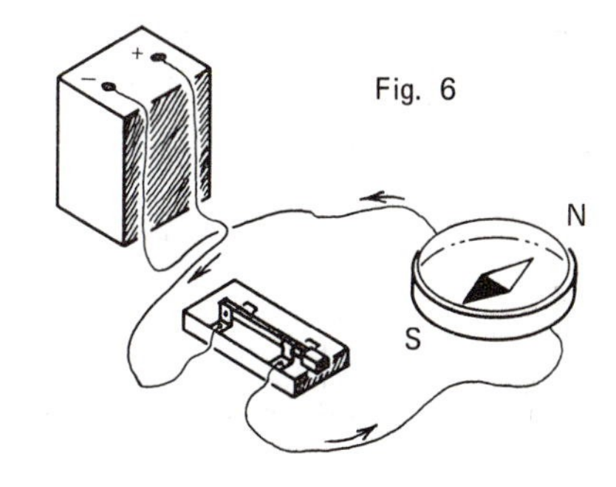

Compass above conductor, SW closed
Battery terminals reversed

The direction of the flux lines actually curve around the straight current-carrying wire. Iron filings on a sheet of paper perpendicular to a current-carrying wire show the direction of the field. The magnetic field is stronger for large currents than for small currents. The direction of the field near a current in a straight wire is given by Ampere's rule:

RULE

Hold the conductor in your right hand, with your thumb extended in the direction of the current. Your fingers circle the wire in the direction of the flux lines.

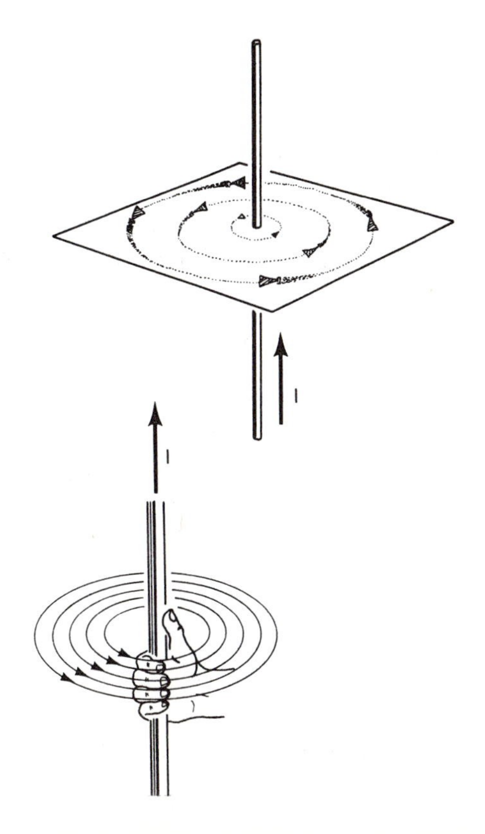

To determine the direction of the flux lines of a current in a loop, use Ampere's rule. If several loops are made into a tight spiral as shown, the flux lines add to form the field shown. A coil of tightly wrapped wire is called a solenoid.

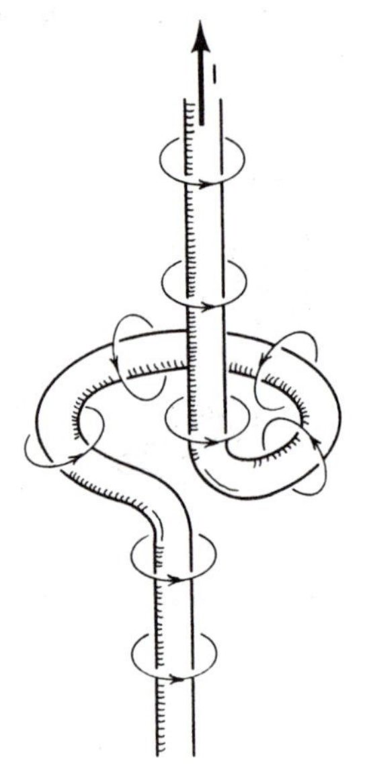

The left side of this solenoid acts like a south magnetic pole. The right side acts like a north magnetic pole. This polarity could be found by using a compass. The rule for finding the polarity of a solenoid is:

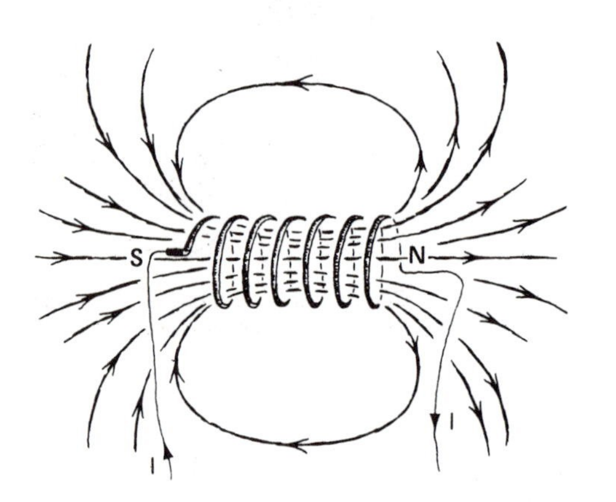

RULE

Hold the solenoid in your right hand so that your fingers circle it in the same direction as the current. Your thumb points to the north pole of the solenoid.

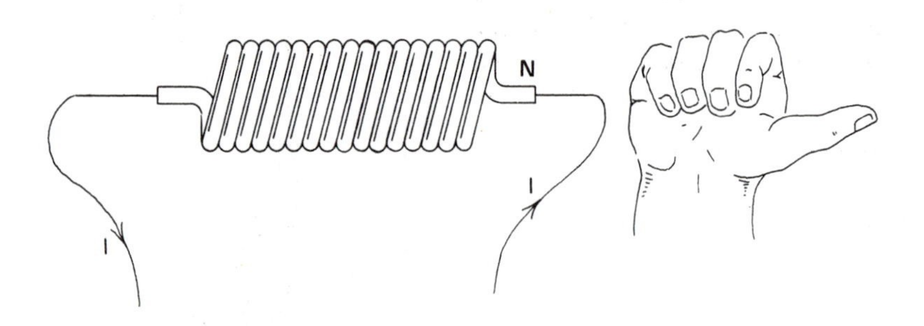

24-7 *INDUCED MAGNETISM AND ELECTROMAGNETS*

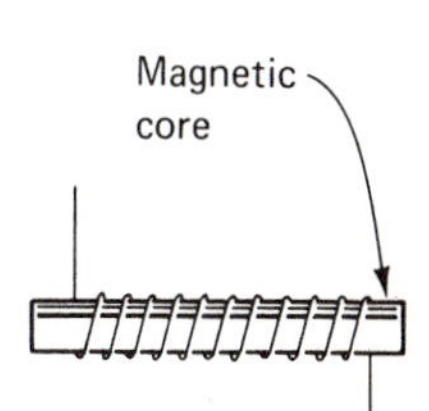

When a magnetic material such as iron is placed in the core of a current-carrying solenoid, the material becomes very strongly magnetized. This is called induced magnetism. The solenoid and magnetic core are called an electromagnet.

When the current through the coil is turned off, the strength of the induced magnet decreases, but some remains. When the core is removed, a magnetic field remains in the core. This is how artificial magnets are made.

A magnet can be thought to consist of many atoms, each behaving like a small magnet. In each atom, the electrons orbit about the nucleus and each electron spins about its own axis. These motions produce a magnetic field in the atom.

Usually the atoms in a material are arranged so that their magnetic fields point in different directions. The result is that the magnetism of one atom is canceled out by its neighbors.

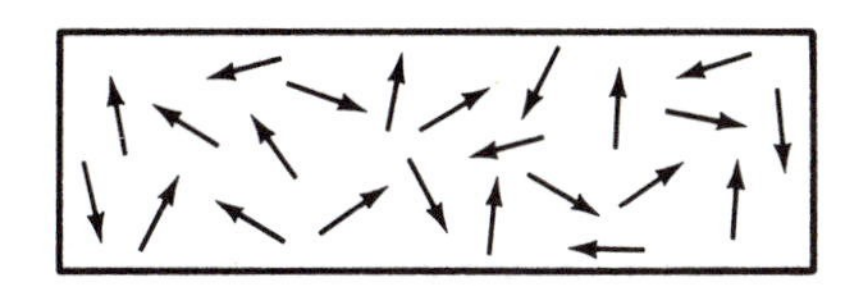

When a magnetic material is magnetized, many of the atoms' magnetic fields line up to point in one direction. The magnetic fields of these atoms add together to give the material a magnetic field.

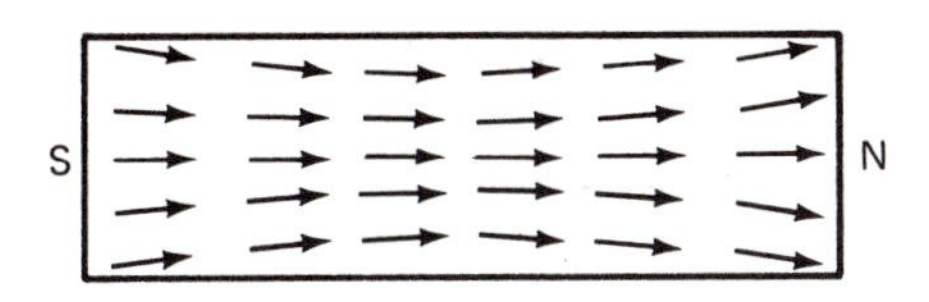

GENERATORS 25

25-1 *INTRODUCTION*

When a magnet is moved so that its flux lines cut across a wire, an emf is induced in the wire. The strength of this induced emf depends on the strength of the magnetic field and on the rate at which the flux lines are cut by moving the magnet or wire. Increasing the strength of the field or increasing the rate at which the flux lines are cut also causes the current to increase.

While the magnet in the diagram is moved downward, the galvanometer indicates that a current flows through the wire.

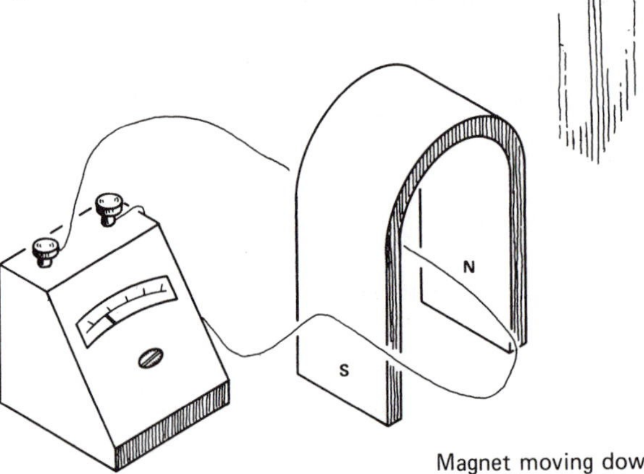

Magnet moving downward.

If the magnet is moved upward, the induced current would be in the opposite direction.

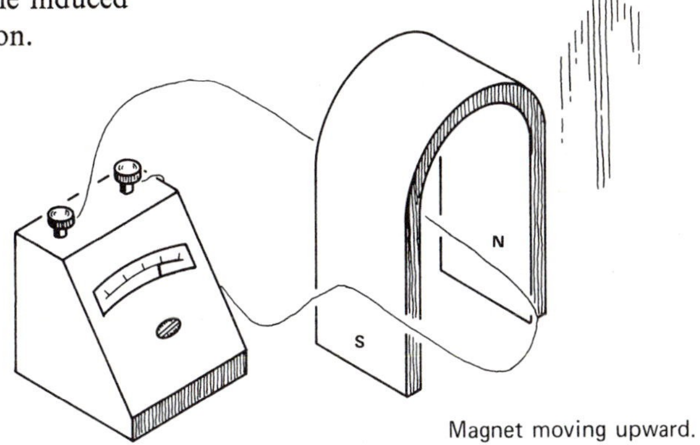

Magnet moving upward.

A current would also flow in the wire if the wire is moved and the magnet is stationary. The current is produced by the relative motion of the magnet and wire. In commercial generators, magnets are spun inside a set of coils of wire.

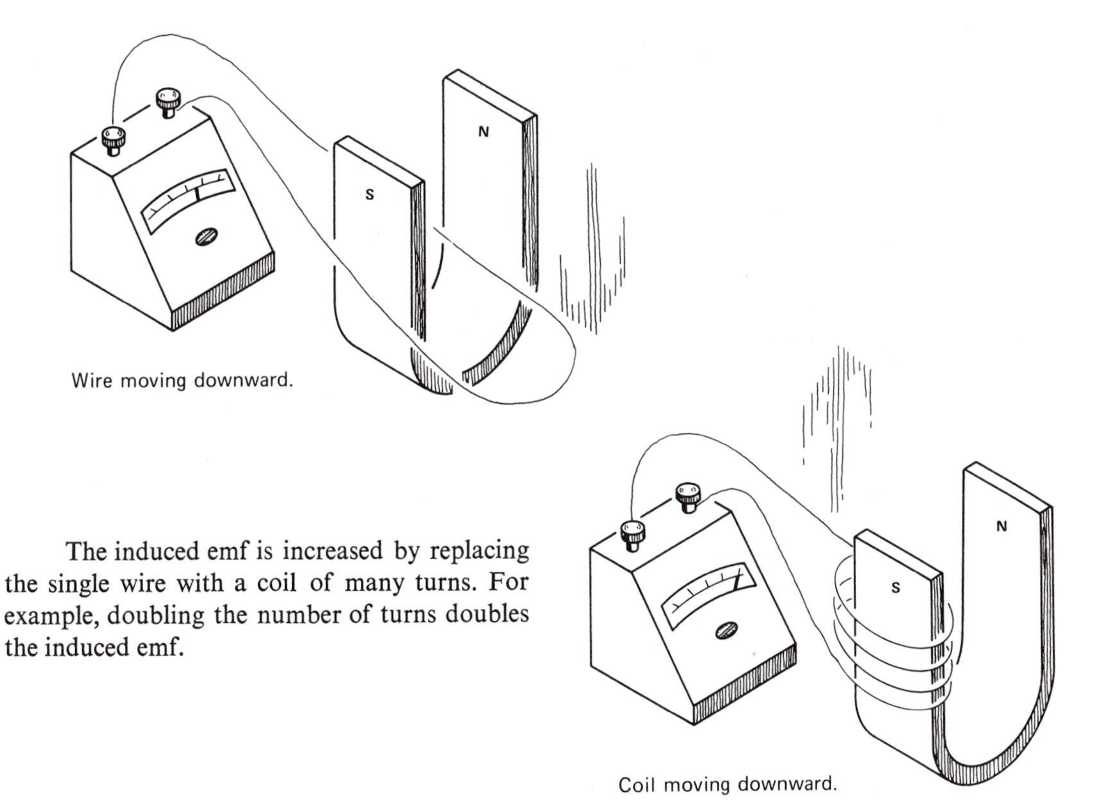

Wire moving downward.

The induced emf is increased by replacing the single wire with a coil of many turns. For example, doubling the number of turns doubles the induced emf.

Coil moving downward.

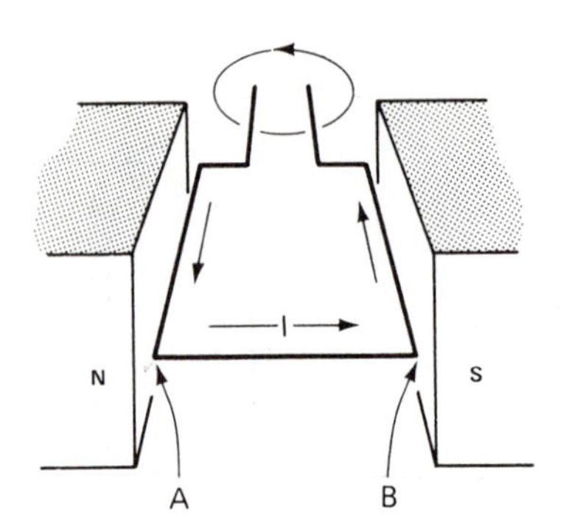

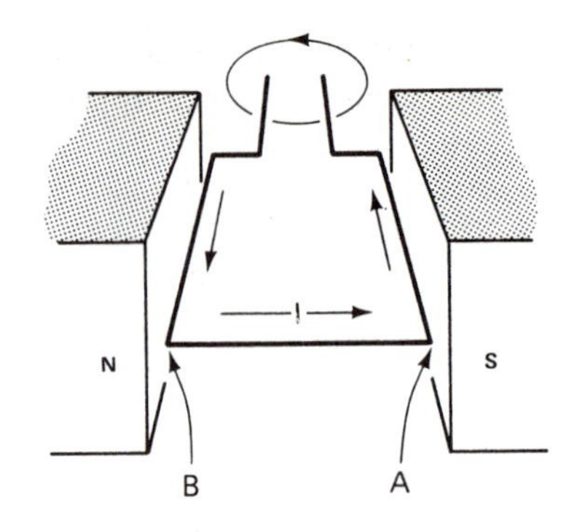

25-2 *AC GENERATORS*

The induction of an emf in a coil of wire can be used to supply electrical power. A coil of wire rotating in a magnetic field produces a fluctuating (changing) emf in the wire. This is the simplest kind of generator to build in the laboratory, so we will study its operation here and compare it to commercial generators where the magnets (electromagnets) are rotated.

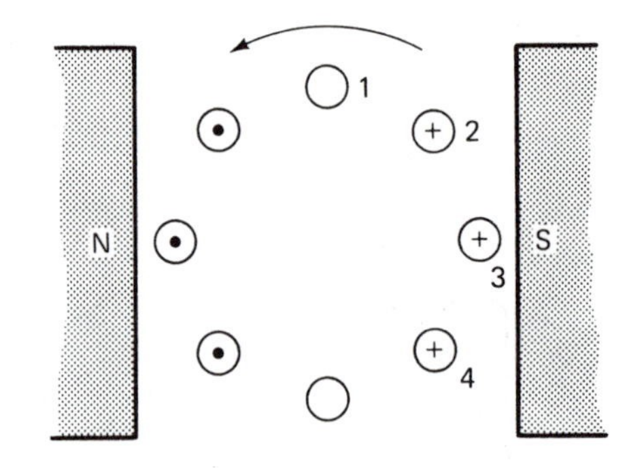

The current produced by rotating the wire through the magnetic field is called an alternating current (ac). As side *A* of the current loop passes downward by the north pole, the induced current is in one direction.

As side *A* (same side of the rotating loop) passes upward by the south pole, the induced current is in the opposite direction. The result is an alternating current induced in the rotating wire. As side *B* (the other side of the rotating loop) passes upward and then downward, the current in it also alternates.

A graph of the induced current is shown below. One cycle is produced by one revolution of the wire. The time required for one cycle depends on the rotational speed of the coil. If the coil rotates 60 times each second, an alternating current of frequency 60 hertz (cycles per second) is produced.

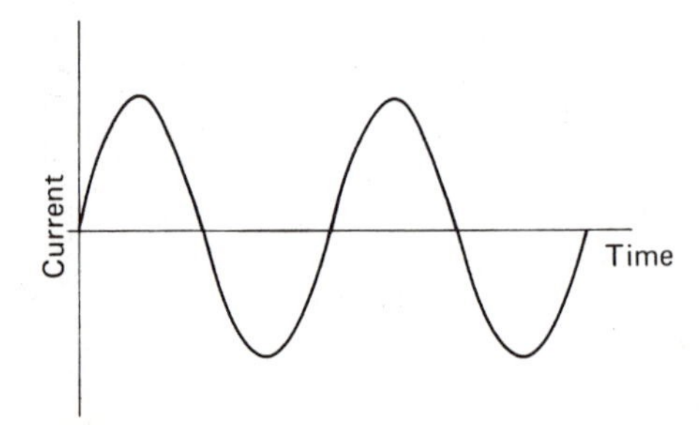

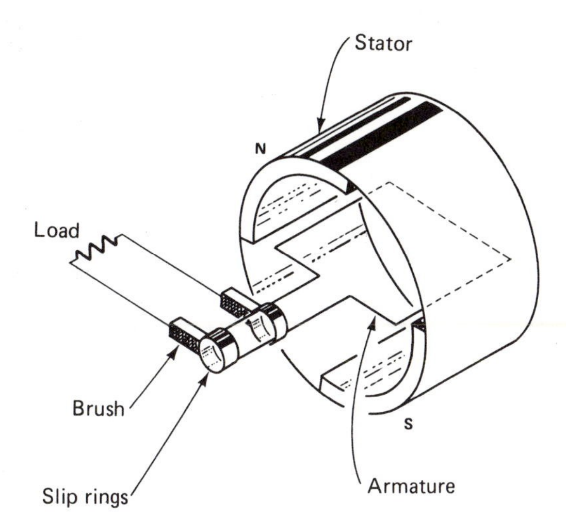

The current produced in the coil is conducted by brushes on slip rings to the external circuit as shown above. The rotating coil is called the rotor or armature, and the field magnets are called the stator.

The generator doesn't actually create electrical energy; it changes the mechanical energy of rotation into electrical energy.

The energy to turn the rotor may be supplied by water falling down a waterfall, a diesel engine, or a steam turbine.

Power companies use large commercial ac generators to produce the current they need to supply to their customers. These generators work in the same manner as the generator discussed here; but they have many coils, and electromagnets are used instead of permanent magnets.

The large generators used by electrical power companies can produce voltages as large as 13,000 volts and currents up to 10 amperes. The alternator used in automobiles is an ac generator which produces about 10 volts and up to 40 amperes.

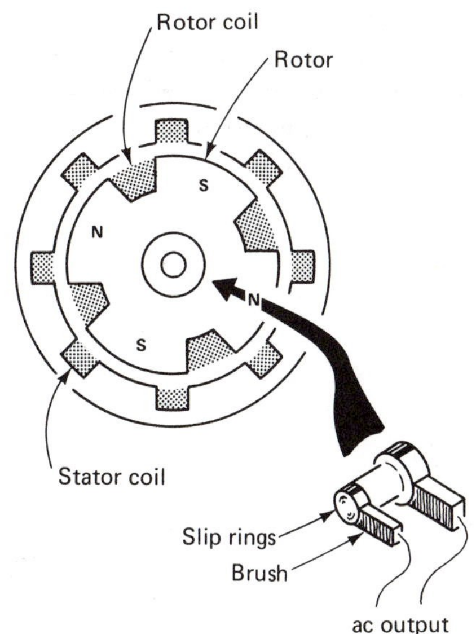

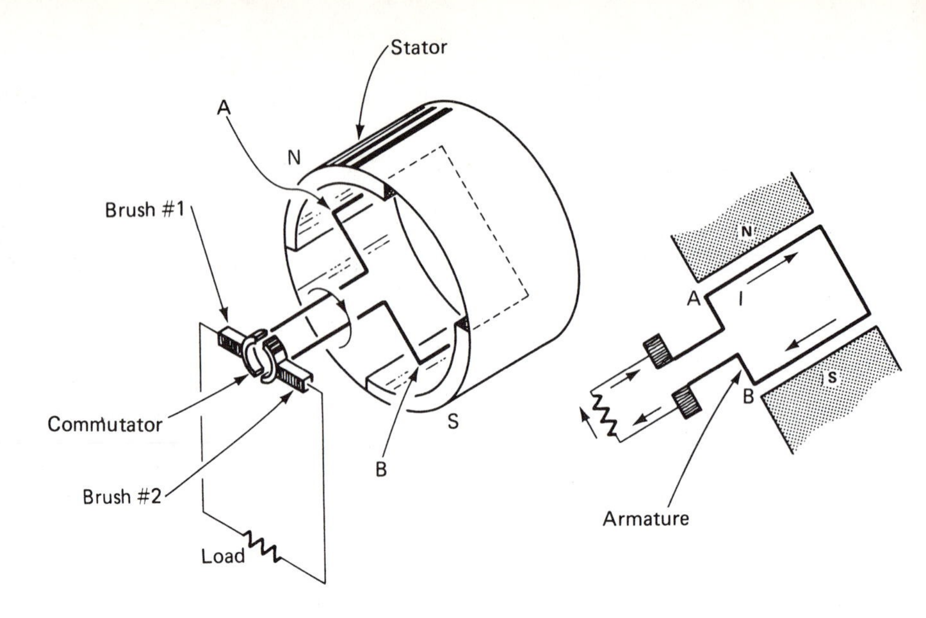

25-3 *DC GENERATORS*

By the use of a special device called a commutator, the ac generator can be used to produce direct current. The commutator is a split ring which replaces the slip rings as shown.

When side *A* of the coil passes upward along the N pole, the induced current flows in the direction shown and is picked up by brush #1. The current in the external circuit is also shown.

When side *B* of the coil passes upward along the N pole, the induced current flows in the direction shown and is picked up by brush #1. The current in the external circuit is in the same direction as it was when *A* passed along the N pole. Thus, this is a direct current.

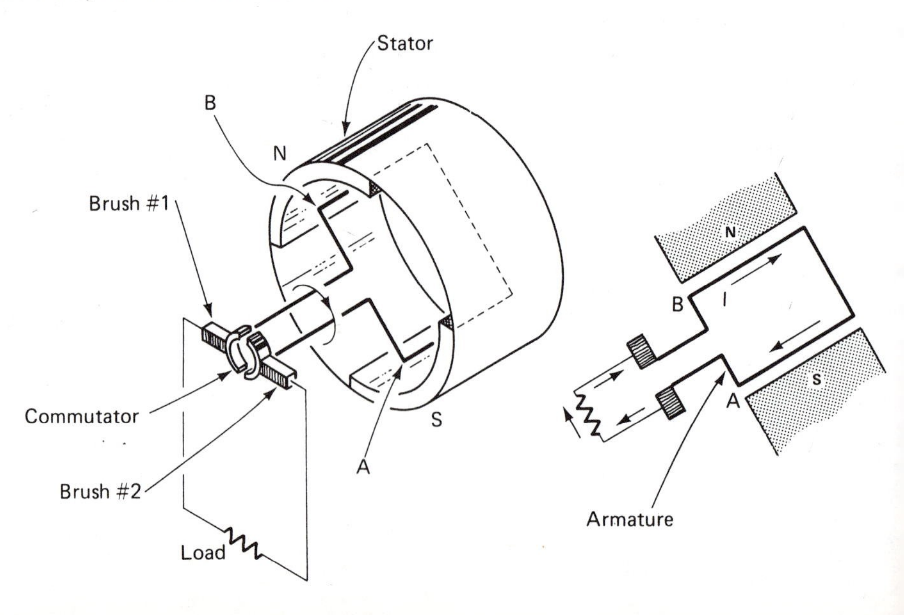

The current produced by this dc generator does not have the same value at all times. A graph of the induced current is shown at the right.

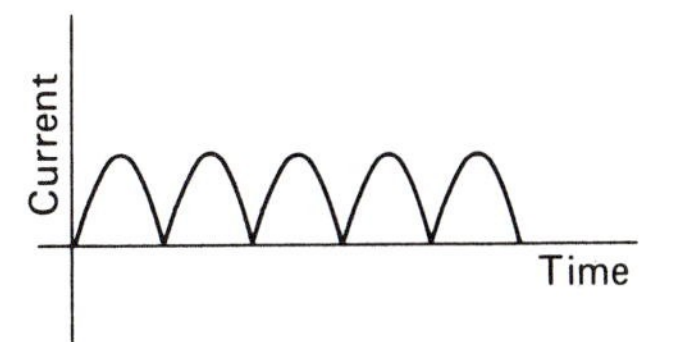

Commercial dc generators that are used for industrial purposes contain many coils. The output current has almost the same value at all times due to the use of the large number of coils.

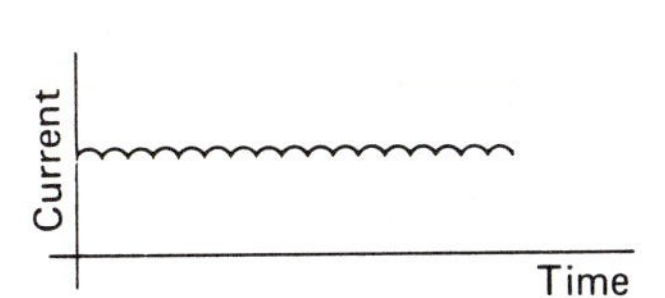

MOTORS 26

26-1 *THE MOTOR PRINCIPLE*

We have seen that like poles of magnets repel each other. A magnet which is pivoted will spin due to the repulsion of another magnet nearby.

We can construct an electromagnet by wrapping wire around an iron core and running a current through the wire.

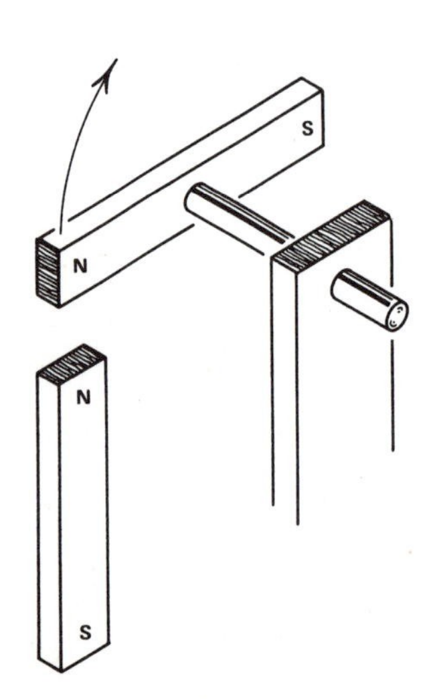

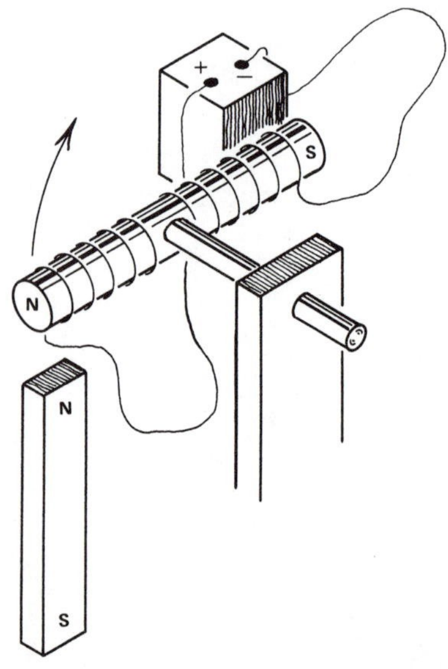

The N pole of the electromagnet will be repelled by a north pole of another magnet. The electromagnet will turn until its S pole is next to the N pole of the permanent magnet.

If we could suddenly change the polarity of the electromagnet (often called the armature), the magnet would repel the N pole, and the electromagnet would continue to spin.

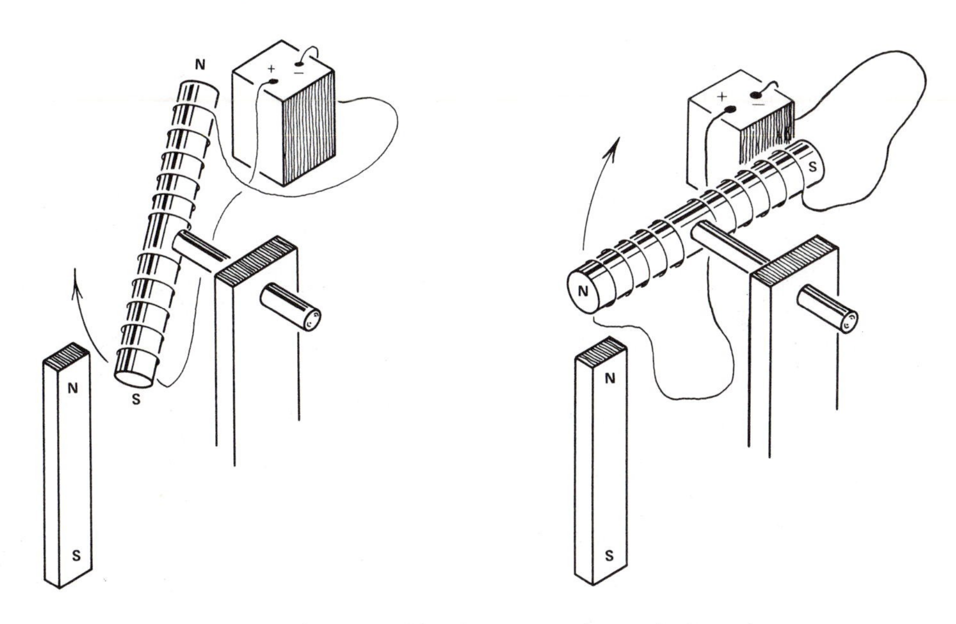

If a dc current supply is used, this change can be made by using a commutator (split ring) to change the direction of the current in the electromagnet. Changing the direction of the current flowing through the coil of the electromagnet changes the poles.

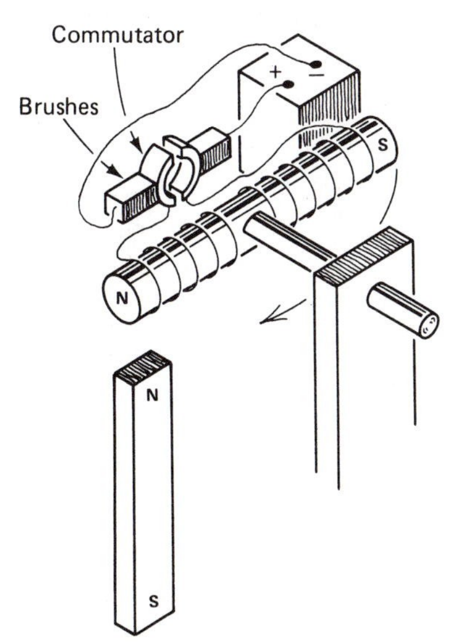

As the current changes direction, the electromagnet spins due to the repulsion of like poles. A shaft may be connected to the electromagnet so that the rotational motion can be used to do work. This device is called a motor. A motor converts electrical energy to mechanical energy and thus performs the reverse function of a generator.

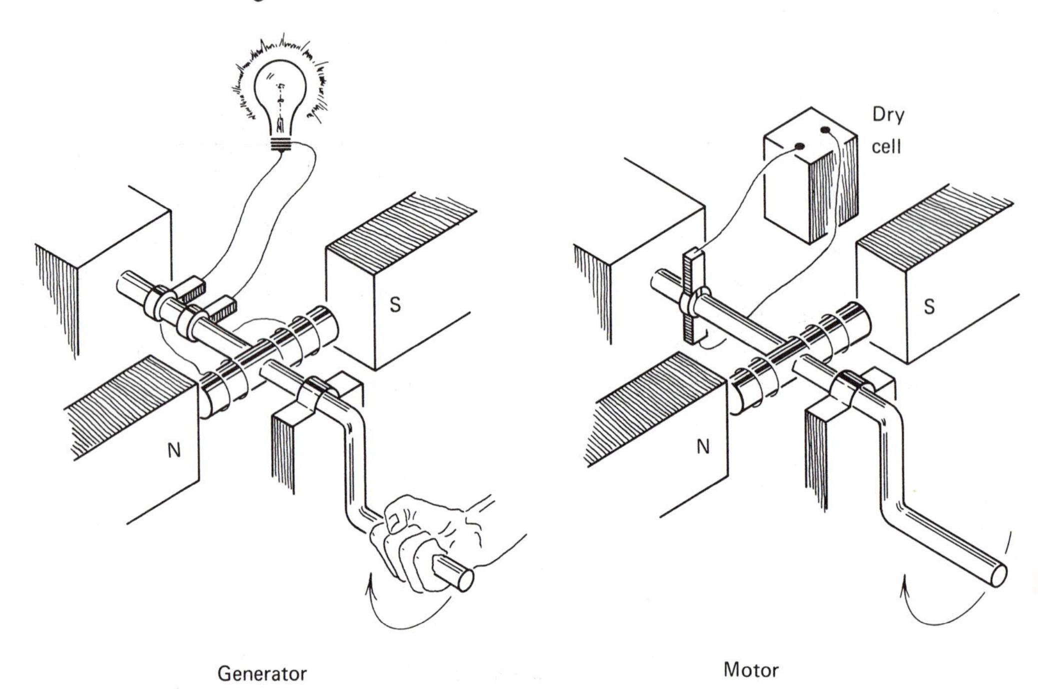

Generator

Motor

If ac current is supplied to the electromagnet, slip rings are used instead of a commutator. The use of alternating current makes the commutator unnecessary. The changes in direction are supplied by the ac current itself.

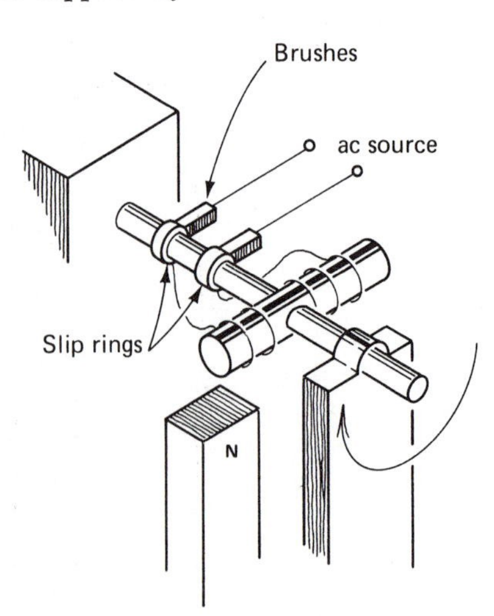

Commercial motors operate in the same way as the motors discussed above. However, they usually use electromagnets in place of the permanent magnets and are much more complex. Slip rings are not necessary in ac motors. The current in the rotating electromagnet can be induced in the same way a current is induced in a generator.

Motors can be designed for many different purposes. Heavily loaded motors need certain types of starters. The torque and power outputs can be greatly varied by differences in design.

Several types of ac motors are discussed below:

1. The universal motor can be run on either ac or dc power. This motor is often used in small hand tools and appliances (see figure at right).
2. The induction motor (see figure below) is the most widely used ac motor. The rotating electromagnet is not connected to a power source by slip rings. Instead, the current in the electromagnet is induced by a moving magnetic field caused by the ac current.

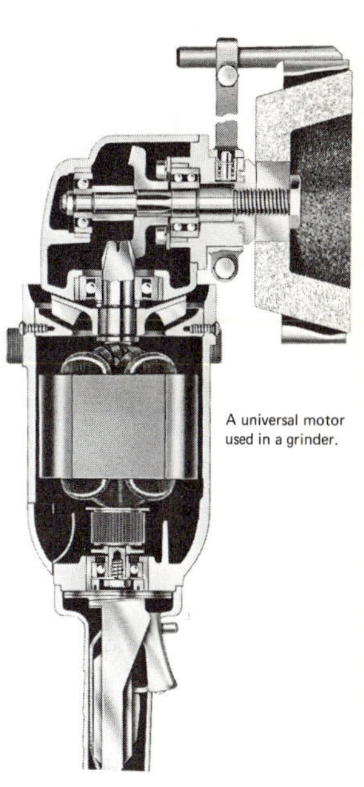

A universal motor used in a grinder.

Photograph courtesy Thor Power Tool Company.

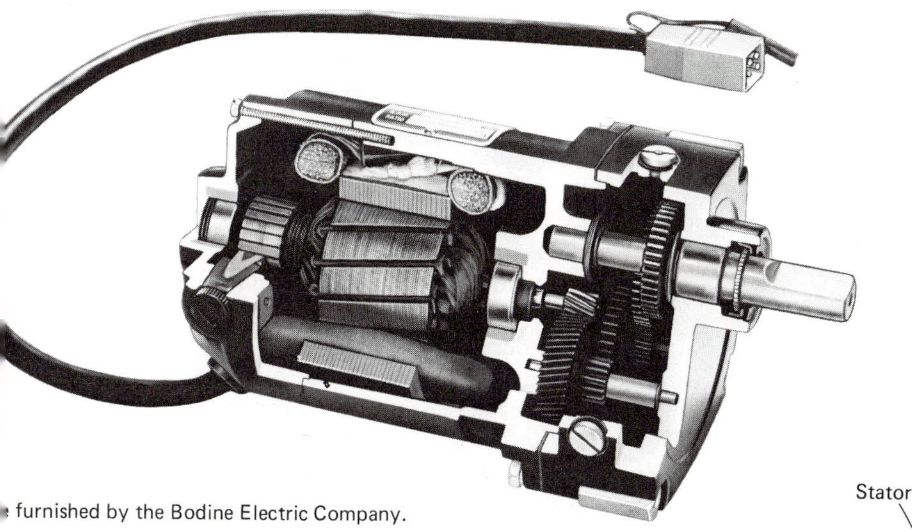

furnished by the Bodine Electric Company.

3. The synchronous motor is very similar to the slip ring ac motor discussed earlier. The rotating electromagnet is supplied with current through slip rings. The speed of rotation of a synchronous motor is constant and depends on the number of coils used. Synchronous motors are used to operate clocks and other devices needing accurate speed control.

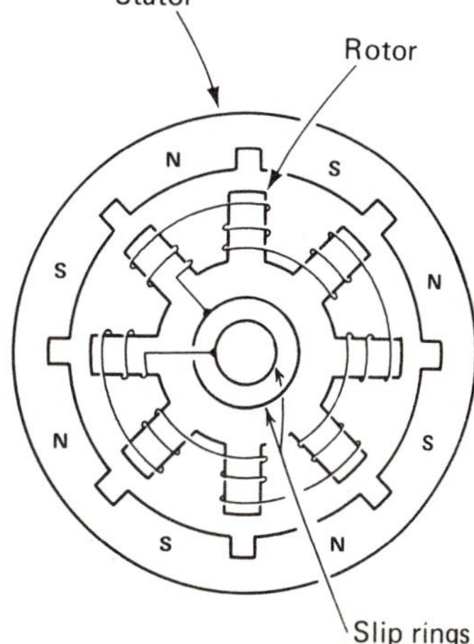

ALTERNATE CURRENT ELECTRICITY 27

27-1 *WHAT IS ALTERNATING CURRENT?*

Alternating current has many more applications in industry and everyday experience than does direct current. We first studied direct current to learn some of the basic ideas of all electricity. Now we will turn our attention to the more common alternating current.

As its name implies, alternating current is current which flows in one direction in a conductor, changes direction, and then flows in the other direction. The direction of flow changes many times in one second. Ordinary household current is 60-Hz current.

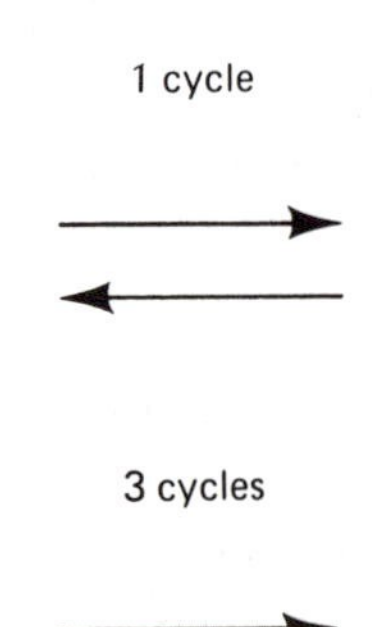

Every time the current repeats itself—flows, changes direction, flows, and changes direction—it goes through one *cycle*. The reason for this alternation is that this is how current comes from electric generators. The emf and current produced by a generator do not alternate instantly between maximum values in each direction, but they build up to maximum values and then decrease, change direction, and build to maximum values in the other direction.

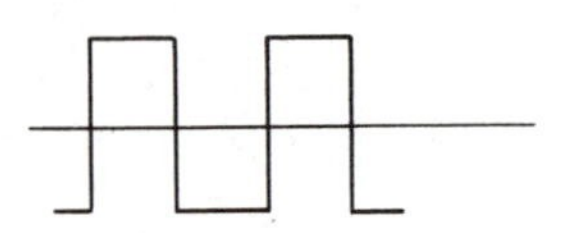

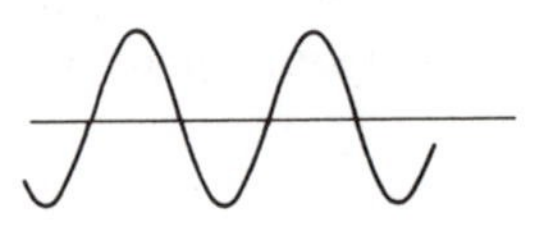

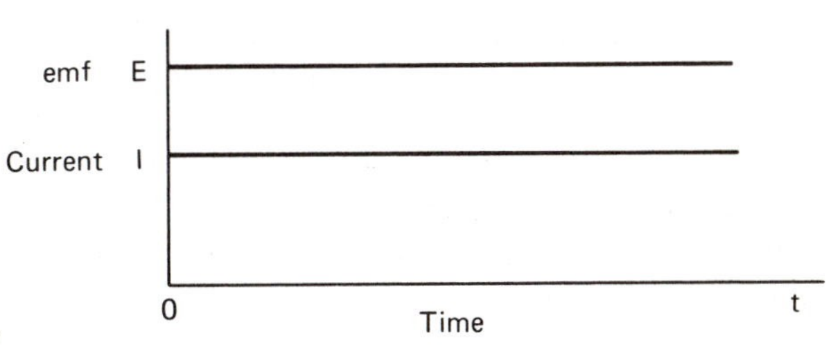

Direct current is usually a steady flow at a fairly constant value. Graphically, it can be represented:

Alternating current, however, is constantly changing. To graphically represent ac, we must show that it builds up and drops off. This can be shown by the curve below called a sine curve. We form the curve by rotating a vector *B* about a point and plotting the vertical components of *B*. Rotating *B* through 360° is graphing one cycle.

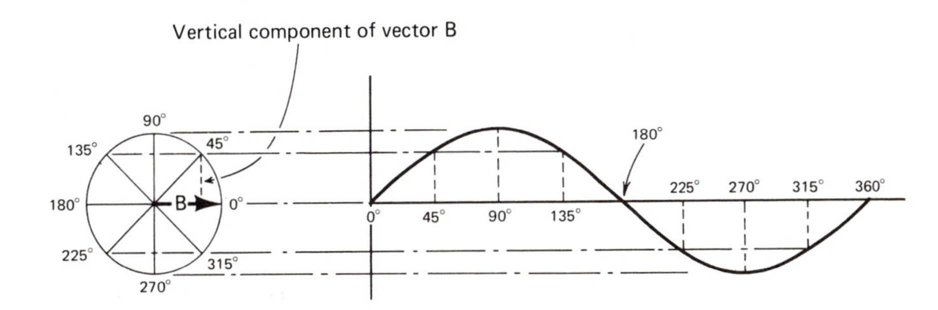

The above graph shows the ac current curve. A graph of ac voltage is also a sine curve.

Ac current and voltage are constantly changing. We can find the value of current or voltage at any instant by using the fact that each makes a sine curve.

$$i = I_{\text{max}} \sin \theta$$
$$e = E_{\text{max}} \sin \theta$$

where: i = instantaneous current (current at any instant),
I_{max} = maximum instantaneous current,
θ = the angle measured from the beginning of the cycle (see graph above),
e = instantaneous voltage,
E_{max} = maximum instantaneous voltage.

Both of the curves at the right show e and i reaching a maximum at the same time and falling to zero at the same time. When this occurs they are said to be "in phase." e and i are in phase in electrical circuits when there is only resistance in the circuit. In a later chapter we will study some other things about ac which will shift the phase.

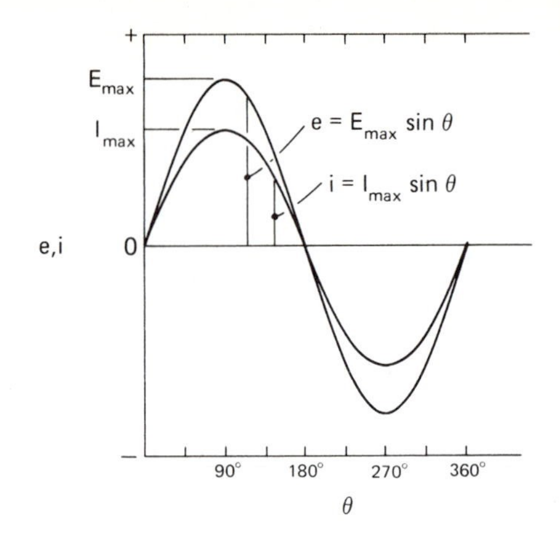

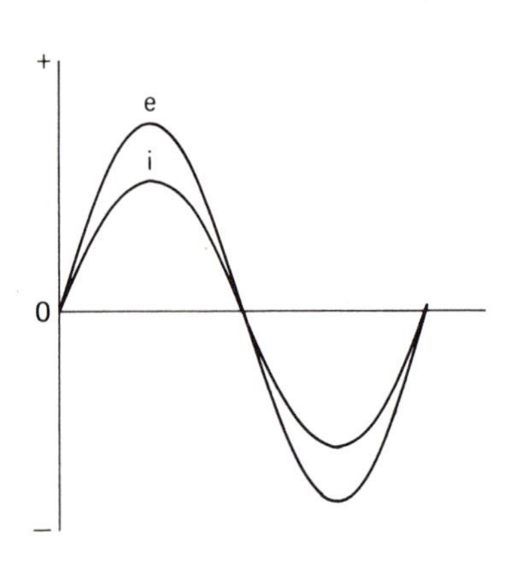

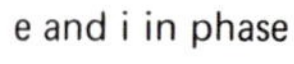
e and i in phase

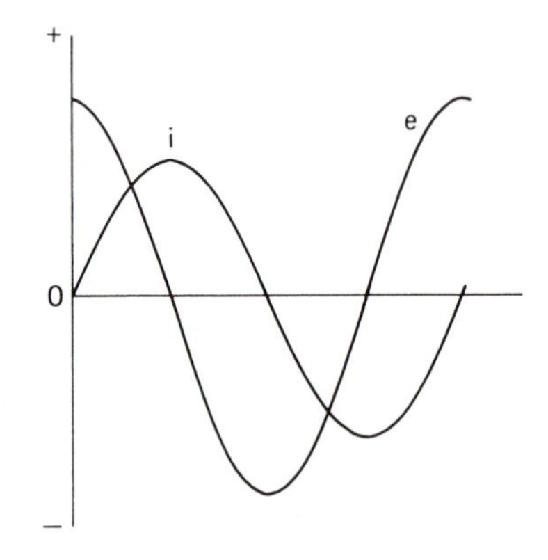

e and i out of phase

27-2 *EFFECTIVE VALUES OF AC*

A direct measurement of ac is difficult because it is constantly changing. The most useful value of ac is based on its heating effect and is called its *effective value*.

The effective value of an alternating current is the number of amperes which produce the same amount of heat in a resistance as an equal number of amperes of a steady direct current.

$$I = 0.707 I_{max}$$
$$I_{max} = 1.41 I$$

where: I = effective value of current (sometimes called rms value),
I_{max} = maximum instantaneous current.

EXAMPLE: The current supplied to a woodworking shop is rated at 10.0 A. What is the maximum value of the current supplied?

DATA: $I = 10.0\text{ A}$

BASIC EQUATION: $I_{max} = 1.41\, I$

WORKING EQUATION: Same

SUBSTITUTION: $I_{max} = 1.41(10.0\text{ A})$
$I_{max} = 14.1\text{ A}$

The effective value for ac voltage may be expressed similarly:

$$E = 0.707E_{max}$$
$$E_{max} = 1.41E$$

where: E = effective value of voltage,
E_{max} = maximum instantaneous voltage.

When we say a house is wired for 120 volts, we are using the effective value of the voltage. Actually the voltage varies between +170 V and −170 V during each cycle. A 10-ampere appliance actually draws between +14.1 A and −14.1 A.

Unless otherwise stated, ac voltage and current are *always* expressed in terms of effective values.

27-3 *AC POWER*

When the load has only resistance, power in ac circuits is determined in the same way as in dc circuits.

$$P = I^2R = VI = \frac{V^2}{R}$$

EXAMPLE 1: What power is expended in a resistance of 37.0 Ω if it has a current of 0.480 A flowing through it?

DATA: $R = 37.0\ \Omega$
$I = 0.48\text{ A}$
$P = ?$

BASIC EQUATION: $P = I^2R$

WORKING EQUATION: Same

SUBSTITUTION: $P = (0.480 \text{ A})^2(37.0 \,\Omega)$
$P = 8.52 \text{ W}$

EXAMPLE 2: What power is expended in a load of 12.0 Ω resistance if the voltage drop across it is 110 V?

DATA: $R = 12.0 \,\Omega$
$V = 110 \text{ V}$
$P = ?$

BASIC EQUATION: $P = \dfrac{V^2}{R}$

WORKING EQUATION: Same

SUBSTITUTION: $P = \dfrac{(110 \text{ V})^2}{12.0 \,\Omega}$
$P = 1010 \text{ W}$

The relationships above are true only when e and i are in phase. Phase differences are produced by capacitance and inductance in ac circuits. These new effects are called impedance. Capacitance, inductance, and impedance will be studied in Chapter 29.

Note that, in the graph comparing dc and ac power, ac power varies but is always positive (+). The sign indicates only the direction of the current. Even so, P is positive in calculations because the product of $-e$ and $-i$ is positive: $P = (-e)(-i) = ei$.

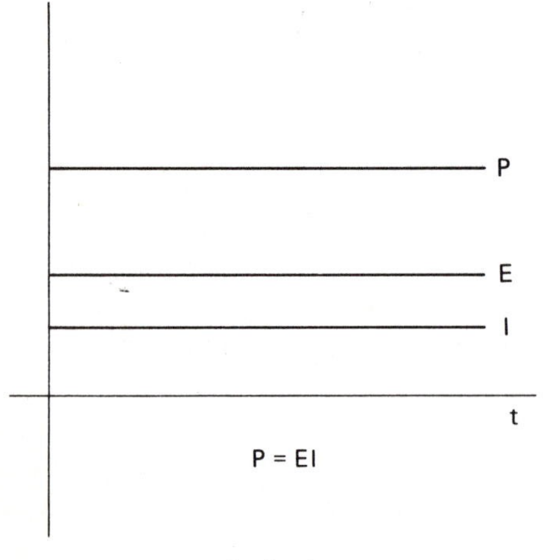

COMPARISON OF POWER IN DC AND AC CIRCUITS

27-4 COMMERCIAL GENERATOR POWER OUTPUT

The power output of the generator is the product of voltage and current. The ac generator converts mechanical energy to electrical energy by performing three functions:

1. The production of voltage—electrical pressure which pushes the current through the loads.
2. The production of power current which is converted into heat, light, and mechanical power.
3. The production of magnetizing current—current transferred back and forth for magnetizing purposes in the generation of electrical power—it is called reactive KVA (kilovolt-amperes).

Apparent Power and Reactive KVA

If the current and voltage are not in phase, the resultant product of current and voltage is *apparent power* instead of actual power. Apparent power is measured in KVA (kilovolt-amperes). Actual power (KW) is the product of KVA and the *power factor* (PF).

$$KW = KVA \cdot PF$$

Mathematically, the power factor is equal to the cosine of the angle by which the current lags behind (or in rare cases leads) the voltage.

The power factor is really a correction factor which must be applied to determine actual power produced. The situation is very similar to finding the amount of work done when a force and the motion are not in the same direction.

EXAMPLE: What is the actual power produced by a generating system which produces 13,600 KVA with a power factor of 0.900?

DATA:

$$KVA = 13{,}600$$
$$PF = 0.900$$
$$KW = ?$$

BASIC EQUATION: $KW = KVA \cdot PF$

WORKING EQUATION: Same

SUBSTITUTION:

$$KW = 13{,}600(0.900)$$
$$= 122{,}400 \text{ KW}$$

27-5 *USES OF AC AND DC*

The uses of direct current in industry are somewhat limited. Primary applications are in charging storage batteries, electroplating, the generation of alternating current, electrolysis, electromagnets, and automobile ignition systems. In all cases, however, ac can be changed to dc by a simple device called a rectifier.

Much more can be done with alternating current. From Mom's mixer to the largest industrial motors, ac finds wide application. There are very practical reasons for this. The voltage of ac can be easily and efficiently changed in transformers to give almost any desired values. Transformers will be studied in Chapter 28. Ac can be used for most purposes just as efficiently as dc. A big advantage of ac is that it can be transmitted over large distances with very little heat loss. (Heat lost in any electrical device is found by: Heat loss (power) $P = I^2R$. The energy wasted as heat can be reduced by making the current smaller. Transformers reduce the current by increasing the voltage.)

PROBLEMS

1. What is the maximum value of the voltage in a circuit where the instantaneous value of the voltage at $\theta = 35°$ is 17.0 V?
2. The instantaneous voltage at $\theta = 50°$ is 82.0 V. What is the maximum voltage?
3. If the maximum ac voltage on a line is 160 V, what is the instantaneous voltage at $\theta = 45°$?
4. If the maximum ac voltage on a line is 2200 V, what is the instantaneous voltage at $\theta = 60°$?
5. What is the effective value of an ac voltage whose maximum voltage is 2200 V?
6. What is the maximum current in a circuit which has a current of 6.00 A in it?
7. What is the effective value of an ac voltage whose maximum voltage is 165 V?
8. What is the maximum current in a circuit which has a current of 4.00 A in it?
9. What is the effective value of a current which reaches a maximum of 17.0 A?
10. A soldering iron is rated at 350 W. If the current in the iron is 4.00 A, what is the resistance of the iron?
11. What power is developed by a motor which draws 6.00 A and has a resistance of 12.0 Ω?

SKETCH
DATA
$a = 1, b = 2,$
$c = ?$

BASIC
EQUATION
$a = bc$

WORKING
EQUATION
$c = \frac{a}{b}$

SUBSTITUTION
$c = \frac{1}{2}$

12. What is the output power of a transformer which has an output voltage of 500 V and current 7.00 A?
13. A heater operates on a 110-V line and is rated at 450 W. What is the resistance of the heater?
14. A heating element draws 6.00 A on a 220-V line. What power is expended in the element?
15. A resistance coil has a resistance of 32.0 Ω. If it expends 375 watts of power, what is the current in the coil?
16. What power is used by a heater which has a resistance of 12.0 Ω and draws a current of 7.00 A?
17. What is the actual power produced by a generating station which produces 14,800 KVA with a power factor of 0.85?
18. A generating station operates with a power factor of 0.87. What actual power is available on the transmission lines if the KVA is 12,800?

TRANSFORMERS 28

28-1 *CHANGING VOLTAGE WITH TRANSFORMERS*

The major advantage of ac over dc is that ac voltage can be easily changed to meet our needs. It can also be transmitted with very little waste of power due to heat loss (I^2R). The amount of the loss is determined by the square of the current.

EXAMPLE 1: Suppose a plant generates 50.0 kW (50,000 W) of power to be sent to a substation on a line with a resistance of 3.00 Ω. We know that some power will be lost as heat during the transmission. The power lost is $P_{\text{lost}} = I^2R$.

(a) How much power is lost if the transmission is at 1100 V?
(b) What percent of the power generated is lost in transmission at 1100 V?
(c) How much power is lost if the transmission is at 11,000 V?
(d) What percent of the power generated is lost in transmission at 11,000 V?

Now compare the power losses at the two different transmission voltages.

(a) At 1100 V:

$$P = VI$$
$$I = \frac{P}{V}$$
$$I = \frac{50{,}000\ \text{W}}{1100\ \text{V}}$$
$$I = 45.5\ \text{A}$$
$$P_{\text{lost}} = I^2R$$
$$P_{\text{lost}} = (45.5\ \text{A})^2(3.00\ \Omega)$$
$$P_{\text{lost}} = 6210\ \text{W}$$
$$P_{\text{lost}} = 6.21\ \text{kW}$$

(b)
$$\%_{\text{lost}} = \frac{\text{power lost}}{\text{power generated}} \times 100\%$$
$$\%_{\text{lost}} = \frac{6.21\ \text{kW}}{50.0\ \text{kW}} \times 100\%$$
$$\%_{\text{lost}} = 0.124 \times 100\%$$
$$\%_{\text{lost}} = 12.4\%$$

(c) At 11,000 V

$$P = VI$$
$$I = \frac{P}{V}$$
$$I = \frac{50{,}000\ \text{W}}{11{,}000\ \text{V}}$$
$$I = 4.55\ \text{A}$$
$$P_{\text{lost}} = I^2R$$
$$P_{\text{lost}} = (4.55\ \text{A})^2(3.00\ \Omega)$$
$$P_{\text{lost}} = 62.1\ \text{W}$$
$$P_{\text{lost}} = 0.0621\ \text{kW}$$

(d)
$$\%_{\text{lost}} = \frac{\text{power lost}}{\text{power generated}} \times 100\%$$
$$\%_{\text{lost}} = \frac{0.0621\ \text{kW}}{50.0\ \text{kW}} \times 100\%$$
$$\%_{\text{lost}} = 0.00124 \times 100\%$$
$$\%_{\text{lost}} = 0.124\%$$

This example shows that where 12.4% of the power would be lost during transmission at 1100 V, only 0.124% would be lost at 11,000 V. So, by increasing the voltage the current is correspondingly lowered, and the power wasted in transmission is greatly reduced. The transformer is a device which can be used to change the voltage to reduce the current and thereby lessen the power loss.

A transformer consists of two coils of wire wrapped on an iron core. It is used to change the voltage of electricity.

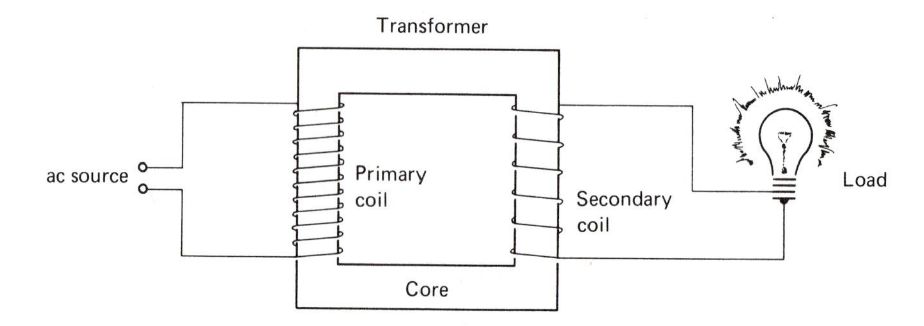

When an alternating current passes through the primary coil, it induces an alternating magnetic field in the core. This magnetic field in turn induces an alternating current in the secondary coil. The magnitude of the voltage

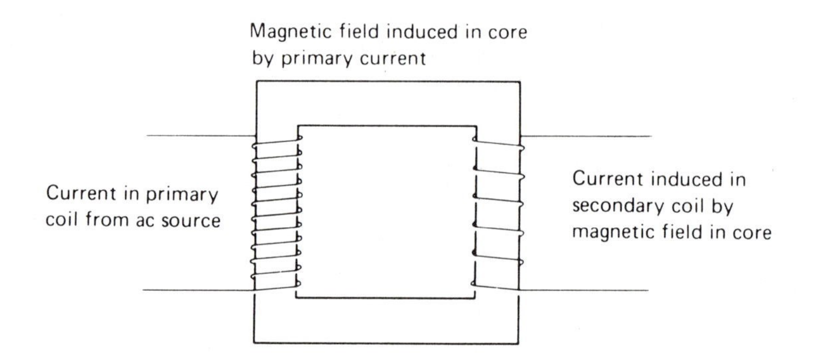

induced in the secondary coil depends on:

1. The voltage applied to the primary coil.
2. The number of turns on the primary coil.
3. The number of turns on the secondary coil.

In equation form:

$$\frac{V_P}{V_S} = \frac{N_P}{N_S}$$

where: V_P = primary voltage
V_S = secondary voltage
N_P = number of primary turns
N_S = number of secondary turns

EXAMPLE: A transformer on a neon sign has 100 turns on its primary coil and 15,000 turns on its secondary coil. If the voltage applied to the primary is 110 V, what is the secondary voltage?

DATA: $V_P = 110\text{ V}$
$N_P = 100\text{ turns}$
$N_S = 15{,}000\text{ turns}$
$V_S = ?$

BASIC EQUATION: $\frac{V_P}{V_S} = \frac{N_P}{N_S}$

WORKING EQUATION: $V_S = V_P \frac{N_S}{N_P}$

SUBSTITUTION: $V_S = \frac{(110\text{ V})(15{,}000\text{ turns})}{100\text{ turns}}$
$V_S = 16{,}500\text{ V}$

Since transformers are used to raise or lower voltage, they are called either step-up or step-down transformers.

Step-up transformers are used when a high voltage is needed to operate X-ray tubes, neon signs, and for transmission of electric power over long distances. A step-up transformer raises the voltage by having more turns on the secondary than on the primary coil.

Step-down transformers are used to lower the voltage from high voltage transmission lines to regular 110 V and 220 V for home and industrial use. Voltage is lowered in the step-down transformer because it has more turns in the primary coil than in the secondary coil.

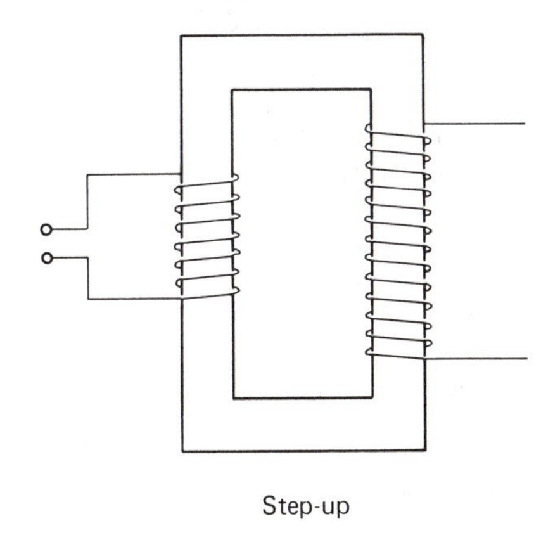

Step-up

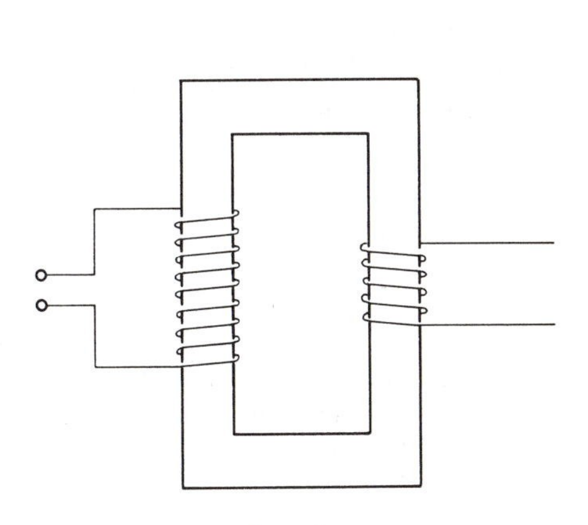

Step-down

Transformers do not create energy. Some energy is lost, however, during the change of voltage. Energy losses in transformers are of three types:

1. Copper losses—these result from the resistance of the copper wires in the coils and are unavoidable.
2. Magnetic losses (called hysteresis losses)—some energy is lost (turned into heat) by reversing the magnetism in the core.
3. Eddy currents—when a mass of metal (the core) is subjected to a changing magnetic field, currents are set up in the metal which do no useful work, waste energy, and produce heat. These losses can be lessened by *laminating* the core. Instead of using a solid block of metal for the core, thin sheets of metal with insulated surfaces are used, reducing these induced currents.

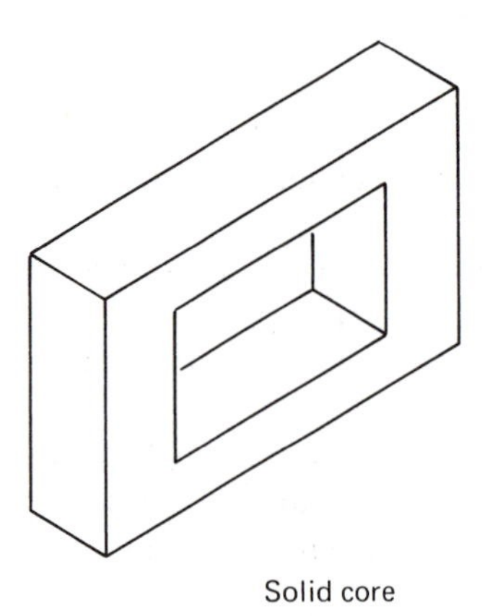

Solid core

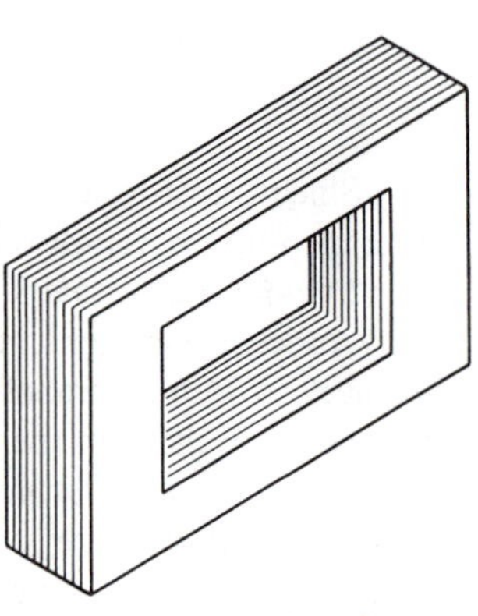

Laminated core

When a transformer steps up the voltage applied to its primary, it reduces the current. Energy is conserved—we cannot get any more electrical energy out of a transformer than we put into it. The relationship between primary and secondary currents is:

$$\frac{I_S}{I_P} = \frac{N_P}{N_S}$$

where: I_S = current in secondary coil
I_P = current in primary coil
N_P = number of turns in primary
N_S = number of turns in secondary

EXAMPLE 1: The primary current of a transformer is 10.0 A. If the primary has 550 turns and the secondary has 2500 turns, what current flows in the secondary coil?

DATA:

$$N_P = 550 \text{ turns}$$
$$I_P = 10.0 \text{ A}$$
$$N_S = 2500 \text{ turns}$$
$$I_S = ?$$

BASIC EQUATION:

$$\frac{I_S}{I_P} = \frac{N_P}{N_S}$$

WORKING EQUATION:

$$I_S = \frac{(I_P)N_P}{N_S}$$

SUBSTITUTION:

$$I_S = \frac{(10.0 \text{ A})(550 \text{ turns})}{2500 \text{ turns}}$$
$$I_S = 2.20 \text{ A}$$

Another way of showing power is conserved is:

$$I_P V_P = I_S V_S \quad \text{or} \quad P_P = P_S$$

where: I_P = current in primary coil
I_S = current in secondary coil
V_P = voltage in primary coil
V_S = voltage in secondary coil

where: P_P = power (watts) in primary coil
P_S = power (watts) in secondary coil

EXAMPLE 2: The power in the primary coil of a transformer is 375 W. If the current in the secondary is 11.4 A, what is the voltage in the secondary?

DATA:

$$P_P = 375 \text{ W}$$
$$I_S = 11.4 \text{ A}$$
$$V_S = ?$$

BASIC EQUATIONS:

$$P_P = P_S \quad \text{and} \quad P_S = V_S I_S$$

WORKING EQUATION:

$$P_P = V_S I_S \quad (\textit{Note}: \text{substitute for } P_S)$$
$$V_S = \frac{P_P}{I_S}$$

SUBSTITUTION:

$$V_S = \frac{375 \text{ W}}{11.4 \text{ A}}$$
$$V_S = 32.9 \text{ V}$$

Good transformers are more than 98 % efficient. This is very important in power transmission. It is impractical to generate electricity at high voltage, but high voltage is desirable for transmission. Therefore, transformers are used to step up the voltage for transmission. High voltage is unsuitable, though, for consumer use. So transformers are used to reduce the voltage.

A simplified diagram of a power distribution system appears below.

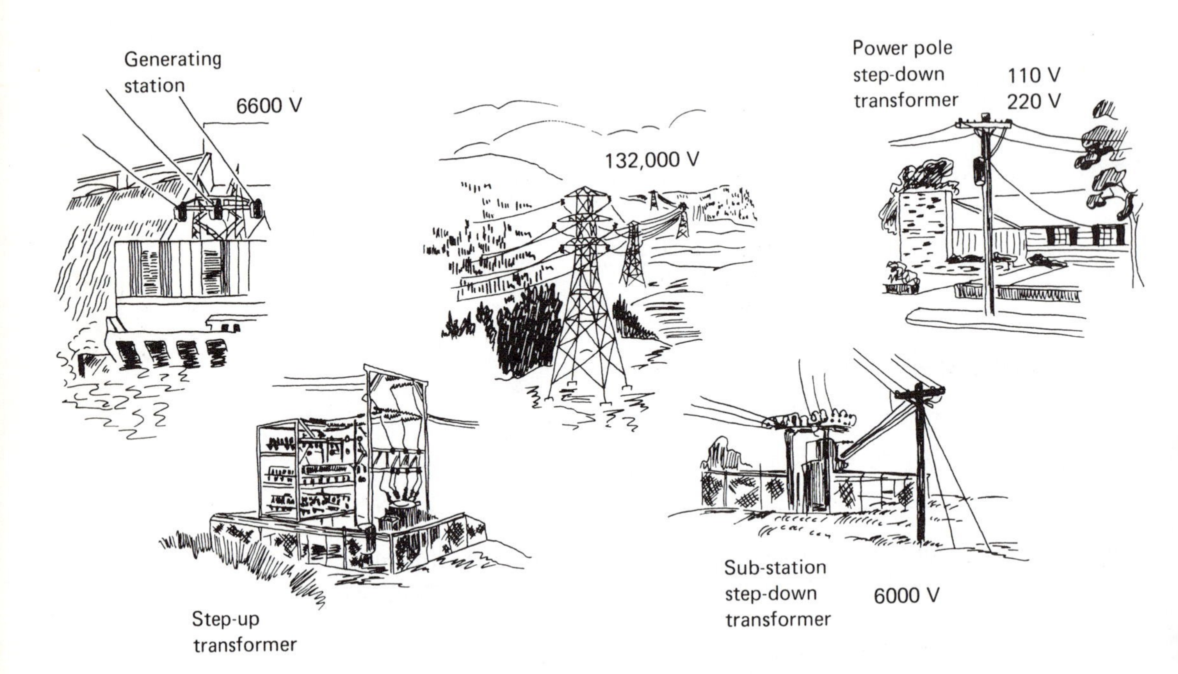

PROBLEMS

1. $V_P = 20.0$ V
   $V_S = 30.0$ V
   $N_S = 15$ turns
   Find N_P.
2. $V_P = 250$ V
   $N_P = 730$ turns
   $N_S = 375$ turns
   Find V_S.
3. $I_P = 6.00$ A
   $I_S = 4.00$ A
   $V_P = 39.0$ V
   Find V_S.

SKETCH
DATA
$a = 1, b = 2$
$c = ?$

BASIC
EQUATION
$a = bc$

WORKING
EQUATION
$c = \frac{a}{b}$

SUBSTITUTION
$c = \frac{1}{2}$

4. A step-up transformer on a 115-V line provides a voltage of 2300 V. If the primary has 65 turns, how many turns does the secondary have?
5. A step-down transformer on a 115-V line provides a voltage of 11.5 V. If the secondary has 35 turns, how many turns does the primary have?
6. A transformer has 20 turns in the primary coil and 2200 in the secondary. If the primary voltage is 12.0 V, what is the secondary voltage?
7. If there is a current of 7.00 A in the primary in problem 3, what is the current in the secondary?
8. If the voltage in the secondary coil of a transformer is 110 V and the current in it is 15.0 A, what power does it supply?
9. A neon sign has a transformer which changes electricity from 110 V to 15,000 V. If the primary current is 10.0 A, what is the current in the secondary?
10. What is the power in the primary in problem 9?
11. A transformer has an output power of 990 W. If the current in the secondary is 0.45 A, what is the voltage in the secondary?
12. The current in the secondary of a transformer is 3.00 A. What is the voltage in the secondary if the power is 775 W?
13. A transformer steps down 6600 V to 120 V. If the secondary current is 14.0 A, what is the primary current?
14. What is the power in the primary in problem 13?

AC CIRCUITS 29

29-1 *INDUCTANCE*

A coil of wire in an ac circuit opposes a change in the value of the current. This is due to the emf induced in the coil itself as the magnetic field of the coil changes. This emf opposes a change in the current.

The unit of inductance (L) is the henry (H). A coil has an inductance of 1 henry if an emf of 1 volt is induced when the current changes at the rate of 1 A/s. A henry can be expressed as Ω s.

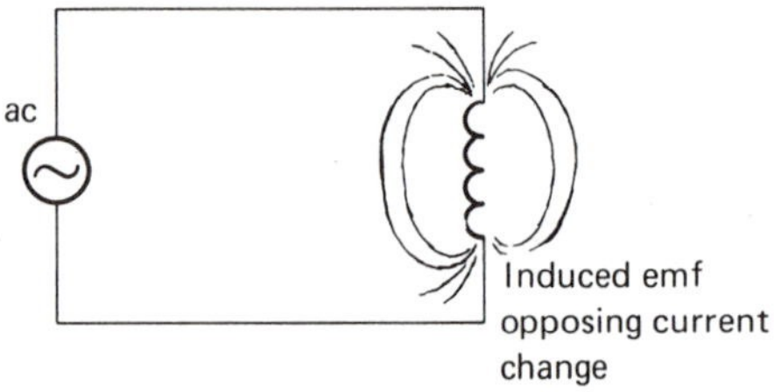

ac

Induced emf opposing current change

$$1 \text{ henry} = 1 \, \Omega \text{ s}$$

The henry is a large unit. A more practical unit is the millihenry (mH) which is one-thousandth of a henry.

Inductance can be illustrated by connecting a coil with a large number of turns and a lamp in series. When connected to a dc source, the lamp burns brightly.

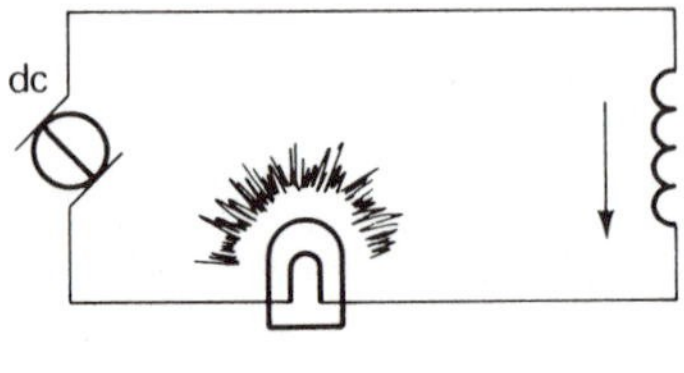

dc

However, when this circuit is connected to an ac power source of the same voltage, the lamp is dimmer due to the inductance of the coil.

The circuit symbol for inductance is shown below.

29-2 *INDUCTIVE REACTANCE*

The opposition to ac current flow in an inductor is called inductive reactance and is measured in ohms. This is usually represented by X_L. The inductive reactance of a coil is found by:

$$X_L = 2\pi f L$$

where: $X_L =$ the inductive reactance
$f =$ the frequency of the ac voltage expressed in hertz (cycles per second), such as 60 Hz or 60/s.
$L =$ the inductance in henries

The current in a circuit which has only an ac voltage source and an inductor is given by:

$$I = \frac{E}{X_L}$$

where: $I =$ the current
$E =$ the voltage
$X_L =$ the inductive reactance

EXAMPLE: A coil with inductance of 0.100 henries is connected to a 60-Hz ac power source of 110 volts. What is the current in the circuit?

SKETCH:

E = 110 V

60/s

DATA:
$$E = 110\text{ V}$$
$$L = 0.100\text{ H}$$
$$f = \frac{60}{\text{s}}$$
$$I = ?$$

BASIC EQUATIONS: $X_L = 2\pi f L$ and $I = \dfrac{E}{X_L}$

WORKING EQUATIONS: Same

SUBSTITUTION:
$$X_L = 2\pi\left(\frac{60}{\text{s}}\right)(0.100\text{ H})$$
$$X_L = 37.7\,\frac{\text{H}}{\text{s}}$$
$$X_L = 37.7\,\frac{\cancel{\text{H}}}{\cancel{\text{s}}}\left(\frac{1\,\Omega\,\cancel{\text{s}}}{1\,\cancel{\text{H}}}\right) \quad \text{(Note conversion factor)}$$
$$X_L = 37.7\,\Omega$$
$$I = \frac{E}{X_L}$$
$$I = \frac{110\text{ V}}{37.7\,\Omega}$$
$$I = 2.92\,\frac{\cancel{\text{V}}}{\cancel{\Omega}}\left(\frac{\text{A}\cancel{\Omega}}{\cancel{\text{V}}}\right)$$
$$I = 2.92\text{ A}$$

In an inductive circuit the current lags behind the voltage by a quarter of a cycle. The maximum voltage in a 60-Hz circuit thus occurs

$$\tfrac{1}{4} \times \tfrac{1}{60}\text{ s} = \tfrac{1}{240}\text{ s}$$

before the maximum current. The current lag is usually measured in degrees. A quarter of a cycle is 90°.

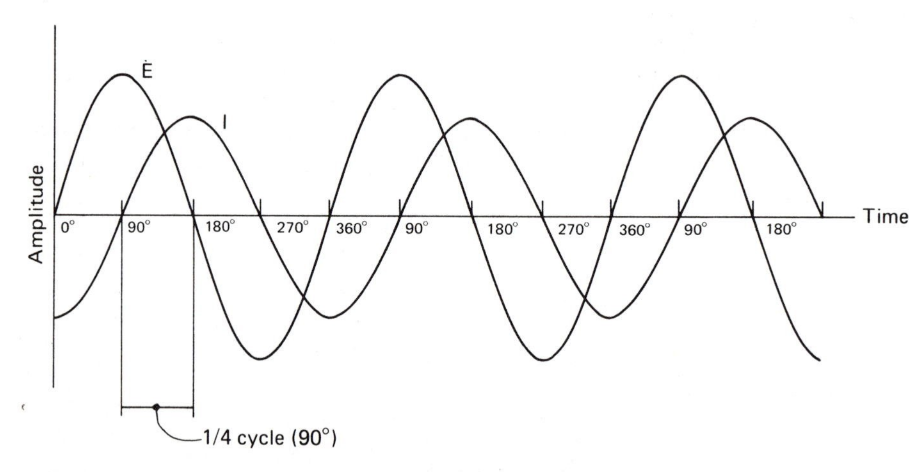

PROBLEMS

Calculate the inductive reactance of the following inductances at the given frequency.

1. $L = 5.00 \times 10^{-3}$ H, $f = 60$/s, $X_L =$ _______ Ω
2. $L = 2.00 \times 10^{-2}$ H, $f = 90$/s, $X_L =$ _______ Ω
3. $L = 7.00 \times 10^{-4}$ H, $f = 10^4$/s, $X_L =$ _______ Ω
4. $L = 8.00 \times 10^{-1}$ H, $f = 10^1$/s, $X_L =$ _______ Ω
5. $L = 4.00 \times 10^{-2}$ H, $f = 10^3$/s, $X_L =$ _______ Ω

Calculate the current in the inductive circuits with the following characteristics:

6. $L = 3.00 \times 10^{-4}$ H, $f = 100$/s, $E = 17.0$ V, $I =$ _______ A
7. $L = 10^{-3}$ H, $f = 10^4$/s, $E = 150$ V, $I =$ _______ A
8. $L = 5.00 \times 10^{-3}$ H, $f = 2.00 \times 10^3$/s, $E = 50.0$ V, $I =$ _______ A
9. $L = 4.00 \times 10^{-2}$ H, $f = 7.00 \times 10^2$/s, $E = 27.0$ V, $I =$ _______ A
10. $L = 7.20 \times 10^{-3}$ H, $f = 10^6$/s, $E = 100$ V, $I =$ _______ A

29-3 *INDUCTANCE AND RESISTANCE IN SERIES*

Most ac circuits have resistance in the form of lights or resistors in addition to inductance. The current lags behind the voltage by any amount of time greater than zero and as large as a quarter cycle.

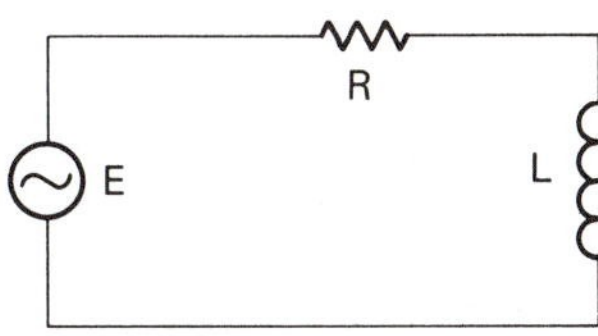

The effect of both the resistance and the inductance on a circuit is called the impedance. It is represented by the letter Z. Ohm's law in an ac circuit can be written as:

$$I = \frac{E}{Z}$$

where: $I =$ the current
$E =$ the voltage
$Z =$ the impedance

The impedance of a series circuit containing a resistance and an inductance is:

$$Z = \sqrt{R^2 + X_L^2}$$
$$Z = \sqrt{R^2 + (2\pi f L)^2}$$

where: $Z =$ the impedance
$R =$ the resistance
$X_L =$ the inductive reactance
$f =$ the frequency
$L =$ the inductance

The impedance can be represented vectorially as the hypotenuse of the right triangle shown. The resistance is always drawn as a vector pointing in the positive x direction. The inductive reactance is drawn as a vector pointing in the positive y direction.

The angle ϕ shown is the phase angle and is equal to the amount by which the current lags behind the voltage. The phase angle is given by:

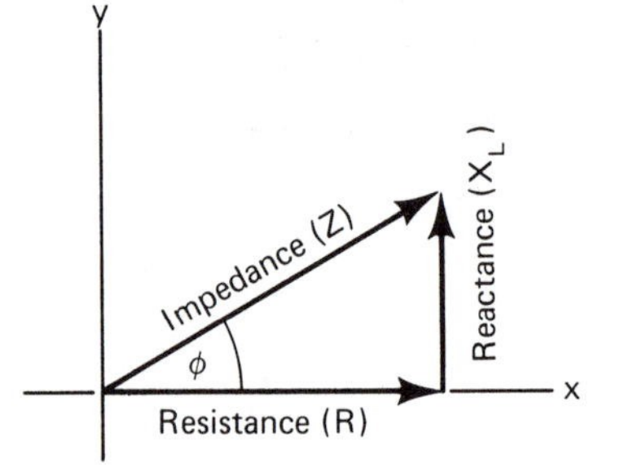

$$\tan \phi = \frac{X_L}{R}$$

EXAMPLE: A lamp of resistance 40.0 Ω is connected in series with an inductance of 0.100 H. This circuit is connected to a 110-V, 60-Hz power supply. What is the current in the circuit? What is the phase angle?

SKETCH:

R
E = 110 V
f = 60/s
L

DATA: $E = 110$ V
$f = 60/\text{s}$
$R = 40.0\ \Omega$
$L = 0.100$ H
$I = ?$
$\phi = ?$

BASIC EQUATIONS: $Z = \sqrt{R^2 + (2\pi f L)^2}$ and $I = \frac{E}{Z}$

WORKING EQUATION: First calculate the impedance.

$$Z = \sqrt{(40.0\ \Omega)^2 + \left[(2\pi)\left(\frac{60}{s}\right)(0.100\ \text{H})\right]^2}$$

$$Z = \sqrt{1600\ \Omega^2 + \left(37.7\ \frac{\text{H}}{\text{s}}\right)^2}$$

$$Z = \sqrt{1600\ \Omega^2 + \left[\left(37.7\ \frac{\cancel{\text{H}}}{\cancel{\text{s}}}\right)\left(\frac{1\ \Omega\ \cancel{\text{s}}}{1\ \cancel{\text{H}}}\right)\right]^2} \quad \text{(Note conversion factor)}$$

$$Z = \sqrt{1600\ \Omega^2 + 1420\ \Omega^2}$$

$$Z = \sqrt{3020\ \Omega^2}$$

$$Z = 55.0\ \Omega$$

$$I = \frac{E}{Z}$$

$$I = \frac{110\ \cancel{\text{V}}}{55.0\ \cancel{\Omega}}\left(\frac{1\ \text{A}\ \cancel{\Omega}}{1\ \cancel{\text{V}}}\right)$$

$$I = 2.00\ \text{A}$$

To find the phase angle (ϕ), we construct the vector right triangle. Then find the angle whose tangent is given by:

$$\tan\phi = \frac{X_L}{R} = \frac{37.7\ \Omega}{40.0\ \Omega} = 0.943$$

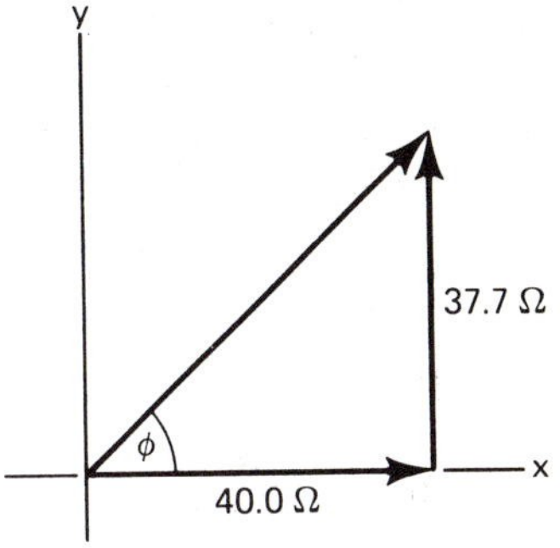

From trigonometric tables the angle whose tangent is 0.943 is most nearly 43°; so

$\phi = 43°$

PROBLEMS

Find the impedance of the circuits with the following characteristics.

1. $R = 200\ \Omega$, $L = 10^{-2}$ H, $f = 1500/\text{s}$, $Z =$ _______ Ω
2. $R = 16.0\ \Omega$, $L = 10^{-3}$ H, $f = 900/\text{s}$, $Z =$ _______ Ω
3. $R = 10^3\ \Omega$, $L \doteq 5.00 \times 10^{-2}$ H, $f = 10^4/\text{s}$, $Z =$ _______ Ω
4. $R = 2.00 \times 10^3\ \Omega$, $L = 7.00 \times 10^{-2}$ H, $f = 5.00 \times 10^3/\text{s}$, $Z =$ _______ Ω
5. $R = 3.00 \times 10^2\ \Omega$, $L = 2.00 \times 10^{-3}$ H, $f = 4000/\text{s}$, $Z =$ _______ Ω
6. Find the phase angle in problem 1.
7. Find the phase angle in problem 2.

8. Find the phase angle in problem 3.
9. Find the phase angle in problem 4.
10. Find the phase angle in problem 5.
11. Find the current in problem 1 if the voltage is 50.0 V.
12. Find the current in problem 2 if the voltage is 10.0 V.
13. Find the current in problem 3 if the voltage is 15.0 V.
14. Find the current in problem 4 if the voltage is 12.0 V.
15. Find the current in problem 5 if the voltage is 8.00 V.

29-4 *CAPACITANCE*

An important component of many ac circuits is the capacitor. A capacitor consists of two conductors which are usually parallel plates separated by a thin insulator. The plates are often made of a metal foil which is rolled in convenient size. Capacitors are represented in circuit diagrams as shown.

The unique property of a capacitor is that it can build up and store charge. When a capacitor is connected to a battery, electrons flow from the negative terminal to one capacitor plate as shown.

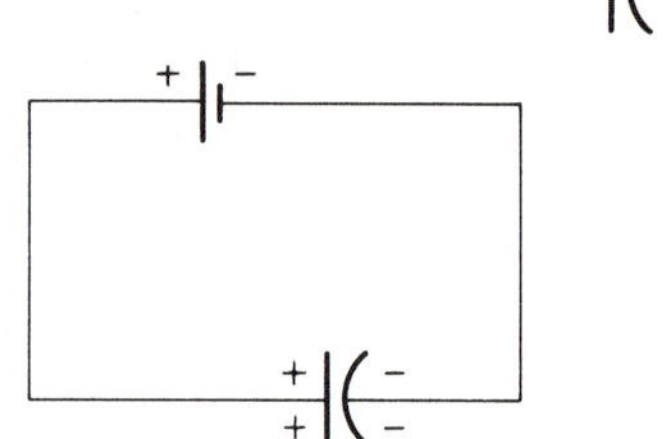

When the capacitor is removed from the battery, the charges remain on the capacitor.

If the capacitor is then connected to a resistor, electrons will flow through the circuit until the capacitor has lost its charge.

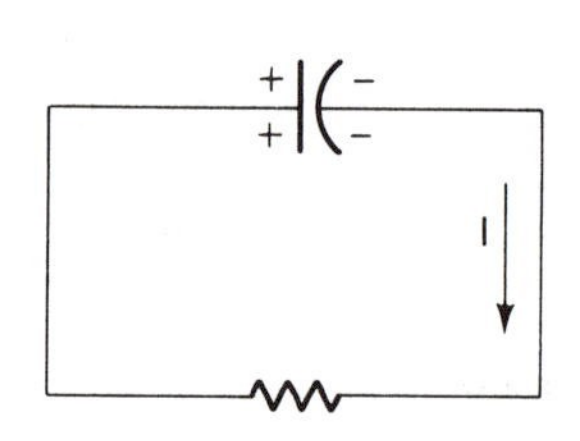

A capacitor in an ac circuit causes a current to flow when the ac voltage is zero. The unit of capacitance is the farad (F). A more practical unit is the μF (10^{-6} F). The effect of a capacitor on a circuit is measured as capacitive reactance and is measured in ohms. It is given by

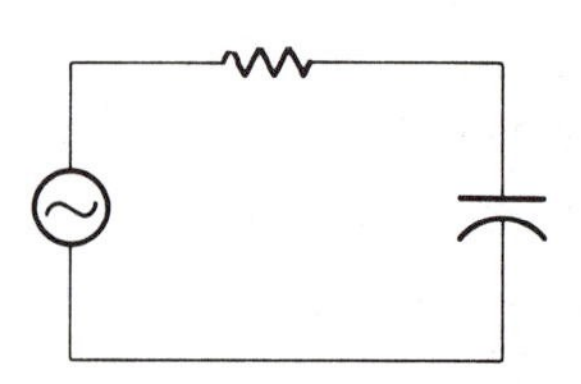

$$X_C = \frac{1}{2\pi f C}$$

where: X_C = the capacitive reactance
f = the frequency
C = the capacitance in farads

$$1F = 1\text{s}/\Omega$$

In a circuit which contains only capacitors, the current *leads* the voltage by 90° (one quarter cycle).

EXAMPLE: Calculate the capacitive reactance of a 10^{-5} F capacitor in a circuit of frequency 100/s.

DATA:
$C = 10^{-5}$ F
$f = 100$/s
$X_C = ?$

BASIC EQUATION: $X_C = \dfrac{1}{2\pi f C}$

WORKING EQUATION: Same

SUBSTITUTION:
$$X_C = \frac{1}{2\pi(100/\text{s})(10^{-5}\text{ F})}$$

$$X_C = \frac{1}{6.28 \times 10^{-3}\frac{\cancel{F}}{\cancel{s}}\left(\frac{1\,\cancel{s}}{1\,\cancel{F}\,\Omega}\right)}$$ (Note conversion factor)

$$X_C = 159\ \Omega$$

PROBLEMS

Find the capacitive reactance of the circuits with the following characteristics.

1. $C = 4.00 \times 10^{-5}$ F, $f = 10^3$/s, X_C = ________ Ω
2. $C = 7.00 \times 10^{-3}$ F, $f = 10^2$/s, X_C = ________ Ω
3. $C = 6.00 \times 10^{-4}$ F, $f = 10^2$/s, X_C = ________ Ω
4. $C = 3.00 \times 10^{-3}$ F, $f = 10^4$/s, X_C = ________ Ω
5. $C = 8.00 \times 10^{-4}$ F, $f = 10^5$/s, X_C = ________ Ω

29-5 *CAPACITANCE AND RESISTANCE IN SERIES*

The combined effect of capacitance and resistance in series is measured by the impedance of the circuit, Z.

$$Z = \sqrt{R^2 + X_C^2}$$

$$Z = \sqrt{R^2 + \left(\frac{1}{2\pi f C}\right)^2}$$

where: Z = the impedance
R = the resistance
X_C = the capacitive reactance
f = the frequency
C = the capacitance

The current is given by Ohm's law

$$I = \frac{E}{Z}$$

where: I = current
E = voltage
Z = impedance

The phase angle can be found by drawing the resistance as a vector in the positive x direction and the capacitive impedance as a vector in the negative y direction as shown. The phase angle gives the amount by which the voltage lags the current.

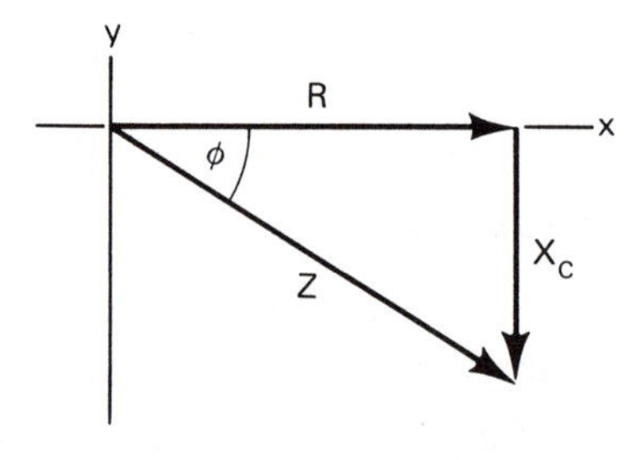

$$\tan \phi = \frac{X_C}{R}$$

EXAMPLE: What current will flow in a 60-Hz ac circuit which includes a 110-V source, a capacitor of 40.0 μF, and a 16.0 Ω resistance in series? Also find the phase angle.

SKETCH:

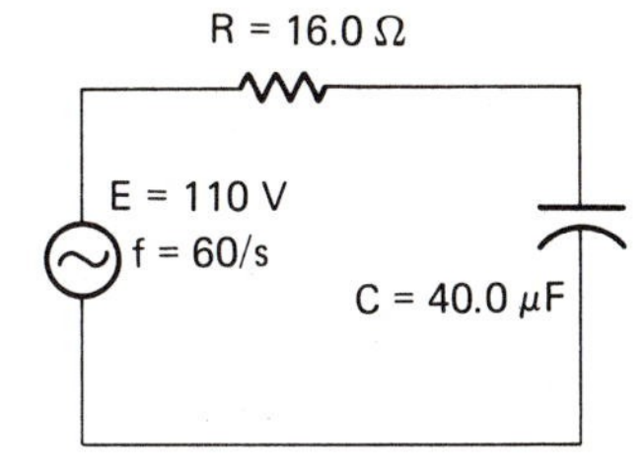

DATA:

$E = 110\ V$
$f = 60/\text{s}$
$R = 16.0\ \Omega$
$C = 40.0\ \mu\text{F}$
$Z = ?$
$I = ?$

BASIC EQUATIONS: $Z = \sqrt{R^2 + \left(\frac{1}{2\pi f C}\right)^2}$ and $I = \frac{E}{Z}$

WORKING EQUATIONS: Same

SUBSTITUTIONS:

$$Z = \sqrt{(16.0\ \Omega)^2 + \left(\frac{1}{2 \times 3.14 \times 60/\text{s} \times 4.00 \times 10^{-5}\ \text{F}}\right)^2}$$

$$Z = \sqrt{256\ \Omega^2 + \left(\frac{1}{0.0151\ \text{F/s}}\right)^2}$$

$$Z = \sqrt{256\ \Omega^2 + \left(66.2\ \frac{\cancel{\text{s}}}{\cancel{\text{F}}} \times \frac{1\ \Omega\ \cancel{\text{F}}}{1\ \cancel{\text{s}}}\right)^2}$$

$$Z = \sqrt{256\ \Omega^2 + 4380\ \Omega^2}$$

$$Z = \sqrt{4636\ \Omega^2}$$

$$Z = 68.1\ \Omega$$

$$I = \frac{E}{Z}$$

$$I = \frac{110\ \text{V}}{68.1\ \Omega}$$

$$I = 1.62\ \frac{\cancel{\text{V}}}{\cancel{\Omega}}\left(\frac{1\ \Omega\ \cancel{\text{A}}}{1\ \cancel{\text{V}}}\right)$$

$$I = 1.62\ \text{A}$$

Then find the phase angle

$$\tan \phi = \frac{X_C}{R}$$

$$\tan \phi = \frac{66.2\ \cancel{\Omega}}{16.0\ \cancel{\Omega}}$$

$$\tan \phi = 4.14$$

so

$$\phi = 76^\circ$$

PROBLEMS

Find the impedance in the following five circuits.

1. $R = 10^3\ \Omega$, $C = 1.00 \times 10^{-6}$ F, $V = 100$ V, $f = 200$/s, $Z =$ ________ Ω
2. $R = 250\ \Omega$, $C = 5.00 \times 10^{-6}$ F, $V = 20.0\ V$, $f = 500$/s, $Z =$ ________ Ω
3. $R = 7.80\ \Omega$, $C = 45.0 \times 10^{-6}$ F, $V = 15.0$ V, $f = 950$/s, $Z =$ ________ Ω
4. $R = 145\ \Omega$, $C = 1.00 \times 10^{-5}$ F, $V = 7.00$ V, $f = 750$/s, $Z =$ ________ Ω
5. $R = 10.0\ \Omega$, $C = 5.00 \times 10^{-5}$ F, $V = 15.0$ V, $F = 10^2$/s, $Z =$ ________ Ω
6. Find the phase angle in problem 1.
7. Find the phase angle in problem 2.
8. Find the phase angle in problem 3.
9. Find the phase angle in problem 4.
10. Find the phase angle in problem 5.
11. Find the current in problem 1.
12. Find the current in problem 2.
13. Find the current in problem 3.
14. Find the current in problem 4.
15. Find the current in problem 5.

29-6 *CAPACITANCE, INDUCTANCE, AND RESISTANCE IN SERIES*

Many circuits which are important in the design of electronic equipment contain all three types of circuit elements discussed in this chapter. The impedance of a circuit containing resistance, capacitance, and inductance in series can be found from the equation:

$$Z = \sqrt{R^2 + (X_L - X_C)^2}$$

where: $Z =$ the impedance
$R =$ the resistance
$X_L =$ the inductive reactance
$X_C =$ the capacitive reactance

The vector diagram for this type of circuit is shown below.
The phase angle is given by:

$$\tan \phi = \frac{X_L - X_C}{R}$$

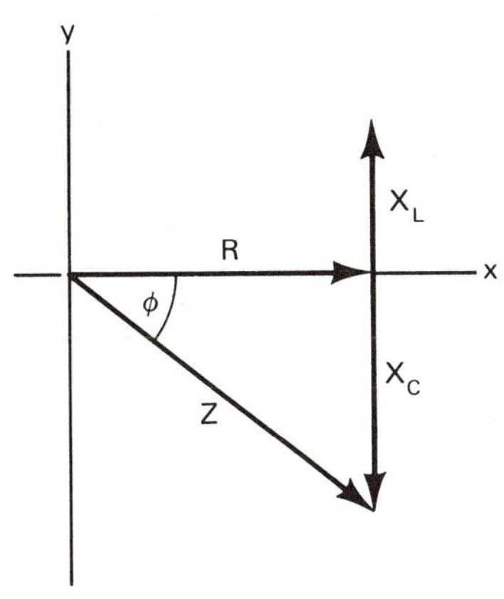

The current in this type of circuit is given by:

$$I = \frac{E}{Z} = \frac{E}{\sqrt{R^2 + \left[2\pi f L - \frac{1}{2\pi f C}\right]^2}}$$

EXAMPLE: A circuit contains a 100-Ω resistance, a 10^{-5}-F capacitor, and a 10^{-2}-H inductance in series with a 25.0-V, 200-Hz/s source. Find the impedance and the current.

SKETCH:

R = 10^2 Ω
E = 25.0 V
L = 10^{-2} H
C = 10^{-5} F

DATA:

$R = 100\ \Omega$
$C = 10^{-5}\ \text{F}$
$L = 10^{-2}\ \text{H}$
$E = 25.0\ \text{V}$
$f = 200/\text{s}$
$Z = ?$
$I = ?$

BASIC EQUATIONS: $X_L = 2\pi fL$, $X_C = \dfrac{1}{2\pi fC}$, $Z = \sqrt{R^2 + (X_L - X_C)^2}$ $I = \dfrac{E}{Z}$

WORKING EQUATIONS: Same

SUBSTITUTIONS: First find X_L:

$$X_L = 2\pi fL$$
$$X_L = 2(3.14)\left(\frac{200}{\text{s}}\right)(10^{-2}\ \text{H})$$
$$X_L = 12.6\ \frac{\text{H}}{\text{s}}\left(\frac{1\ \Omega\ \text{s}}{1\ \text{H}}\right)$$
$$X_L = 12.6\ \Omega$$

NOW FIND X_C:

$$X_C = \frac{1}{2\pi fC}$$
$$X_C = \frac{1}{(2)(3.14)(200/\text{s})(10^{-5}\ \text{F})}$$
$$X_C = 79.4\ \Omega$$

NOW FIND THE IMPEDANCE:

$$Z = \sqrt{R^2 + (X_L - X_C)^2}$$
$$Z = \sqrt{(10^2\ \Omega)^2 + (12.6\ \Omega - 79.4\ \Omega)^2}$$
$$Z = \sqrt{10^4\ \Omega^2 + (-66.8\ \Omega)^2}$$
$$Z = \sqrt{10^4\ \Omega^2 + 4460\ \Omega^2}$$
$$Z = \sqrt{14{,}460\ \Omega^2}$$
$$Z = 120\ \Omega$$

NOW FIND THE CURRENT:

$$I = \frac{E}{Z}$$
$$I = \frac{120\ \Omega}{25\ \text{V}}$$
$$I = 0.208\ \frac{\cancel{\text{V}}}{\cancel{\Omega}}\left(\frac{1\ \text{A}\ \cancel{\Omega}}{1\ \cancel{\text{V}}}\right)$$
$$I = 0.208\ \text{A}$$

PROBLEMS

Find the impedance and current in the following circuits.

1. $R = 40.0\ \Omega$, $L = 0.500$ H, $C = 50.0 \times 10^{-6}$ F, $f = 60$/s, $V = 5.00$ V.
2. $R = 200\ \Omega$, $L = 10^{-2}$ H, $C = 2.00 \times 10^{-7}$ F, $f = 10^3$/s, $V = 10.0$ V.
3. $R = 10^2\ \Omega$, $L = 10^{-2}$ H, $C = 3.00 \times 10^{-4}$ F, $f = 10^4$/s, $V = 15.0$ V.
4. $R = 60.0\ \Omega$, $L = 0.700$ H, $C = 30.0 \times 10^{-6}$ F, $f = 60$/s, $V = 8.00$ V.

29-7 RESONANCE

The current in a circuit containing resistance, capacitance, and inductance is given by the equation:

$$I = \frac{E}{\sqrt{R^2 + (X_L - X_C)^2}}$$

When the inductive reactance equals the capacitive reactance, they nullify each other and the current is given by:

$$I = \frac{E}{R}$$

which is its maximum possible value. When this condition exists, the circuit is in resonance with the applied voltage.

Resonant circuits are used in radios and televisions. The frequency of a certain station is tuned in when a resonant circuit (antenna circuit) is adjusted to that frequency. This is accomplished by changing the capacitance until the capacitive reactance equals the inductive reactance. The applied voltage is the radio signal picked up by the antenna.

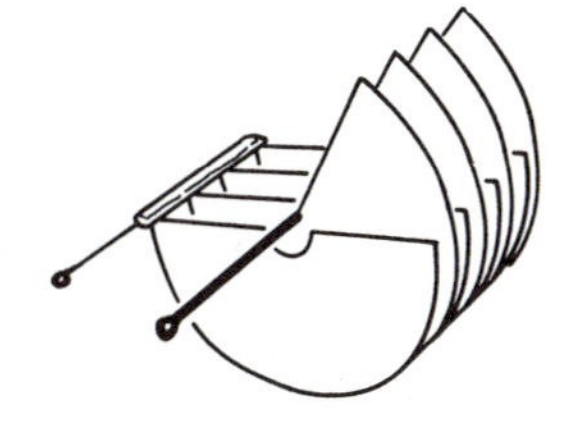

The resonant frequency occurs when $X_L = X_C$. To find this frequency we write:

$$X_L = X_C$$

$$2\pi f L = \frac{1}{2\pi f C}$$

$$f^2 = \frac{1}{4\pi^2 LC}$$

$$f = \frac{1}{\sqrt{4\pi^2 LC}}$$

$$f = \frac{1}{2\pi\sqrt{LC}}$$

The circuit can be adjusted to any frequency by varying the capacitance or the inductance.

EXAMPLE: Find the resonant frequency of a circuit containing a 5.00×10^{-9} F capacitor in series with a 2.60×10^{-6} H inductor.

SKETCH:

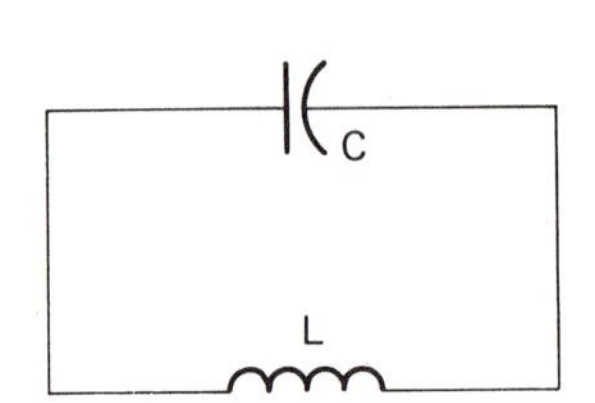

DATA: $C = 5.00 \times 10^{-9}$ F
$L = 2.60 \times 10^{-6}$ H

BASIC EQUATION: $f = \frac{1}{2\pi\sqrt{LC}}$

WORKING EQUATION: Same

SUBSTITUTION:

$$f = \frac{1}{2(3.14)\sqrt{(2.60 \times 10^{-6}\ \text{H})(5.00 \times 10^{-9}\ \text{F})}}$$

$$f = \frac{1}{6.28\sqrt{1.30 \times 10^{-14}\ \not{H}\not{F}\left(\frac{1\ \not{\Omega}\ \text{s}}{1\ \not{H}}\right)\left(\frac{1\ \text{s}}{1\ \not{F}\not{\Omega}}\right)}}$$

$$f = \frac{1}{6.28(1.14 \times 10^{-7})\text{s}}$$

$$f = \frac{1}{(7.16 \times 10^{-7})\text{s}}$$

$$f = 1.4 \times 10^{6}\ \frac{\text{cycle}}{\text{s}} \quad \text{or} \quad 1400\ \frac{\text{kilocycles}}{\text{s}} \quad \text{or} \quad 1400\ \text{kHz}$$

This frequency is in the AM radio band.

PROBLEMS

1. Find the resonant frequency of a circuit with $L = 10^{-5}$ H and $C = 5.00 \times 10^{-6}$ F.
2. Find the resonant frequency of a circuit with $L = 2.00 \times 10^{-4}$ H and $C = 4.00 \times 10^{-5}$ F.
3. Find the resonant frequency of a circuit with $L = 3.00 \times 10^{-2}$ H and $C = 6.00 \times 10^{-4}$ F.

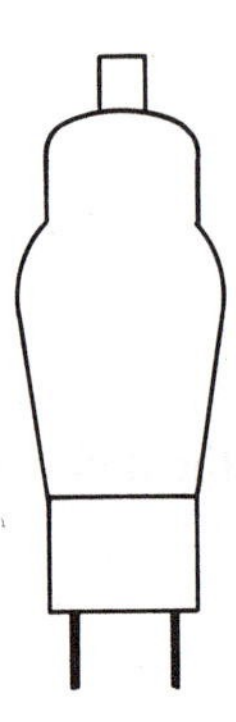

29-8 *RECTIFICATION*

Technicians find it is often necessary to change ac into dc. This process is called rectification. A device which accomplishes this is called a diode. Early diodes were constructed as vacuum tubes.

Many diodes today are constructed out of a small piece of semiconductor material. These diodes are usually less than an eighth of an inch long.

A diode allows current to flow only in one direction. It is similar to a turnstile that revolves only in one direction.

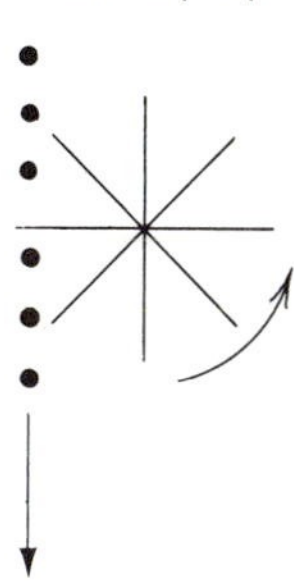

People may pass the turnstile in one direction but are blocked when attempting to pass in the opposite direction.

A diode allows electrons to pass in only one direction and not in the other.

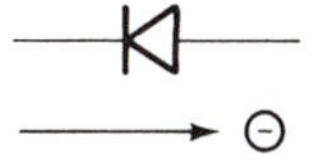

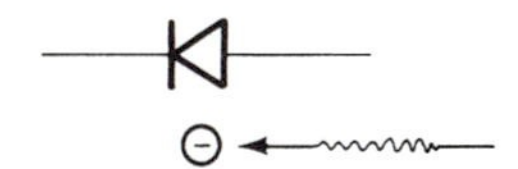

An alternating current is allowed to pass only in one direction and is thus changed to a direct current.

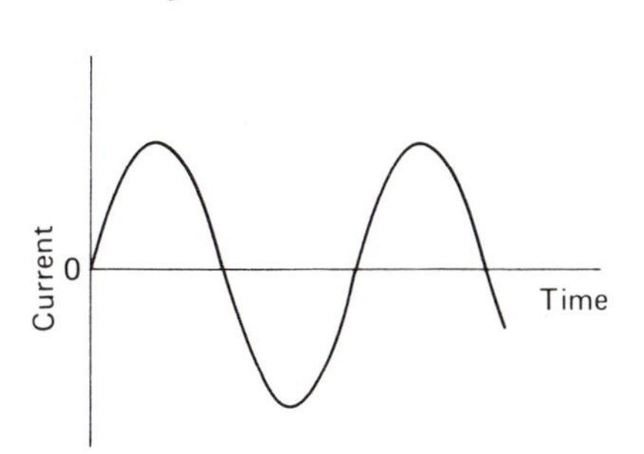

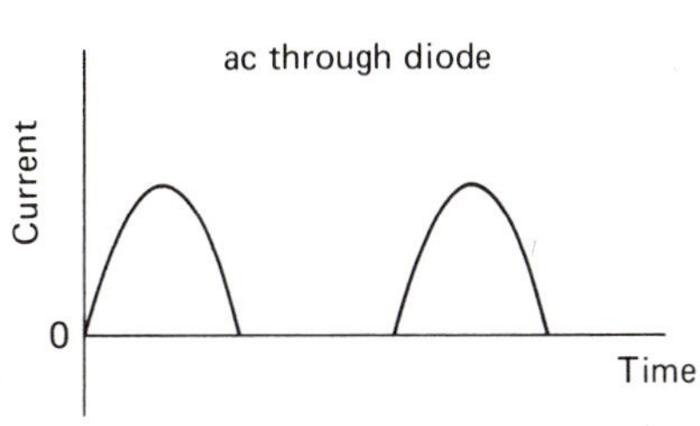

Additional circuit devices can be added to the rectifier which will smooth out the direct current so that it appears as shown.

Rectifiers are used in automobiles to change the alternating current produced by the alternator into direct current. They are used in all ac radios and televisions to produce dc power for the tubes and transistors.

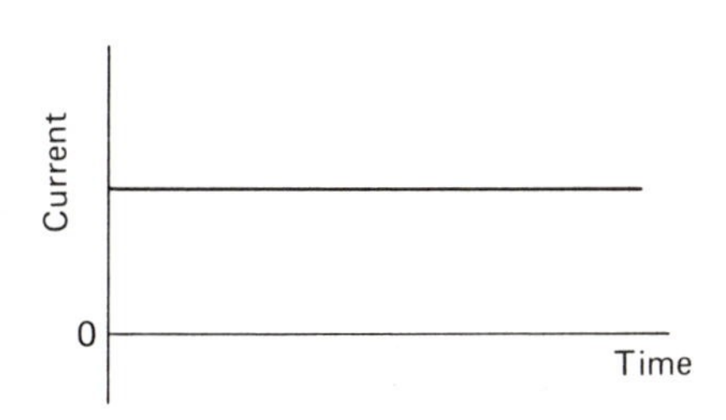

LIGHT 30

30-1 *THE NATURE OF LIGHT*

There are two distinct ways of transporting energy from one point to another. The first is to transport particles of matter which carry energy with them. Examples of this method are electron conduction in a wire, shooting a bullet, natural gas pipelines, and gasoline transport. The second method is the propagation of a wave disturbance through the medium between two points. Sound and water waves are examples of this method of energy transport. Light is unusual in that it appears to combine characteristics of each method. When light is traveling through a medium it appears to behave like a wave with the following characteristics: (1) reflection at the surface of a medium; (2) refraction when passing from one medium to another; (3) interference (cancellation) when two waves are properly superimposed; and (4) diffraction (bending) when the waves pass the corners of an obstacle. When light interacts with matter, such as when it is absorbed or emitted, it behaves as if it were a massless particle.

30-2 *THE SPEED OF LIGHT*

Light is one form of a class of radiation called electromagnetic radiation, which also includes radio waves, TV, infrared, gamma rays, and X rays. A chart of the entire electromagnetic spectrum is shown in the table below.

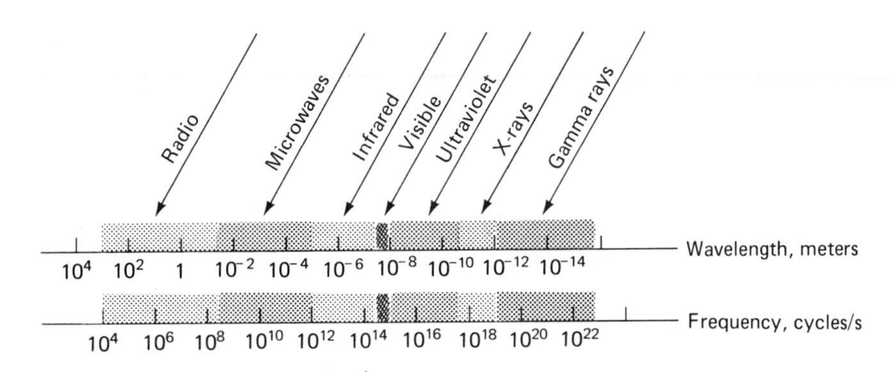

The Electromagnetic Spectrum

All the forms of electromagnetic radiation travel at the same velocity which is 186,000 mi/s or 2.99×10^8 m/s. This is often referred to as the speed of light and is represented by the letter c.

$$c = 186{,}000 \text{ mi/s} = 2.99 \times 10^8 \text{ m/s}$$

The distance traveled by any form of electromagnetic radiation can be found by combining this velocity with the equation relating distance traveled, velocity, and time: $x = vt$. Since the velocity is c, this equation becomes:

$$x = ct$$

EXAMPLE: Find the distance traveled by an X ray in 0.100 s.

DATA: $c = 186{,}000$ mi/s
$t = 0.100$ s
$x = ?$

BASIC EQUATION: $x = ct$

WORKING EQUATION: Same

SUBSTITUTION: $x = (186{,}000 \text{ mi/s})(0.100 \text{ s})$
$= 18{,}600$ mi

PROBLEMS

1. Find the distance (in meters) traveled by a radio wave in 5.00 seconds.
2. Find the distance (in miles) traveled by a light wave in 6.40 seconds.
3. A television signal is sent to a communications satellite which is 20,000 miles above a relay station. How long does it take for the signal to reach the satellite?
4. How long does it take for a radio signal from earth to reach an astronaut on the moon? The distance from the earth to the moon is 2.4×10^5 miles.
5. The sun is 9.3×10^7 miles from the earth. How long does it take light to travel from the sun to the earth?
6. A radar wave which is bounced off an airplane returns to the radar receiver in 2.5×10^{-5} s. How far (in km) is the airplane from the radar receiver?

30-3 *LIGHT AS A WAVE*

Light and the other forms of electromagnetic radiation are composed of oscillations in the electric and magnetic fields that exist in space. These oscillations are set up by rapid movement of charged particles such as electrons in radio antennas and electrons in a hot object such as a light bulb filament. All waves are characterized by the distance which separates two points on the wave which are at the same point of vibration. This distance is called the wavelength and is denoted by the Greek letter lambda, λ. The wavelength of visible light ranges from about 4×10^{-7} m to 7.6×10^{-7} m. The wavelengths of other electromagnetic radiations can be found in the table at the beginning of this chapter.

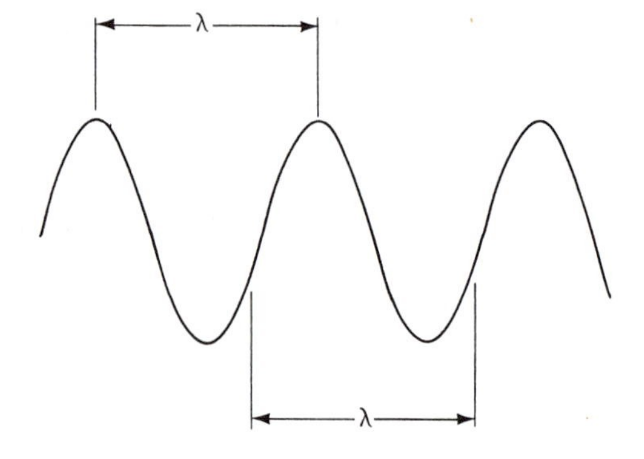

Another characteristic of waves is the frequency, f. Frequency is the number of complete oscillations per second. A basic relationship exists for all waves which relate the frequency, wavelength, and the velocity of the wave. The velocity is equal to the product of the frequency and the wavelength.

$$c = f\lambda$$

EXAMPLE: Find the frequency of a light wave which has a wavelength of 5×10^{-7} m.

DATA: $\lambda = 5 \times 10^{-7}$ m
$c = 2.99 \times 10^{8}$ m/s
$f = ?$

BASIC EQUATION: $c = f\lambda$

WORKING EQUATION: $f = \frac{c}{\lambda}$

SUBSTITUTION: $f = \frac{2.99 \times 10^{8} \text{ m/s}}{5 \times 10^{-7} \text{ m}}$
$= 5.98 \times 10^{14}$ Hz (cycles/s)

PROBLEMS

1. $c = 2.99 \times 10^{8}$ m/s
   $\lambda = 4.00 \times 10^{-5}$ m
   $f = ?$
2. $c = 2.99 \times 10^{8}$ m/s
   $\lambda = 7.80 \times 10^{-10}$ m
   $f = ?$
3. $c = 2.99 \times 10^{8}$ m/s
   $f = 9.70 \times 10^{11}$ Hz
   $\lambda = ?$
4. $c = 2.99 \times 10^{8}$ m/s
   $f = 2.42 \times 10^{7}$ Hz
   $\lambda = ?$
5. Find the wavelength of a radio wave from an AM station broadcasting at a frequency of 1400 kHz.
6. Find the wavelength at a radio wave from a FM station broadcasting at a frequency of 94.5 M Hz.

30-4 *LIGHT AS A PARTICLE*

As mentioned earlier, light sometimes behaves as if it were a massless particle. These particles are called photons and each carries a portion of the energy of the wave. This energy is given by:

$$E = hf$$

where

$$f = \text{frequency}$$
$$h = 6.62 \times 10^{-34} \text{ J-s (Planck's constant)}$$

The particle theory of light must be used to explain some interactions between light and matter such as the photoelectric effect. The photoelectric effect is the ejection of electrons when light strikes certain metal surfaces.

30-5 *LUMINOUS INTENSITY OF LIGHT SOURCES*

The determination of the necessary light sources for proper illumination in homes, business, and industry is often a matter of concern for engineering technicians. We will consider here some simple problems of this type.

The intensity (strength), I, of a light source is measured in terms of candlepower (cp). The early use of certain candles for standards of illumination led to the name of the unit. We now use a platinum source at a certain temperature as the standard for comparison. Another unit, the lumen, is often used for the measurement of the intensity of a source. The conversion factor between candlepower and lumens is:

$$1 \text{ cp} = 4\pi \text{ lumens}$$

Thus a certain 40-W light bulb which is rated at 35 cp would have a rating of 441 lumens.

$$35 \text{ cp} \times \frac{4\pi \text{ lumens}}{1 \text{ cp}} = 441 \text{ lumens}$$

PROBLEMS

1. $I = 48.0$ cp
   $I =$ _______ lumens
2. $I = 197$ cp
   $I =$ _______ lumens
3. $I = 543$ lumens
   $I =$ _______ cp
4. $I = 432$ lumens
   $I =$ _______ cp
5. $I = 75.0$ cp
   $I =$ _______ lumens

30-6 *INTENSITY OF ILLUMINATION*

When a surface is illuminated by a light source, the intensity of the illumination decreases as the distance between the source and the surface increases. If the source radiates light uniformly in all directions, the light is uniformly distributed over a spherical surface centered at the source. Since the surface area of a sphere is $4\pi r^2$, the intensity of illumination, E, at the surface is given by:

$$E = \frac{I}{4\pi r^2}$$

where I is the intensity of the source in lumens and r is the distance between the source and the illuminated surface.

EXAMPLE: Find the intensity of illumination on a surface located 2.00 meters from a source with an intensity of 400 lumens.

DATA:
$I = 400$ lumens
$r = 2.00$ m
$E = ?$

BASIC EQUATION: $E = \dfrac{I}{4\pi r^2}$

WORKING EQUATION: Same

SUBSTITUTION:
$$E = \frac{400 \text{ lumens}}{4\pi(2.00 \text{ m})^2}$$
$$= 7.91 \frac{\text{lumens}}{\text{m}^2}$$

The unit used for intensity of illumination is the lumen/m² in the mks system and the lumen/ft² in the English system. Another unit often used is the foot-candle which is equal to one lumen/ft²

1 ft-candle = 1 lumen/ft²

EXAMPLE 2: Find the intensity of illumination 4 ft from a source with an intensity of 600 lumens.

DATA:
$I = 600$ lumens
$r = 4.00$ ft
$E = ?$

BASIC EQUATION: $E = \dfrac{I}{4\pi r^2}$

WORKING EQUATION: Same

SUBSTITUTION: $$E = \frac{600 \text{ lumens}}{4\pi(4.00 \text{ ft})^2}$$

$$= 2.91 \frac{\text{lumens}}{\text{ft}^2} \cdot \frac{1 \text{ foot-candle}}{1 \text{ lumen/ft}^2}$$

$$= 2.91 \text{ foot-candles}$$

PROBLEMS

1. $I = 900$ lumens
   $r = 7.00$ ft
   $E = ?$
2. $I = 741$ lumens
   $r = 4.00$ ft
   $E = ?$
3. $I = 893$ lumens
   $r = 4.50$ ft
   $E = ?$
4. $E = 4.32$ ft-candles
   $r = 9.00$ ft
   $I = ?$
5. $E = 10.5$ ft-candles
   $r = 6.00$ ft
   $I = ?$
6. Find the intensity of the light source necessary to produce an illumination of 5.50 ft-candles a distance of 9.85 ft from the source.
7. Find the intensity of the light source necessary to produce an illumination of 2.39 ft-candles a distance of 4.50 ft from the source.
8. Find the intensity of the light source necessary to produce an illumination of 5.28 ft-candles a distance of 6.50 ft from the source.

appendix

SLIDE RULE

A-1 *INTRODUCTION*

The slide rule is a very important tool of the technician. The slide rule can be used to multiply, divide, square, cube, find square roots, find cube roots, find reciprocals, and find sines, cosines, and tangents of angles.

There are three parts to a slide rule: the *body*, *slide*, and the *cursor with hairline*.

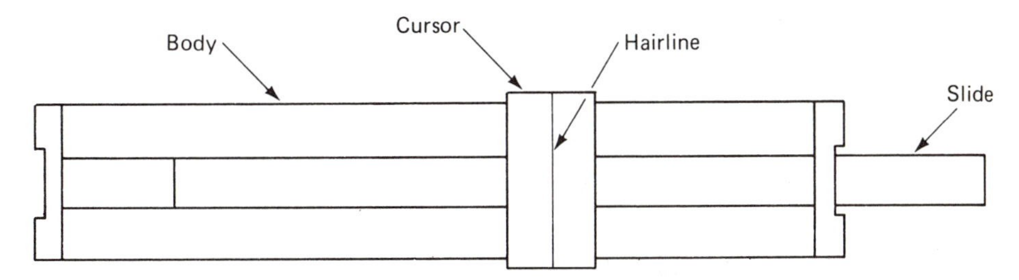

The slide and cursor can be moved across the body. To move the slide and cursor easily, you should hold the rule on the end with the thumb and forefinger of one hand and operate the slide and cursor with the other hand.

There are graduated scales on both body and the slide. The most important scales (most used) are the C and D scales. These scales are used most often for multiplying numbers.

A-2 *READING C AND D SCALES*

Since the C and D scales are identical, you now need only be concerned with studying one of them. Consider the D scale. It is divided into nine divisions by ten graduations called *primary graduations*. The primary graduations start with 1 and end with 1. The 1 on the left of the D scale is called the *left index* (L_i), and the 1 on the right is called the *right index* (R_i). The first digit of the number to be located is always at or to the right of a primary graduation. Each of the ten primary graduations is divided into ten divisions by nine *secondary graduations*. The second digit of the number to be located is always at or to the right of a secondary graduation. Each of the ten secondary divisions is divided further into *tertiary graduations*. The third digit of the number to be located is at or between tertiary graduations of the D scale.

There are three separate considerations that have to be made for the tertiary graduations.

1. Between L_i and 2 there are nine tertiary graduations between the secondary graduations forming ten divisions. Each of these tertiary graduations represents 1.

2. Between 2 and 4 there are four tertiary graduations between the secondary graduations forming five divisions. Each of these tertiary graduations represents 2.

3. Between 4 and R_i there is one tertiary graduation between the secondary graduations forming two divisions. Each tertiary graduation represents 5.

EXAMPLES: Read the digits under the hairline on the D scale shown in the pictures below.

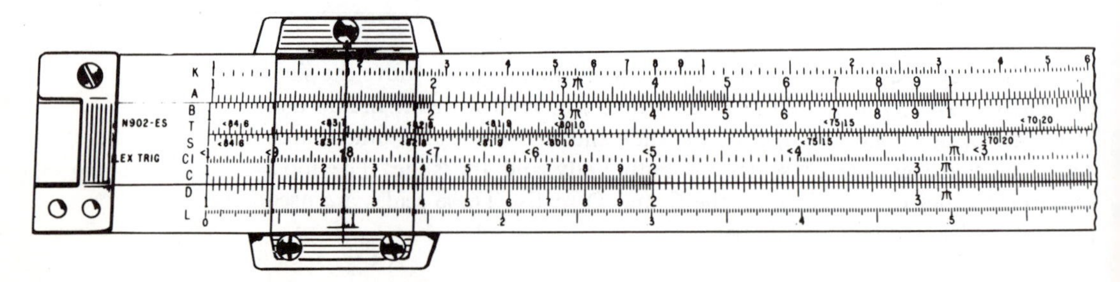

The hairline is to the right of primary 1 and secondary 2. It is on the fourth tertiary mark. Each tertiary mark on this part of the scale represents 1. The reading is 124.

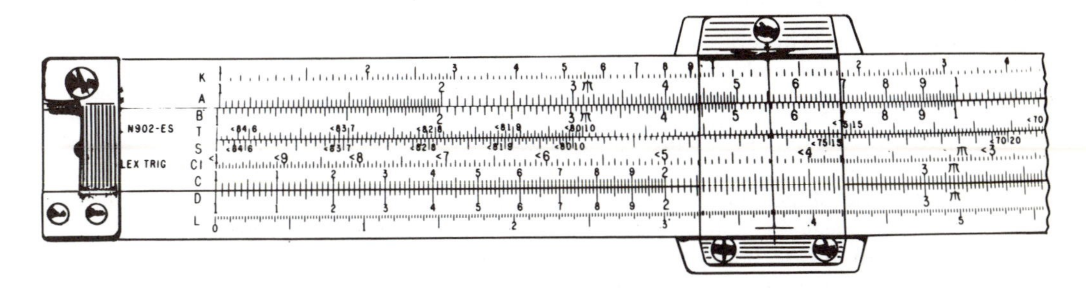

The hairline is to the right of primary 2 and secondary 3. It is on the third tertiary mark to the right of 3. Each tertiary mark on this part of scale represents 2. The reading is 236.

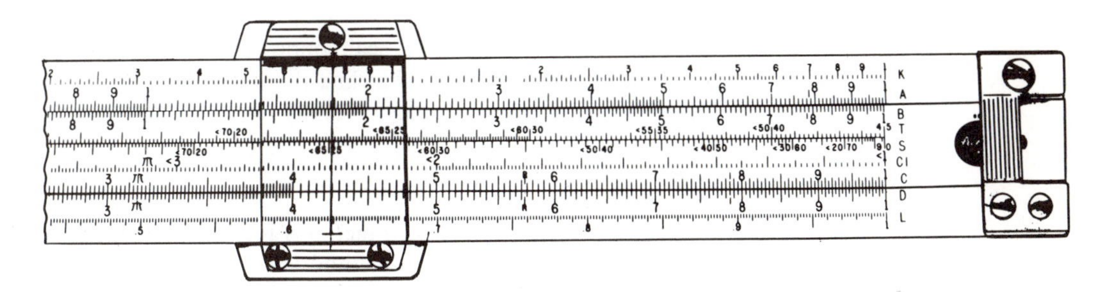

The hairline is to the right of primary 4 and secondary 2. It is on the tertiary mark which represents 5. The reading is 425.

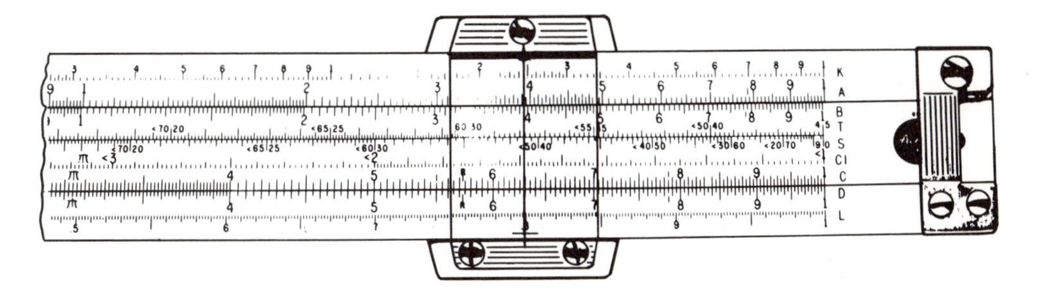

The hairline is to the right of primary 6 and to the left of secondary 3. It is to the right of the tertiary mark which represents 5. It must be greater than 5 but less than 10. Since it is much closer to 10 than 5, we estimate the tertiary mark as 9. The reading is 629.

A-3 *SLIDE RULE CALCULATIONS*

Before studying the different operations on the slide rule, you need to know the general procedure to follow when doing a calculation on the slide rule.

1. The slide rule is accurate to only three significant digits. Therefore, all numbers must first be rounded to three significant digits.
2. The slide rule does not allow for decimal points. For example, the numbers 3.24, 324, 0.00324, 32,400, and 32.4 are all indicated by the same mark.
3. Therefore, each calculation with a slide rule involves two parts:
   (a) Use the slide rule to get the three significant digits in the answer.
   (b) Locate the decimal point in the answer by scientific notation and laws of exponents or by estimation.

A-4 *MULTIPLICATION*

EXAMPLE 1: 3.46×1.92

(a) Place L_i (left index) of the C scale opposite 346 on the D scale. Slide the cursor so that the hairline is over 192 on the C scale. Read the digits of the product under the hairline on the D scale, 664.

(b) Locating the decimal point.

Note: You will be shown two methods for locating the decimal point. It will be your choice which method you use.

METHOD 1: Approximate the product by rounding each number to one significant digit and then multiply: $3 \times 2 = 6$. Place the decimal point in 664 so that the number is "nearest" 6, which is 6.64. Therefore, $3.46 \times 1.92 = 6.64$.

METHOD 2:

RULE

When multiplying two numbers between 1 and 10 *on the slide rule, there is one digit to the left of the decimal point if the left index* (L_i) *is used, and there are two digits to the left of the decimal point if the right index* (R_i) *is used.*

Since the left index of the C scale is used, there is only one digit to the left of the decimal point, 6.64

EXAMPLE 2: 7.23×5.76

(a) Place L_i of the C scale opposite 723 on the D scale. Slide the cursor so the hairline is over 576 on the C scale, which is impossible. Therefore, place R_i of the C scale opposite 723 on the D scale. Slide the cursor so the hairline is over 576 on the C scale. Read the digits of the product under the hairline on the D scale, 416.

(b) Locating the decimal point.

METHOD 1: Approximate the product by rounding each number to one significant digit and then multiply: $7 \times 6 = 42$. Place the decimal point in 416 so that the number is "nearest" 42, which is 41.6. Therefore, $7.23 \times 5.76 = 41.6$.

METHOD 2: Since R_i of the C scale is used and both factors are between 1 and 10, there are two digits to the left of the decimal point, 41.6.

Note: As in the example above, whenever using one index results in not being able to complete an operation on the slide rule, use the other index.

PROBLEMS

Find the products using a slide rule.

1. 1.73×3.12
2. 4.91×5.62
3. 8.43×9.77
4. 6.13×5.02
5. 3.06×1.17
6. 7.90×3.67
7. 2.06×6.44
8. 5.12×8.14
9. 8.33×9.01
10. 9.24×1.02

EXAMPLE 3: 35.1×81.6

(a) Place R_i of the C scale opposite 351 on the D scale. Slide the cursor so that the hairline is over 816 on the C scale. Read the digits of the product under the hairline on the D scale, 286.

(b) Locating the decimal point.

METHOD 1: Approximate the product by rounding each number to one significant digit and then multiply: $40 \times 80 = 3200$. Place the decimal point in 286 so that the number is "nearest" 3200, which is 2860. Therefore $35.1 \times 81.6 = 2860$.

METHOD 2: Write the problem with the factors in scientific notation, $35.1 \times 81.6 = 3.51 \times 10^1 \times 8.16 \times 10^1 = 3.51 \times 8.16 \times 10^1 \times 10^1 = 3.51 \times 8.16 \times 10^2$. We have found the digits 286 on the slide rule, and since the right index was used, there are two digits to the left of the decimal point so, $3.51 \times 8.16 \times 10^2 = 28.6 \times 10^2$. Now move the decimal point two places to the right to multiply by 10^2, and we have $35.1 \times 81.6 = 2860$.

EXAMPLE 4: 152.7×0.000376.

Since only three places can be read on a slide rule, the numbers must be rounded to three significant digits.

(a) Place L_i of the C scale opposite 153 on the D scale. Slide the cursor so the hairline is over 376 on the C scale. Read the digits of the product under the hairline on the D scale, 575.

(b) Locating the decimal point.

METHOD 1: Approximate the product by rounding each number to one significant digit and multiply: $200 \times 0.0004 = 0.08$. Place the decimal point in 575 so that the number is "nearest" 0.08, which is 0.0575. Therefore, $152.7 \times 0.000376 = 0.0575$.

METHOD 2: Write the problem with the factors in scientific notation, $153 \times 0.000376 = 1.53 \times 10^2 \times 3.76 \times 10^{-4} = 1.53 \times 3.76 \times 10^{-2}$. We have found the digits 575 while using the left index on the slide rule, so there is one digit to the left of the decimal point, $5.75 \times 10^{-2} = 0.0575$.

PROBLEMS

Use a slide rule to multiply the following.

1. 93.1×6.42
2. 0.817×13.4
3. 433×31.13
4. 0.614×0.01421
5. 91.4×16.7
6. 8310×61.3
7. 0.00144×0.369
8. 0.0543×63100
9. 19×0.01614
10. 8×91732

Recall from Chapter 2 the formula for the area of a rectangle is $A = lw$.

11. If $l = 21.7$ ft and $w = 13.6$ ft, find A.
12. If $l = 5.71$ m and $w = 3.12$ m, find A.
13. Find the area of a rectangular metal plate which measures 5.61 in. by 3.37 in.
14. A machinist is paid \$7.16 per hour. How much does he earn working an 8-hour day?
15. How much does the machinist in problem 14 earn working a 40-hour week?

A-5 *RECIPROCALS*

The reciprocal of a number, n, is $1/n$. As examples of reciprocals: the reciprocal of 16 is $\frac{1}{16}$ and the reciprocal of 0.01 is 100.

The CI and C scales are used to find the reciprocal of a number. The graduations on the CI scale are the same as those on the C and D scales except the numbers are in *reverse order*. There is a symbol “$<$” to remind you in which direction to read the scale. The CI scale is read from right to left.

EXAMPLE 1: Find the reciprocal of 4 in decimal form.

(a) Move the cursor so the hairline is over 4 on the CI scale. The digits of the reciprocal are on the C scale, 250.

(b) Locating the decimal point.

METHOD 1: *Approximation:* Round the number to one significant digit and then divide 1 by it.

$$\begin{array}{r} 0.2 \\ 4\overline{)1.00} \end{array}$$

Therefore, the reciprocal of 4 is 0.25.

RULE

A number between 1 *and* 10 *always has a reciprocal such that the decimal point is before the first significant digit.*

METHOD 2: The digits of the reciprocal on the C scale are 250; so by the rule, the reciprocal of 4 is 0.25.

EXAMPLE 2: Find the reciprocal of 16.5 in decimal form.

(a) Move the cursor so the hairline is over 165 on the CI scale. The digits of the reciprocal are read on the C scale, 606.

(b) Locating the decimal point.

METHOD 1: *Approximation*: Round 16.5 to one significant digit and divide 1 by it.

$$\begin{array}{r} 0.05 \\ 20\overline{)1.00} \end{array}$$

The approximation is 0.05. Place the decimal point in 606 so that the number is “nearest” 0.05, which is 0.0606. Therefore,

$$\frac{1}{16.5} = 0.0606$$

METHOD 2: Write the problem with the given number in scientific notation. The reciprocal of 16.5 is:

$$\frac{1}{16.5} = \frac{1}{1.65 \times 10^1} = \frac{1}{1.65} \times \frac{1}{10^1} = \frac{1}{1.65} \times 10^{-1} = 0.606 \times 10^{-1}$$

$$= 0.0606$$

Therefore, the reciprocal of 16.5 is 0.0606.

EXAMPLE 3: Find the reciprocal of 0.00651 in decimal form.

(a) Move the cursor so the hairline is over 651 on the CI scale. The digits of the reciprocal are read on the C scale, 154.

(b) Locating the decimal point.

METHOD 1: *Approximation*: Round 0.00651 to 0.007 and divide into 1.

$$0.007.\overline{)1.000.} \quad \text{quotient } 100.$$

The approximation is 100. Place the decimal point in 154 so that the number is "nearest" 100, which is 154. Therefore,

$$\frac{1}{0.00651} = 154$$

METHOD 2: Write the problem with the given number in scientific notation. The reciprocal of 0.00651 is:

$$\frac{1}{0.00651} = \frac{1}{6.51 \times 10^{-3}} = \frac{1}{6.51} \times \frac{1}{10^{-3}} = \frac{1}{6.51} \times 10^3 = 0.154 \times 10^3$$

$$= 154$$

Therefore, the reciprocal of 0.00651 = 154.

Note: $\frac{1}{10^{-3}} = \frac{1}{1/10^3} = 10^3$

PROBLEMS

Find the reciprocals of each of the following in decimal form.

1. 14.6	2. 5.17	3. 37.4
4. 183	5. 0.172	6. 0.0432
7. 37	8. 50	9. 0.712
10. 18.9	11. 3.42	12. 63.8
13. 0.00111	14. 8430	15. 5000

A-6 *DIVISION*

5 divided by 2 gives the same result as 5 times $\frac{1}{2}$ (notice that $\frac{1}{2}$ is the reciprocal of 2). In general, we can think of dividing one number by a second in terms of multiplying the first by the reciprocal of the second. The CI and D scales are used for division.

EXAMPLE 1: $6.22 \div 8.35$.

(a) Set L_i of CI scale opposite 622 on the D scale. Move the cursor until the hairline is over 835 on the CI scale. Read the digits of the quotient under the hairline on the D scale, 745.

(b) Locating the decimal point.

METHOD 1: *Approximation*: Round each number to one significant digit and divide to locate the decimal point.

$$8\overline{)6.0}\quad = 0.7$$

The approximation is 0.7. Place the decimal point in 745 so that the number is "nearest" 0.7, which is 0.745. Therefore,

$$\frac{6.22}{8.35} = 0.745$$

RULE

When the divisor and the number being divided are both between 1 *and* 10 *and the* left *index of the slide rule is used, the decimal point is before the first significant digit of the quotient.*

METHOD 2: Since both the numbers are between 1 and 10 and the left index is used, $6.22 \div 8.35 = 0.745$ by the rule above.

EXAMPLE 2: $\frac{8.23}{1.73}$.

(a) Set R_i of CI scale opposite 823 on the D scale. Move the cursor so that the hairline is over 173 on the CI scale. Read the digits of the quotient under the hairline on the D scale, 476.

(b) Locating the decimal point.

METHOD 1: *Approximation*: Round each number to one significant digit and divide to locate the decimal point.

$$2\overline{)8.0}\quad = 4.0$$

The approximation is 4. Place the decimal point in 476 so that the number is "nearest" 4, which is 4.76. Therefore,

$$\frac{8.23}{1.73} = 4.76$$

RULE

When the divisor and the number being divided are both between 1 *and* 10 *and the* right *index of the slide rule is used, there is one digit to the left of the decimal point.*

METHOD 2: Since both the numbers are between 1 and 10 and the right index is used, $8.23 \div 1.73 = 4.76$ by the rule above.

EXAMPLE 3: $0.231 \div 560$.

(a) Set L_i of the CI scale opposite 231 on the D scale. Move the cursor until the hairline is over 560 on the CI scale. Read 413 under the hairline on the D scale.

(b) Locating the decimal point.

METHOD 1: *Approximation*: Round and divide.

$$600 \overline{\big) 0.2000}^{\;0.0003}$$

Place the decimal point in 413 so that the number is "nearest" 0.0003, which is 0.000413. Therefore,

$$\frac{0.231}{560} = 0.000413$$

METHOD 2: Write the problem with the numbers in scientific notation:

$$\frac{2.31 \times 10^{-1}}{5.60 \times 10^{2}} = \frac{2.31}{5.60} \times \frac{10^{-1}}{10^{2}}$$

$$= 0.413 \times \frac{10^{-1}}{10^{2}} \qquad \text{(the } L_i \text{ was used)}$$

$$= 0.413 \times 10^{-1-2} = 0.413 \times 10^{-3} = 0.000413$$

EXAMPLE 4: $560 \div 0.231$.

(a) Set R_i of the CI scale opposite 560 on the D scale. Move the cursor until the hairline is over 231 on the CI scale. Read 242 under the hairline on the D scale.

(b) Locating the decimal point.

METHOD 1: *Approximation*: Round and divide.

$$0.2\overline{)600} = 0.2.\overline{)600.0.}\quad \text{quotient } 3000$$

Place the decimal point in 242 so that the number is "nearest" 3000, which is 2420. Therefore,

$$\frac{560}{0.231} = 2420$$

METHOD 2:

$$\frac{5.60 \times 10^2}{2.31 \times 10^{-1}} = \frac{5.60}{2.31} \times \frac{10^2}{10^{-1}}$$

$$= 2.42 \times \frac{10^2}{10^{-1}} \qquad \text{(the } R_i \text{ was used)}$$

$$= 2.42 \times 10^{2-(-1)} = 2.42 \times 10^3 = 2420$$

PROBLEMS

Divide using a slide rule.

1. $\frac{352}{16}$
2. $\frac{94.3}{8.72}$
3. $\frac{378}{572}$
4. $\frac{26.4}{5.2}$
5. $843 \div 9.11$
6. $24.8 \div 0.612$
7. $18.9 \div 0.00122$
8. $0.044 \div 0.0171$
9. $71.3 \div 0.694$
10. $\frac{159}{63.2}$
11. $\frac{18.7}{94}$
12. $\frac{1430}{64.1}$
13. $\frac{18,000}{6.3}$
14. $\frac{496}{35,000}$
15. $\frac{48}{0.0012}$
16. $\frac{321}{0.00242}$
17. $\frac{31942}{614.912}$
18. $\frac{1864}{0.0397}$
19. $\frac{466}{0.0125}$
20. $\frac{0.00473}{0.0009265}$

Use the formula $V = d/t$ in each of the following:

21. If $d = 630$ and $t = 9$, find V.
22. If $d = 150$ and $t = 30$, find V.
23. If $d = 5170$ and $t = 5.7$, find V.

PROBLEMS

Use a slide rule and the conversion tables in the Appendix to convert the following (express all answers in decimal rather than in fractional form):

1. 6.2 m = _______ft
2. 15,270 ft = _______mi
3. 847 in. = _______ft
4. 210 yd = _______ft
5. 24.1 kg = _______g
6. 48.3 oz = _______lb
7. 1.7 mi = _______ft
8. 430 l = _______qt
9. 61.1 m = _______cm
10. 18.4 in. = _______cm
11. 736 ft = _______m
12. 43,700 g = _______kg
13. 496 oz = _______lb
14. 63.8 cm = _______in.
15. 4700 mg = _______g
16. 87.5 ft = _______yd
17. 16 mi = _______m
18. 83.7 cm = _______m
19. 142 ft = _______in.
20. 614 qt = _______l

A-7 *COMBINED OPERATIONS USING MULTIPLICATION AND DIVISION*

The previous rules for multiplication and division to locate the decimal point using method 2 are used when *only two numbers* are multiplied or divided. Therefore, these rules may not be used when more than two numbers are being used.

To multiply more than two numbers on the slide rule: multiply the first two numbers; move the correct index of the C scale to this result and multiply by the third number; move the correct index of the C scale to this result and multiply by the fourth number; and so on until all the numbers have been used. Read the digits of the final result under the hairline on the D scale.

EXAMPLE 1: $0.887 \times 378 \times 57 \times 0.00241$.

(a) Set R_i of the C scale opposite 887 on the D scale. Move the cursor so that the hairline is over 378 on the C scale. Move R_i of the C scale under the hairline and then move the cursor so that the hairline is over 570 on C. Move L_i of the C scale under the hairline and then move the cursor so that the hairline is over 241 on the C scale. Read 461 under the hairline on the D scale.

(b) Locating the decimal point.

METHOD 1: *Approximation*: Round the numbers to one significant digit and multiply.

$0.9 \times 400 \times 60 \times 0.002 = 43.2$

Place the decimal point in 461 so that the number is "nearest" 43.2, which is 46.1. Therefore,

$0.887 \times 378 \times 57 \times 0.00241 = 46.1$

METHOD 2: Write the numbers in scientific notation.

$8.87 \times 10^{-1} \times 3.78 \times 10^{2} \times 5.7 \times 10^{1} \times 2.41 \times 10^{-3}$

Combine decimals and powers of 10:

$8.87 \times 3.78 \times 5.7 \times 2.41 \times 10^{-1} \times 10^{2} \times 10^{1} \times 10^{-3}$

Round the decimal numbers to one significant digit and multiply them. Add the exponents of the powers of 10.

$$8.87 \times 3.78 \times 5.7 \times 2.41 \times 10^{-1+2+1+(-3)}$$
$$= 9 \times 4 \times 6 \times 2 \times 10^{-1} = 432 \times 10^{-1}$$
$$= 43.2$$

Place the decimal point in 461 so that the number is "nearest" 43.2, which is 46.1. Therefore, $0.887 \times 378 \times 57 \times 0.00241 = 46.1$.

PROBLEMS

1. $3.16 \times 9.65 \times 1.50$
2. $10.1 \times 73.1 \times 0.361$
3. $2.47 \times 66.7 \times 587$
4. $2080 \times 0.0123 \times 6080 \times 5$
5. $572 \times 86.1 \times 2930 \times 25$
6. $0.363 \times 2330 \times 0.0637 \times 2.5$
7. $0.031 \times 0.061 \times 0.044 \times 0.030$
8. $83.4 \times 0.440 \times 577 \times 0.0945 \times 0.00515 \times 86.9$

To solve a problem which has both multiplications and divisions: multiply all the numbers in the numerator; divide that result by the first number in the denominator; then divide that result by the second number in the denominator; and so on until you have divided by all the numbers in the denominator. (On the slide rule it is *not advisable* to multiply all the numbers in the numerator, multiply all the numbers in the denominator, and then divide the two results. This method leads to errors and takes more time.)

EXAMPLE 2: $\dfrac{14.1 \times 0.0595 \times 8.15}{0.222 \times 84.9}$.

(a) Set L_i of the C scale opposite 141 on the D scale. Move the cursor until the hairline is over 595 on the C scale. Set R_i of the C scale under the hairline; then move the cursor until the hairline is over 815 on C scale. Set R_i of CI scale under the hairline; then move the cursor until the hairline is over 222 on the CI scale. Set L_i of CI scale under the hairline, then move the cursor until the hairline is over 849 on the CI scale. Read the digits of the result on the D scale, 363.

(b) Locating the decimal point.

METHOD 1: *Approximation*: Round the numbers to one significant digit and multiply and divide.

$$\frac{10 \times 0.06 \times 8}{0.2 \times 80} = \frac{4.8}{16} = 0.3$$

Place the decimal point in 363 so that the number is "nearest" 0.3, which is 0.363. Therefore,

$$\frac{14.1 \times 0.0595 \times 8.15}{0.222 \times 84.9} = 0.363$$

METHOD 2: Write the numbers in scientific notation.

$$\frac{1.41 \times 10^{1} \times 5.95 \times 10^{-2} \times 8.15 \times 10^{0}}{2.22 \times 10^{-1} \times 8.49 \times 10^{1}}$$

Combine decimals and powers of 10

$$\frac{1.41 \times 5.95 \times 8.15 \times 10^{1} \times 10^{-2} \times 10^{0}}{2.22 \times 8.49 \times 10^{-1} \times 10^{1}}$$

Round the decimal numbers to one significant digit.

$$\frac{1 \times 6 \times 8 \times 10^{1} \times 10^{-2} \times 10^{0}}{2 \times 8 \times 10^{-1} \times 10^{1}}$$

In the numerator and then in the denominator: multiply the numbers and add the exponents of the powers of 10.

$$\frac{48 \times 10^{1+(-2)+0}}{16 \times 10^{(-1)+1}} = \frac{48 \times 10^{-1}}{16 \times 10^{0}}$$

Divide the numerator by the denominator and subtract the powers of 10.

$$16\overline{)48}^{\,3.} \qquad \frac{48 \times 10^{-1}}{16 \times 10^{0}} = 3 \times 10^{-1-0} = 3 \times 10^{-1} = 0.3$$

Place the decimal point in 363 so that the number is "nearest" 0.3, which is 0.363. Therefore,

$$\frac{14.1 \times 0.0595 \times 8.15}{0.222 \times 84.9} = 0.363$$

PROBLEMS

Perform the indicated operations with a slide rule.

1. $\dfrac{2.02 \times 39.3}{97}$

2. $\dfrac{45.8 \times 38.4}{78.1}$

3. $\dfrac{726 \times 513}{12}$

4. $\dfrac{0.0342 \times 33.2}{0.0466}$

5. $\dfrac{21.1 \times 81.1}{35.8 \times 9.65}$

6. $\dfrac{8.74 \times 53}{2.11 \times 72.6}$

7. $\dfrac{1560 \times 345}{23.9 \times 262}$

8. $\dfrac{1860 \times 65400}{703 \times 0.0524}$

9. $\dfrac{341}{8.31 \times 2.42}$

10. $\dfrac{39.1}{3.66 \times 5.3}$

11. $\dfrac{80.5}{3.81 \times 0.0735}$

12. $\dfrac{56100}{8710 \times 0.05}$

13. $\dfrac{751.5 \times 2860 \times 612}{926}$

14. $\dfrac{853 \times 0.324 \times 0.0259}{0.0679}$

15. $\dfrac{47900}{835 \times 47.7 \times 1.74}$

16. $\dfrac{347}{0.0855 \times 0.0466 \times 1360}$

17. $\dfrac{42.4 \times 33.1 \times 3470}{650 \times 0.315}$

18. $\dfrac{971 \times 0.00348 \times 0.543}{87 \times 0.227}$

19. $\dfrac{39 \times 56}{67 \times 12 \times 82}$

20. $\dfrac{5.39 \times 40.5}{0.478 \times 78.4 \times 0.0084}$

A-8 *SQUARING*

To find squares and square roots of numbers on the slide rule we use the A and D scales. Notice that the A scale looks like two half size D scales laid end to end. Also notice that there is a "1" at each end and in the middle of the A scale. The middle "1" divides the A scale into the left half and the right half. When you read the A scale, read the division marks carefully.

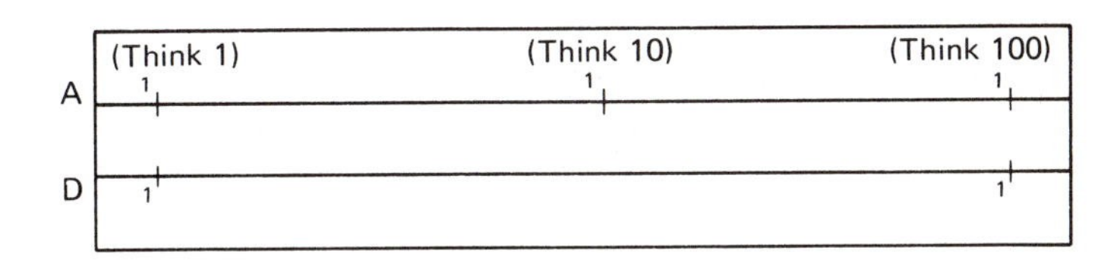

Let's learn to square numbers first. Write the number in scientific notation in the form $M \times 10^n$, where n is an integer and M is between 1 and 10. Then, $(M \times 10^n)^2 = M^2 \times 10^{2n}$, where M^2 is between 1 and 100.

On your slide rule, set the hairline on the number M on the D scale and read M^2 under the hairline on the A scale. If the hairline is on the left half of the A scale, M^2 is between 1 and 10, and if the hairline is on the right half of the A scale, M^2 is between 10 and 100. Then multiply the exponent of the power of 10 by 2 and multiply $M^2 \times 10^{2n}$.

EXAMPLE 1: $(46.4)^2 = (4.64 \times 10^1)^2 = 21.5 \times 10^2 = 2150.$

Solution. Set the hairline on 464 on the D scale, and on the A scale read the digits of the square of 464 which are 215. The hairline is on the *right* half of the A scale; therefore, M^2 is between 10 and 100. Hence, $M^2 = 21.5$. Multiply the exponent of the power of 10 by 2 and multiply $21.5 \times 10^2 = 2150$.

EXAMPLE 2: $1370^2 = (1.37 \times 10^3)^2 = 1.88 \times 10^6 = 1{,}880{,}000.$

Solution. Set the hairline on 137 on the D scale, and on the A scale read the digits of the square of 137 which are 188. The hairline is on the *left* half of the A scale; therefore, M^2 is between 1 and 10. Hence, $M^2 = 1.88$. Multiply the exponent of the power of 10 by 2 and multiply $1.88 \times 10^6 = 1{,}880{,}000$.

EXAMPLE 3: $0.0578^2 = (5.78 \times 10^{-2})^2 = 33.4 \times 10^{-4} = 0.00334.$

Solution. Set the hairline on 578 on the D scale, and on the A scale read the digits of the square of 578 which are 334. The hairline is on the *right* half of the A scale; therefore, M^2 is between 10 and 100. Hence, $M^2 = 33.4$. Multiply the exponent of the power of 10 by 2 and multiply $33.4 \times 10^{-4} = 0.00334$.

EXAMPLE 4: $0.119^2 = (1.19 \times 10^{-1})^2 = 1.42 \times 10^{-2} = 0.0142.$

Solution. Set the hairline on 119 on the D scale, and on the A scale read the digits of the square of 119 which are 142. The hairline is on the *left* half of A scale; therefore, M^2 is between 1 and 10. Hence, $M^2 = 1.42$. Multiply the exponent of the power of 10 by 2 and multiply $1.42 \times 10^{-2} = 0.0142$.

PROBLEMS

Use a slide rule to square.

1. $(1.93)^2$	2. $(2.61)^2$	3. $(0.512)^2$	4. $(0.719)^2$
5. $(83.1)^2$	6. $(96.5)^2$	7. $(0.061)^2$	8. $(0.0132)^2$
9. $(23.2)^2$	10. $(12.3)^2$	11. $(145)^2$	12. $(203)^2$
13. $(0.566)^2$	14. $(0.00771)^2$	15. $(17.7)^2$	16. $(44.2)^2$
17. $(0.706)^2$	18. $(0.102)^2$	19. $(5{,}280)^2$	20. $(16{,}273)^2$

A-9 *SQUARE ROOT*

Finding the square root of a number is the inverse of squaring a number; therefore, the operation of the slide rule is just the reverse of squaring.

Write the number in the form $R \times 10^n$, where n is an *even* integer and R is between 1 and 100. Then

$$\sqrt{R \times 10^n} = \sqrt{R} \times 10^{n/2}, \qquad \sqrt{R} \text{ is between 1 and 10}$$

If R is between 1 and 10, set the hairline on the *left* half of the A scale and read the square root of R on the D scale. Then divide the exponent of the power of 10 by 2 and multiply $\sqrt{R}$ and $10^{n/2}$.

If R is between 10 and 100, set the hairline on the *right* half of the A scale and read the square root of R on the D scale. Then divide the exponent of the power of 10 by 2 and multiply $\sqrt{R}$ and $10^{n/2}$.

Note: $\sqrt{R}$ is *always* between 1 and 10, since R is between 1 and 100.

EXAMPLE 1: $\sqrt{513} = \sqrt{5.13 \times 10^2} = \sqrt{5.13} \times 10^1 = 2.26 \times 10 = 22.6.$

Solution. Because R is between 1 and 10, set the hairline over 513 on the *left* half of the A scale. Under the hairline on the D scale, read the digits of the square root of 513, which are 226. Hence, $\sqrt{R} = 2.26$. Then divide the exponent of the power of 10 by 2 and multiply $2.26 \times 10 = 22.6$.

EXAMPLE 2: $\sqrt{3670} = \sqrt{36.7 \times 10^2} = 6.06 \times 10^1 = 60.6.$

Solution. Because R is between 10 and 100, set the hairline over 367 on the *right* half of the A scale. Under the hairline on the D scale, read the digits of the square root of 367, which are 606. Hence, $\sqrt{R} = 6.06$. Then divide the exponent of the power of 10 by 2 and multiply $6.06 \times 10 = 60.6$.

EXAMPLE 3: $\sqrt{833{,}000} = \sqrt{83.3 \times 10^4} = 9.13 \times 10^2 = 913.$

Solution. Because R is between 10 and 100, set the hairline over 833 on the *right* side of the A scale. Under the hairline on the D scale, read the digits of the square root of 833, which are 913. Hence, $\sqrt{R} = 9.13$ .Then divide the exponent of the power of 10 by 2 and multiply $9.13 \times 10^2 = 913$.

EXAMPLE 4: $\sqrt{0.0415} = \sqrt{4.15 \times 10^{-2}} = 2.04 \times 10^{-1} = 0.204.$

Solution. Because R is between 1 and 10, set the hairline over 415 on the *left* half of the A scale. Under the hairline on the D scale, read the digits of the square root of 415, which are 204. Hence, $\sqrt{R} = 2.04$. Then divide the exponent of the power of 10 by 2 and multiply $2.04 \times 10^{-1} = 0.204$.

EXAMPLE 5: $\sqrt{0.00717} = \sqrt{71.7 \times 10^{-4}} = 8.47 \times 10^{-2} = 0.0847.$

Solution. Because R is between 10 and 100, set the hairline over 717 on the *right* half of the A scale. Under the hairline on the D scale, read the digits of the square root of 717, which are 847. Hence, $\sqrt{R} = 8.47$. Then divide the exponent of the power of 10 by 2 and multiply $8.47 \times 10^{-2} = 0.0847$.

EXAMPLE 6: $\sqrt{0.00000621} = \sqrt{6.21 \times 10^{-6}} = 2.49 \times 10^{-3} = 0.00249.$

Solution. Because R is between 1 and 10, set the hairline over 621 on the *left* of the A scale. Under the hairline on the D scale, read the digits of the square root of 621, which are 249. Hence, $\sqrt{R} = 2.49$. Then divide the exponent of the power of 10 by 2 and multiply $2.49 \times 10^{-3} = 0.00249$.

PROBLEMS

Use a slide rule to find the square root.

1. $\sqrt{272}$
2. $\sqrt{507}$
3. $\sqrt{0.0714}$
4. $\sqrt{0.0000572}$
5. $\sqrt{5720}$
6. $\sqrt{6350}$
7. $\sqrt{0.0921}$
8. $\sqrt{0.000372}$
9. $\sqrt{15.9}$
10. $\sqrt{69}$
11. $\sqrt{157{,}600}$
12. $\sqrt{47{,}200}$
13. $\sqrt{0.000612}$
14. $\sqrt{0.0715}$
15. $\sqrt{0.41}$
16. $\sqrt{0.924}$
17. $\sqrt{7.4}$
18. $\sqrt{1.2}$
19. $\sqrt{1600}$
20. $\sqrt{625}$
21. $\sqrt{0.472}$
22. $\sqrt{63.5}$
23. $\sqrt{870}$
24. $\sqrt{0.919}$
25. $\sqrt{0.000000666}$

PROBLEMS

Use a slide rule to perform the indicated operations.

1. $(0.173)^2$
2. $(0.0625)^2$
3. $\sqrt{84{,}000}$
4. $\sqrt{827}$
5. $(0.161)^2$
6. $\sqrt{8320}$
7. $\sqrt{39.2}$
8. $(81.2)^2$
9. $(172)^2$
10. $\sqrt{0.0617}$
11. $\sqrt{0.478}$
12. $\sqrt{530{,}000}$
13. $(1024)^2$
14. $(5.71)^2$
15. 8.2^2
16. $\sqrt{0.000137}$
17. $\sqrt{0.00170}$
18. $\sqrt{863}$
19. 82.5^2
20. 107^2

A-10 *CUBING*

To find cubes and cube roots of numbers on the slide rule, we use the D and K scales. Notice that the K scale looks like three one third size D scales laid end to end. Also notice that there are four 1's on the K scale which divide it into thirds—left third, middle third, and right third. When reading the K scale, be sure to read the division marks carefully.

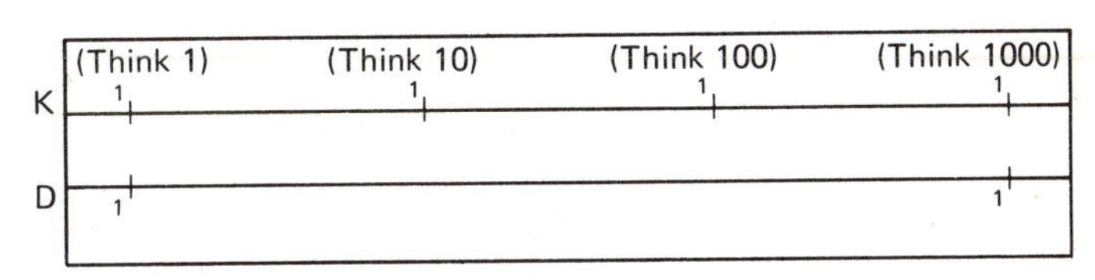

Let's learn to cube numbers first. Write the number in scientific notation in the form $M \times 10^n$, where n is an integer and M is between 1 and 10. Then, $(M \times 10^n)^3 = M^3 \times 10^{3n}$, where M^3 is between 1 and 1000.

On your slide rule, set the hairline on the number M on the D scale, and read M^3 under the hairline on the K scale. Then:

(a) if the hairline is on the *left* third of the K scale, M^3 is between 1 and 10;
(b) if the hairline is on the *middle* third of the K scale, M^3 is between 10 and 100; and
(c) if the hairline is on the *right* third of the K scale, M^3 is between 100 and 1000.

Then multiply the exponent of the power of 10 by 3 and multiply $M^3 \times 10^{3n}$.

EXAMPLE 1: $30.4^3 = (3.04 \times 10^1)^3 = 28.1 \times 10^3 = 28{,}100$.

Solution. Set the hairline on 304 on the D scale, and on the K scale read the digits of the cube of 304, which are 281. The hairline is on the *middle* third of the K scale; therefore, M^3 is between 10 and 100. Hence, $M^3 = 28.1$. Multiply the exponent of the power of 10 by 3 and multiply $28.1 \times 10^3 = 28{,}100$.

EXAMPLE 2: $153^3 = (1.53 \times 10^2)^3 = 3.58 \times 10^6 = 3{,}580{,}000$.

Solution. Set the hairline on 153 on the D scale, and on the K scale read the digits of the cube of 153, which are 358. The hairline is on the *left* third of the K scale; therefore, M^3 is between 1 and 10. Hence, $M^3 = 3.58$. Multiply the exponent of the power of 10 by 3 and multiply $3.58 \times 10^6 = 3{,}580{,}000$.

EXAMPLE 3: $7.24^3 = (7.24 \times 10^0)^3 = 380 \times 10^0 = 380.$

Solution. Set the hairline on 724 on the D scale, and on the K scale read the digits of the cube of 724, which are 380. The hairline is on the *right* third of the K scale; therefore, M^3 is between 100 and 1000. Hence, $M^3 = 380$. Multiply the exponent of the power of 10 by 3 and multiply $380 \times 10^0 = 380$.

EXAMPLE 4: $0.278^3 = (2.78 \times 10^{-1})^3 = 21.5 \times 10^{-3} = 0.0215.$

Solution. Set the hairline on 278 on the D scale, and on the K scale read the digits of the cube of 278, which are 215. The hairline is on the *middle* third of the K scale; therefore, M^3 is between 10 and 100. Hence, $M^3 = 21.5$. Multiply the exponent of the power of 10 by 3 and multiply $21.5 \times 10^{-3} = 0.0215$.

EXAMPLE 5: $0.0655^3 = (6.55 \times 10^{-2})^3 = 281 \times 10^{-6} = 0.000281.$

Solution. Set the hairline on 655 on the D scale, and on the K scale read the digits of the cube of 655, which are 281. The hairline is on the *right* third of the K scale; therefore, M^3 is between 100 and 1000. Hence, $M^3 = 281$. Multiply the exponent of the power of 10 by 3 and multiply $281 \times 10^{-6} = 0.000281$.

PROBLEMS

Use the slide rule to cube.

1. $(6.35)^3$
2. $(17.5)^3$
3. $(73.1)^3$
4. $(2.53)^3$
5. $(0.172)^3$
6. $(52.7)^3$
7. $(63.9)^3$
8. $(430)^3$
9. $(1.17)^3$
10. $(4.25)^3$

A-11 *CUBE ROOT*

Finding the cube root of a number is the inverse of cubing a number; therefore, the operation of the slide rule for finding cube roots is just the reverse of cubing.

Write the number in the form

$$S \times 10^n$$

where n is an integer divisible by 3 and S is between 1 and 1000. Then,

$$\sqrt[3]{S \times 10^n} = \sqrt[3]{S} \times 10^{n/3}, \qquad \sqrt[3]{S} \text{ is always between 1 and 10}$$

(a) If S is between 1 and 10, set the hairline on the *left* third of the K scale and read the cube root of S on the D scale. Then divide the exponent of the power of 10 by 3 and multiply

$$\sqrt[3]{S} \quad \text{and} \quad 10^{n/3}.$$

(b) If S is between 10 and 100, set the hairline on the *middle* third of the K scale and read the cube root of S on the D scale. Then divide the exponent of the power of 10 by 3 and multiply

$$\sqrt[3]{S} \quad \text{and} \quad 10^{n/3}.$$

(c) If S is between 100 and 1000, set the hairline on the *right* third of the K scale and read the cube root of S on the D scale. Then divide the exponent of the power of 10 by 3 and multiply

$$\sqrt[3]{S} \quad \text{and} \quad 10^{n/3}.$$

Note: $\sqrt[3]{S}$ is *always* between 1 and 10, since S is between 1 and 1000.

EXAMPLE 1: $\sqrt[3]{5630} = \sqrt[3]{5.63 \times 10^3} = 1.78 \times 10^1 = 17.8.$

Solution. Because S is between 1 and 10, set the hairline over 563 on the *left* third of the K scale. Under the hairline on the D scale, read the digits of the cube root of 563, which are 178. Hence, $\sqrt[3]{S} = 1.78$. Then divide the exponent of the power of 10 by 3 and multiply $1.78 \times 10^1 = 17.8$.

EXAMPLE 2: $\sqrt[3]{15{,}800} = \sqrt[3]{15.8 \times 10^3} = 2.51 \times 10^1 = 25.1.$

Solution. Because S is between 10 and 100, set the hairline over 158 on the *middle* third of the K scale. Under the hairline on the D scale, read the digits of the cube root of 158, which are 251. Hence, $\sqrt[3]{S} = 2.51$. Then divide the exponent of the power of 10 by 3 and multiply $2.51 \times 10^1 = 25.1$.

EXAMPLE 3: $\sqrt[3]{832{,}000{,}000} = \sqrt[3]{832 \times 10^6} = 9.41 \times 10^2 = 941.$

Solution. Because S is between 100 and 1000, set the hairline over 832 on the *right* third of the K scale. Under the hairline on the D scale, read the digits of the cube root of 832, which are 941. Hence, $\sqrt[3]{S} = 9.41$. Then divide the exponent of the power of 10 by 3 and multiply $9.41 \times 10^2 = 941$.

EXAMPLE 4: $\sqrt[3]{0.0172} = \sqrt[3]{17.2 \times 10^{-3}} = 2.58 \times 10^{-1} = 0.258.$

Solution. Because S is between 10 and 100, set the hairline over 172 on the *middle* third of the K scale. Under the hairline on the D scale, read the digits of the cube root of 172, which are 258. Hence, $\sqrt[3]{S} = 2.58$. Then divide the exponent of the power of 10 by 3 and multiply $2.58 = 10^{-1} = 0.258$.

EXAMPLE 5: $\sqrt[3]{0.00322} = \sqrt[3]{3.22 \times 10^{-3}} = 1.48 \times 10^{-1} = 0.148$.

Solution. Because S is between 1 and 10, set the hairline over 322 on the *left* third of the K scale. Under the hairline on the D scale, read the digits of the cube root of 322, which are 148. Hence, $\sqrt[3]{S} = 1.48$. Then divide the exponent of the power of 10 by 3 and multiply $1.48 \times 10^{-1} = 0.148$.

EXAMPLE 6: $\sqrt[3]{0.00052} = \sqrt[3]{520 \times 10^{-6}} = 8.04 \times 10^{-2} = 0.0804$.

Solution. Because S is between 100 and 1000, set the hairline over 520 on the *right* third of the K scale. Under the hairline on the D scale, read the digits of the cube root of 520, which are 804. Hence, $\sqrt[3]{S} = 8.04$. Then divide the exponent of the power of 10 by 3 and multiply $8.04 \times 10^{-2} = 0.0804$.

PROBLEMS

Use a slide rule to find cube root.

1. $\sqrt[3]{6.51}$	2. $\sqrt[3]{41.7}$	3. $\sqrt[3]{420}$	4. $\sqrt[3]{2570}$
5. $\sqrt[3]{0.614}$	6. $\sqrt[3]{0.0129}$	7. $\sqrt[3]{0.00712}$	8. $\sqrt[3]{0.000426}$
9. $\sqrt[3]{0.527}$	10. $\sqrt[3]{827,000}$		

PROBLEMS

Use a slide rule to perform the indicated operations.

1. $\sqrt[3]{268,000}$	2. $(2.72)^3$	3. $(34.4)^3$	4. $\sqrt[3]{61,700}$
5. $\sqrt[3]{573}$	6. $(0.225)^3$	7. $(0.037)^3$	8. $\sqrt[3]{1620}$
9. $\sqrt[3]{84,800}$	10. $(3.21)^3$	11. $(11.425)^3$	12. $\sqrt[3]{1.03}$
13. $\sqrt[3]{327}$	14. $(0.966)^3$	15. $\sqrt[3]{88.4}$	16. $(29.9)^3$
17. $\sqrt[3]{629,000}$	18. $\sqrt[3]{59}$	19. $(0.071)^3$	20. $(0.00652)^3$

A-12 *COMBINED OPERATIONS WITH POWERS AND ROOTS*

To evaluate a combined operation with a root as a factor:

1. Find the root using the methods in the previous sections.
2. Replace the root with its decimal equivalent in the original problem.
3. Evaluate the resulting problem using the combined operations methods in Section 7.

EXAMPLE 1: $\dfrac{43\sqrt{269}}{24.2}$

1. $\sqrt{269} = 16.4$
2. $\dfrac{43 \times 16.4}{24.2} = 29.1$

EXAMPLE 2: $\dfrac{36.8 \times \pi}{\sqrt[3]{44.3}}$.

1. $\sqrt[3]{44.3} = 3.54$
2. $\dfrac{36.8 \times \pi}{3.54} = 32.7$

To evaluate a combined operation with a power as a factor, use the number being raised to a power as a factor

(a) twice if being squared or
(b) three times if being cubed

and follow the combined operations methods in Section 7.

EXAMPLE 3: $\dfrac{48.1 \times 69.6^2}{30.8}$.

Evaluate $\dfrac{48.1 \times 69.6 \times 69.6}{30.8} = 7570$

EXAMPLE 4: $\dfrac{75.2}{4.16^3}$.

Evaluate $\dfrac{75.2}{4.16 \times 4.16 \times 4.16} = 1.04$

PROBLEMS

Use a slide rule to perform the indicated operations.

1. $\dfrac{30.1^2}{37.6}$
2. $\dfrac{\sqrt{46.4}}{67.1}$
3. $32.8\sqrt[3]{33.4}$
4. $42.4(71.1)^3$
5. $\dfrac{83 \times 41}{\sqrt{23}}$
6. $\dfrac{84.7 \times 69.9}{\sqrt[3]{699}}$
7. $\dfrac{84.3 \times 61}{\sqrt{0.00672}}$
8. $\dfrac{19 \times 29^3}{2810}$
9. $(\sqrt{59.1})(\sqrt[3]{26.2})$
10. $\dfrac{77.6^2 \times 93.6}{\sqrt{68}}$
11. $\dfrac{29 \times \sqrt{15}}{\sqrt[3]{82}}$
12. $\sqrt{(16)(91)}$
13. $\sqrt{(29.4)(45.9)}$
14. $\sqrt[3]{\dfrac{43.4}{29.2}}$
15. $\dfrac{6.1^2 \times 1.8^3}{\sqrt{5100}}$

PROBLEMS

(Review) Use a slide rule to perform the indicated operations.

1. 3.72×6.19
2. $52.7 \div 9.71$
3. 27.2^2
4. 6.17^3
5. $\sqrt{9.17}$
6. $\sqrt[3]{16.4}$
7. $\dfrac{15.4 \times 6.17}{9.4}$
8. $\dfrac{5\sqrt{7}}{3}$
9. $\dfrac{4\pi}{9}$
10. 13.7×16.4
11. $\dfrac{0.037}{0.0054}$
12. $\dfrac{57 \times 67}{\sqrt[3]{17}}$
13. 115^2
14. 0.374^3
15. $\dfrac{2\pi}{3}$
16. $\dfrac{2\sqrt{3}}{5}$
17. $\dfrac{15.7 \times 6.3 \times 971}{48.1 \times 0.763}$
18. 0.00173×580
19. $\dfrac{0.00711}{0.0619}$
20. $\sqrt{0.00535}$
21. $\sqrt[3]{0.0974}$
22. $\dfrac{1}{9.41 \times 6.71}$
23. $\dfrac{2}{3\pi}$
24. $\dfrac{5}{6\sqrt{3}}$
25. $\dfrac{19.4 \times 6.71}{0.714}$
26. $\dfrac{18.9 \times 543}{0.713}$
27. $\dfrac{9.11 \times 7.65^2}{3.1^3}$
28. $\dfrac{0.92 \times 0.73}{0.16 \times 0.37}$
29. $\dfrac{8 \times 67}{34}$
30. $\dfrac{18 \times 63}{47 \times 93}$
31. $\dfrac{3\sqrt{5}}{4}$
32. $\dfrac{7\pi}{3}$
33. $\dfrac{3\sqrt{6}}{2\pi}$
34. $\sqrt[3]{17{,}400}$
35. $\sqrt{9680}$
36. 3740×6800
37. $\dfrac{19{,}740}{2{,}643}$
38. $\sqrt[3]{0.0000172}$
39. $\sqrt{10.4}$
40. $\sqrt{(19.7)(1.7)}$
41. $\dfrac{149 \times 681}{7.92}$
42. $\dfrac{811 \times 1.77}{6.71}$
43. $\dfrac{3\pi}{4}$
44. $\dfrac{6\pi}{11}$
45. $\dfrac{9\sqrt{10}}{4}$
46. $4\pi(1.3)^2$
47. $2\pi(31)^2$
48. $\dfrac{3\sqrt{16.7}}{8.21}$
49. $\sqrt{\dfrac{6.17 \times 8.1}{19.4}}$
50. $\sqrt{\dfrac{18.4}{0.671 \times 0.811}}$

TABLES

TABLE A
ENGLISH WEIGHTS AND MEASURES

UNITS OF LENGTH

Standard unit—inch (in. or ″)
12 inches = 1 foot (ft or ′)
3 feet = 1 yard (yd)
$5\frac{1}{2}$ yards or $16\frac{1}{2}$ feet = 1 rod (rd)
5280 feet = 1 mile (mi)

UNITS OF WEIGHT

Standard unit—pound (lb)
16 ounces (oz) = 1 pound
2000 pounds = 1 ton (T)

VOLUME MEASURE

Liquid

16 ounces (fl oz) = 1 pint (pt)
2 pints = 1 quart (qt)
4 quarts = 1 gallon (gal)

Dry

2 pints (pt) = 1 quart (qt)
8 quarts = 1 peck (pk)
4 pecks = 1 bushel (bu)

*To facilitate use of these tables with slide rule, all data have been rounded to three significant digits.

TABLE B
METRIC MASS AND MEASURES

UNITS OF LENGTH

Standard unit—meter (m)

1 micrometer (μm)	=	0.000001 m	=	10^{-6} m
1 millimeter (mm)	=	0.001 m	=	10^{-3} m
1 centimeter (cm)	=	0.01 m	=	10^{-2} m
1 decimeter (dm)	=	0.1 m	=	10^{-1} m
1 dekameter (dam)	=	10 m	=	10 m
1 hectometer (hm)	=	100 m	=	10^{2} m
1 kilometer (km)	=	1,000 m	=	10^{3} m
1 megameter (Mm)	=	1,000,000 m	=	10^{6} m

UNITS OF MASS

Standard unit—gram (g)

1 microgram (μg)	=	0.000001 g	=	10^{-6} g
1 milligram (mg)	=	0.001 g	=	10^{-3} g
1 centigram (cg)	=	0.01 g	=	10^{-2} g
1 decigram (dg)	=	0.1 g	=	10^{-1} g
1 dekagram (dag)	=	10 g	=	10 g
1 hectogram (hg)	=	100 g	=	10^{2} g
1 kilogram (kg)	=	1,000 g	=	10^{3} g
1 megagram (Mg)	=	1,000,000 g	=	10^{6} g

UNITS OF CAPACITY

Standard unit—liter (1)

1 microliter (μl)	=	0.000001 l	=	10^{-6} liter
1 milliliter (ml)	=	0.001 l	=	10^{-3} liter
1 centiliter (cl)	=	0.01 l	=	10^{-2} liter
1 deciliter (dl)	=	0.1 l	=	10^{-1} liter
1 dekaliter (dal)	=	10 l	=	10 liters
1 hectoliter (hl)	=	100 l	=	10^{2} liters
1 kiloliter (kl)	=	1,000 l	=	10^{3} liters
1 megaliter (Ml)	=	1,000,000 l	=	10^{6} liters

CONVERSION FACTORS

Conversion factors may be made directly from the tables. For example, 1 cm = 0.394 in. For further instructions on the use of this table, see page 3 of the text.

TABLE C
CONVERSION TABLE FOR LENGTH

	cm	METER	km	in.	ft	mile
1 centimeter =	1	10^{-2}	10^{-5}	0.394	3.28×10^{-2}	6.21×10^{-6}
1 METER =	100	1	10^{-3}	39.4	3.28	6.21×10^{-4}
1 kilometer =	10^{5}	1000	1	3.94×10^{4}	3280	0.621
1 inch =	2.54	2.54×10^{-2}	2.54×10^{-5}	1	8.33×10^{-2}	1.58×10^{-5}
1 foot =	30.5	0.305	3.05×10^{-4}	12	1	1.89×10^{-4}
1 mile =	1.61×10^{5}	1610	1.61	6.34×10^{4}	5280	1

TABLE D
CONVERSION TABLE FOR AREA

English

1 sq ft = 144 sq in.
1 sq yd = 9 sq ft
1 sq rd = 30.25 sq yd
1 acre = 160 sq rd
= 4840 sq yd
= 43,560 sq ft
1 sq mi = 640 acres

Metric

1 sq m = 10,000 sq cm
= 1,000,000 sq mm
1 sq cm = 100 sq mm
= 0.0001 sq m
1 sq km = 1,000,000 sq m

	METER^2	cm^2	ft^2	in^2
1 SQUARE METER =	1	10^4	10.8	1550
1 square centimeter =	10^{-4}	1	1.08×10^{-3}	0.155
1 square foot =	9.29×10^{-2}	929	1	144
1 square inch =	6.45×10^{-4}	6.45	6.94×10^{-3}	1

1 square mile = 2.79×10^8 ft^2 = 640 acres; 1 acre = 43,560 ft^2;
1 circular mil = 5.07×10^{-6} cm^2 = 7.85×10^{-7} in^2

TABLE E
CONVERSION TABLE FOR VOLUME

English

1 cu ft = 1728 cu in.
1 cu yd = 27 cu ft

Metric

1 cu m = 10^6 cu cm
1 cu cm = 10^{-6} cu m
= 10^3 cu mm

	METER^3	cm^3	liter	ft^3	in^3
1 CU M =	1	10^6	1000	35.3	6.10×10^4
1 cu cm =	10^{-6}	1	1.00×10^{-3}	3.53×10^{-5}	6.10×10^{-2}
1 liter =	1.00×10^{-3}	1000	1	3.53×10^{-2}	61.0
1 cu ft =	2.83×10^{-2}	2.83×10^4	28.3	1	1730
1 cu in. =	1.64×10^{-5}	16.4	1.64×10^{-2}	5.79×10^{-4}	1

1 U. S. fluid gallon = 4 U. S. fluid quarts = 8 U. S. pints = 128 U. S. fluid ounces = 231 in^3 = 0.134 ft^3.

1 liter = 1000 cm^3 = 1.06 qt

EXAMPLE: 1 cu m = 35.3 ft^3.

TABLE F
CONVERSION TABLE FOR MASS

Quantities in the shaded areas are not mass units. When we write, for example, 1 kg "=" 2.21 lb, this means that a kilogram is a mass that weighs 2.21 pounds under standard conditions of gravity ($g = 9.81\ m/s^2 = 32.2\ ft/s^2$).

	g	KG	slug	oz	lb	ton
1 gram =	1	0.001	6.85×10^{-5}	3.53×10^{-2}	2.21×10^{-3}	1.10×10^{-6}
1 KILOGRAM =	1000	1	6.85×10^{-2}	35.3	2.21	1.10×10^{-3}
1 slug =	1.46×10^{4}	14.6	1	515	32.2	1.61×10^{-2}
1 ounce =	28.4	2.84×10^{-2}	1.94×10^{-3}	1	6.25×10^{-2}	3.13×10^{-5}
1 pound =	454	0.454	3.11×10^{-2}	16	1	0.0005
1 ton =	9.07×10^{5}	907	62.2	3.2×10^{4}	2000	1

EXAMPLE: 1 gram = 6.85×10^{-5} slug.

TABLE G
CONVERSION TABLE FOR DENSITY

Quantities in the shaded areas are weight densities and, as such, are dimensionally different from mass densities. See note for mass table.

	slug/ft³	KG/ METER³	g/cm³	lb/ft³	lb/in³
1 slug per ft³ =	1	515.4	0.515	32.2	1.86 $\times 10^{-2}$
1 KILOGRAM per M³ =	1.94 $\times 10^{-3}$	1	0.001	6.24 $\times 10^{-2}$	3.61 $\times 10^{-5}$
1 gram per cm³ =	1.94	1000	1	62.4	3.61 $\times 10^{-2}$
1 pound per ft³ =	3.11 $\times 10^{-2}$	16.0	1.60 $\times 10^{-2}$	1	5.79 $\times 10^{-4}$
1 pound per in³ =	53.7	2.77 $\times 10^{4}$	27.7	1730	1

TABLE H
CONVERSION TABLE FOR TIME

	yr	day	hr	min	S
1 year =	1	365	8.77×10^{3}	5.26×10^{5}	3.16×10^{7}
1 day =	2.74×10^{-3}	1	24	1440	8.64×10^{4}
1 hour =	1.14×10^{-4}	4.17×10^{-2}	1	60	3600
1 minute =	1.90×10^{-6}	6.94×10^{-4}	1.67×10^{-2}	1	60
1 SECOND =	3.17×10^{-8}	1.16×10^{-5}	2.78×10^{-4}	1.67×10^{-2}	1

EXAMPLE: 1 year = 3.16×10^{7} s.

TABLE I
CONVERSION TABLE FOR SPEED

	ft/s	km/hr	METER/S	mi/hr	cm/s	knot
1 foot per second =	1	1.10	0.305	0.682	30.5	0.593
1 kilometer per hour =	0.911	1	0.278	0.621	27.8	0.540
1 METER per SECOND =	3.28	3.6	1	2.24	100	1.94
1 mile per hour =	1.47	1.61	0.447	1	44.7	0.869
1 centimeter per second =	3.28×10^{-2}	3.6×10^{-2}	0.01	2.24×10^{-2}	1	1.94×10^{-2}
1 knot =	1.69	1.85	0.514	1.15	51.4	1
1 knot = 1 naut mi/hr	1 mi/min = 88.0 ft/s = 60.0 mi/hr					

EXAMPLE: 1 ft/s = 1.10 km/hr.

TABLE J
CONVERSION TABLE FOR FORCE

	N	lb
1 NEWTON =	1	0.225
1 pound =	4.45	1

TABLE K
CONVERSION TABLE FOR PRESSURE

	atm	inch of water	cm-Hg	N/METER^2	lb/in^2	lb/ft^2
1 atmosphere =	1	407	76	1.01×10^5	14.7	2120
1 inch of water* at 4°C =	2.46×10^{-3}	1	0.187	249	3.61×10^{-2}	5.20
1 centimeter of mercury* at 0°C =	1.32×10^{-2}	5.35	1	1330	0.193	27.9
1 NEWTON per METER^2 =	9.87×10^{-6}	4.02×10^{-3}	7.50×10^{-4}	1	1.45×10^{-4}	2.09×10^{-2}
1 pound per in^2 =	6.81×10^{-2}	27.7	5.17	6.90×10^3	1	144
1 pound per ft^2 =	4.73×10^{-4}	0.192	3.59×10^{-2}	47.9	6.94×10^{-3}	1

*Where the acceleration of gravity has the standard value, $9.81 \text{ m/s}^2 = 32.2 \text{ ft/s}^2$.

TABLE L

CONVERSION TABLE FOR ENERGY, WORK, HEAT

	Btu	ft lb	hp-hr	JOULE	cal	kW-hr
1 British thermal unit =	1	778	3.93×10^{-4}	1060	252	2.93×10^{-4}
1 foot pound =	1.29×10^{-3}	1	5.05×10^{-7}	1.36	0.324	3.77×10^{-7}
1 horsepower-hour =	2550	1.98×10^{6}	1	2.69×10^{6}	6.41×10^{5}	0.746
1 JOULE =	9.48×10^{-4}	0.738	3.73×10^{-7}	1	0.239	2.78×10^{-7}
1 calorie =	3.97×10^{-3}	3.09	1.56×10^{-6}	4.19	1	1.16×10^{-6}
1 kilowatt-hour =	3410	2.66×10^{6}	1.34	3.60×10^{6}	8.60×10^{5}	1

EXAMPLE: 1 BTU = 778 ft lb.

TABLE M

CONVERSION TABLE FOR POWER

	Btu/hr	ft lb/s	hp	cal/s	kW	WATT
1 British thermal unit per hour =	1	0.216	3.93×10^{-4}	7.00×10^{-2}	2.93×10^{-4}	0.293
1 foot pound per second =	4.63	1	1.82×10^{-3}	0.324	1.36×10^{-3}	1.36
1 horsepower =	2550	550	1	178	0.746	746
1 calorie per second =	14.3	3.09	5.61×10^{-3}	1	4.19×10^{-3}	4.19
1 kilowatt =	3410	738	1.34	239	1	1000
1 WATT =	3.41	0.738	1.34×10^{-3}	0.239	0.001	1

TABLE N
CONVERSION TABLE FOR PLANE ANGLES

	°	′	″	RADIAN	rev
1 degree =	1	60	3600	1.75×10^{-2}	2.78×10^{-3}
1 minute =	1.67×10^{-2}	1	60	2.91×10^{-4}	4.63×10^{-5}
1 second =	2.78×10^{-4}	1.67×10^{-2}	1	4.85×10^{-6}	7.72×10^{-7}
1 RADIAN =	57.3	3440	2.06×10^{5}	1	0.159
1 revolution =	360	2.16×10^{4}	1.30×10^{6}	6.28	1

TABLE O
CONVERSION TABLE FOR CHARGE

1 electronic charge = 1.60×10^{-19} coulomb
1 ampere-hour = 3600 coulomb

TABLE P
COPPER WIRE TABLE

Gage No.	Diameter in mils	Diameter in mm	Cross Section		Ohms per 1,000 ft		Weight per 1000 ft lb
			Cir mils	Sq In.	25°C (77°F)	65°C (149°F)	
0000	460.0		212,000	0.166	0.0500	0.0577	641.0
000	410.0		168,000	0.132	0.0630	0.0727	508.0
00	365.0		133,000	0.105	0.0795	0.0917	403.0
0	325.0		106,000	0.0829	0.100	0.116	319.0
1	289.0	7.35	83,700	0.0657	0.126	0.146	253.0
2	258.0	6.54	66,400	0.0521	0.159	0.184	201.0
3	229.0	5.83	52,600	0.0413	0.201	0.232	159.0
4	204.0	5.19	41,700	0.0328	0.253	0.292	126.0
5	182.0	4.62	33,100	0.0260	0.319	0.369	100.0
6	162.0	4.12	26,300	0.0206	0.403	0.465	79.5
7	144.0	3.67	20,800	0.0164	0.508	0.586	63.0
8	128.0	3.26	16,500	0.0130	0.641	0.739	50.0
9	114.0	2.91	13,100	0.0103	0.808	0.932	39.6
10	102.0	2.59	10,400	0.00815	1.02	1.18	31.4
11	91.0	2.31	8,230	0.00647	1.28	1.48	24.9
12	81.0	2.05	6,530	0.00513	1.62	1.87	19.8
13	72.0	1.83	5,180	0.00407	2.04	2.36	15.7
14	64.0	1.63	4,110	0.00323	2.58	2.97	12.4
15	57.0	1.45	3,260	0.00256	3.25	3.75	9.86
16	51.0	1.29	2,580	0.00203	4.09	4.73	7.82
17	45.0	1.15	2,050	0.00161	5.16	5.96	6.20
18	40.0	1.02	1,620	0.00128	6.51	7.51	4.92
19	36.0	0.91	1,290	0.00101	8.21	9.48	3.90
20	32.0	0.81	1,020	0.000802	10.4	11.9	3.09
21	28.5	0.72	810	0.000636	13.1	15.1	2.45
22	25.3	0.64	642	0.000505	16.5	19.0	1.94
23	22.6	0.57	509	0.000400	20.8	24.0	1.54
24	20.1	0.51	404	0.000317	26.2	30.2	1.22
25	17.9	0.46	320	0.000252	33.0	38.1	0.970
26	15.9	0.41	254	0.000200	41.6	48.0	0.769
27	14.2	0.36	202	0.000158	52.5	60.6	0.610
28	12.6	0.32	160	0.000126	66.2	76.4	0.484
29	11.3	0.29	127	0.0000995	83.4	96.3	0.384
30	10.0	0.26	101	0.0000789	105	121	0.304
31	8.9	0.23	79.7	0.0000626	133	153	0.241
32	8.0	0.20	63.2	0.0000496	167	193	0.191
33	7.1	0.18	50.1	0.0000394	211	243	0.152
34	6.3	0.16	39.8	0.0000312	266	307	0.120
35	5.6	0.14	31.5	0.0000248	335	387	0.0954
36	5.0	0.13	25.0	0.0000196	423	488	0.0757
37	4.5	0.11	19.8	0.0000156	533	616	0.0600
38	4.0	0.10	15.7	0.0000123	673	776	0.0476
39	3.5	0.09	12.5	0.0000098	848	979	0.0377
40	3.1	0.08	9.9	0.0000078	1070	1230	0.0200

TABLE Q
DECIMAL FRACTION AND COMMON FRACTION EQUIVALENTS

$\frac{1}{64} = 0.015625$	$\frac{11}{32} = 0.34375$	$\frac{43}{64} = 0.671875$
$\frac{1}{32} = 0.03125$	$\frac{23}{64} = 0.359375$	$\frac{11}{16} = 0.6875$
$\frac{3}{64} = 0.046875$	$\frac{3}{8} = 0.375$	$\frac{45}{64} = 0.703125$
$\frac{1}{16} = 0.0625$	$\frac{25}{64} = 0.390625$	$\frac{23}{32} = 0.71875$
$\frac{5}{64} = 0.078125$	$\frac{13}{32} = 0.40625$	$\frac{47}{64} = 0.734375$
$\frac{3}{32} = 0.09375$	$\frac{27}{64} = 0.421875$	$\frac{3}{4} = 0.75$
$\frac{7}{64} = 0.109375$	$\frac{7}{16} = 0.4375$	$\frac{49}{64} = 0.765625$
$\frac{1}{8} = 0.125$	$\frac{29}{64} = 0.453125$	$\frac{25}{32} = 0.78125$
$\frac{9}{64} = 0.140625$	$\frac{15}{32} = 0.46875$	$\frac{51}{64} = 0.796875$
$\frac{5}{32} = 0.15625$	$\frac{31}{64} = 0.484375$	$\frac{13}{16} = 0.8125$
$\frac{11}{64} = 0.171875$	$\frac{1}{2} = 0.5$	$\frac{53}{64} = 0.828125$
$\frac{3}{16} = 0.1875$	$\frac{33}{64} = 0.515625$	$\frac{27}{32} = 0.84375$
$\frac{13}{64} = 0.203125$	$\frac{17}{32} = 0.53125$	$\frac{55}{64} = 0.859375$
$\frac{7}{32} = 0.21875$	$\frac{35}{64} = 0.546875$	$\frac{7}{8} = 0.875$
$\frac{15}{64} = 0.234375$	$\frac{9}{16} = 0.5625$	$\frac{57}{64} = 0.890625$
$\frac{1}{4} = 0.25$	$\frac{37}{64} = 0.578125$	$\frac{29}{32} = 0.90625$
$\frac{17}{64} = 0.265625$	$\frac{19}{32} = 0.59375$	$\frac{59}{64} = 0.921875$
$\frac{9}{32} = 0.28125$	$\frac{39}{64} = 0.609375$	$\frac{15}{16} = 0.9375$
$\frac{19}{64} = 0.296875$	$\frac{5}{8} = 0.625$	$\frac{61}{64} = 0.953125$
$\frac{5}{16} = 0.3125$	$\frac{41}{64} = 0.640625$	$\frac{31}{32} = 0.96875$
$\frac{21}{64} = 0.328125$	$\frac{21}{32} = 0.65625$	$\frac{63}{64} = 0.984375$

$\frac{1}{12} = 0.083\bar{3}$	$\frac{5}{12} = 0.416\bar{6}$	$\frac{3}{4} = 0.75$	$\frac{1}{5} = 0.2$	$\frac{3}{5} = 0.6$
$\frac{1}{6} = 0.16\bar{6}$	$\frac{1}{2} = 0.5$	$\frac{5}{6} = 0.83\bar{3}$	$\frac{2}{5} = 0.4$	$\frac{4}{5} = 0.8$
$\frac{1}{4} = 0.25$	$\frac{7}{12} = 0.583\bar{3}$	$\frac{11}{12} = 0.916\bar{6}$		
$\frac{1}{3} = 0.3\bar{3}$	$\frac{2}{3} = 0.6\bar{6}$			

TABLE R
FORMULAS FROM GEOMETRY

PLANE FIGURES

In the following, a, b, c, d, and h are lengths of sides and altitudes, respectively.

		Perimeter	*Area*
Rectangle		$P = 2(a + b)$	$A = ab$
Square	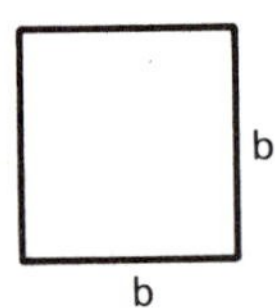	$P = 4b$	$A = b^2$
Parallelogram	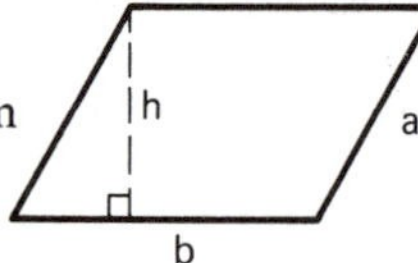	$P = 2(a + b)$	$A = bh$
Rhombus	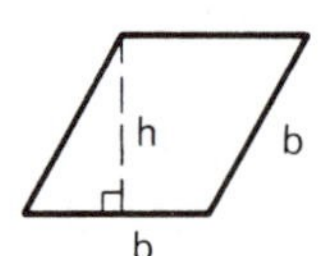	$P = 4b$	$A = bh$
Trapezoid	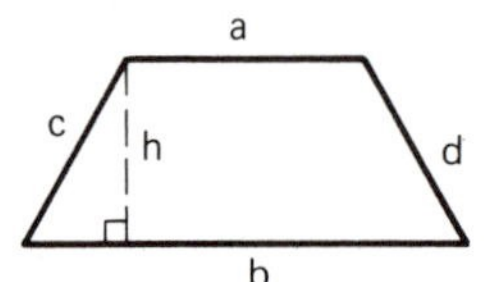	$P = a + b + c + d$	$A = \left(\frac{a + b}{2}\right)h$
Triangle	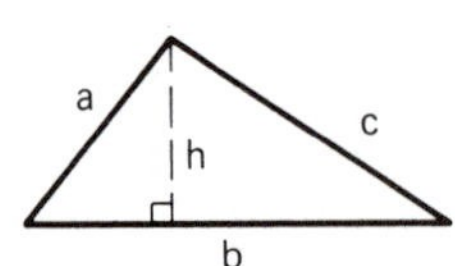	$P = a + b + c$	$A = \frac{1}{2}bh$

The sum of the measures of the angles of a triangle = 180°. In a right triangle:

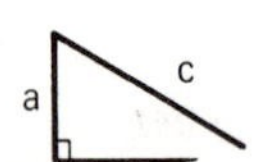

$c^2 = a^2 + b^2$ or $c = \sqrt{a^2 + b^2}$.

	Circumference	Area	
Circle	$C = \pi d$ $C = 2\pi r$	$A = \pi r^2$	$d = 2r$

The sum of the measures of the central angles of a circle $= 360°$.

GEOMETRIC SOLIDS

In the following, B, r, and h are the area of base, length of radius, and height, respectively.

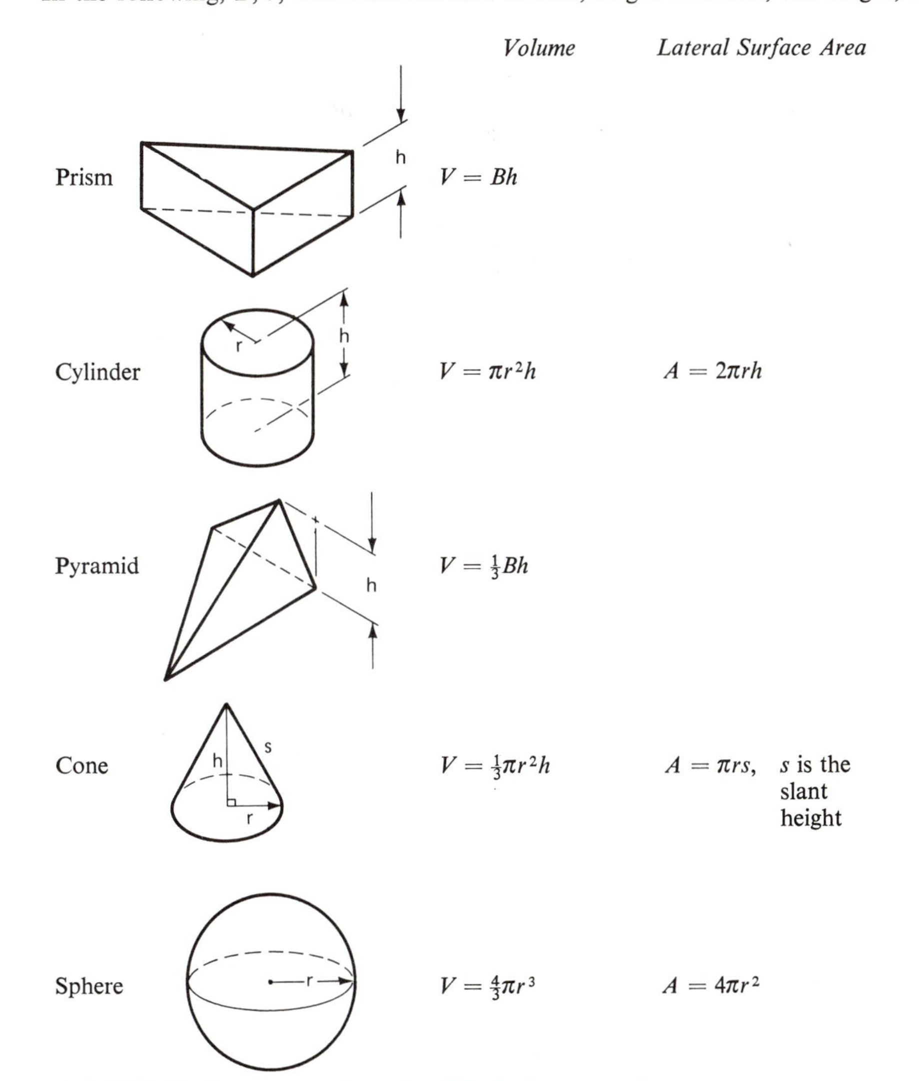

	Volume	Lateral Surface Area
Prism	$V = Bh$	
Cylinder	$V = \pi r^2 h$	$A = 2\pi rh$
Pyramid	$V = \frac{1}{3}Bh$	
Cone	$V = \frac{1}{3}\pi r^2 h$	$A = \pi rs$, s is the slant height
Sphere	$V = \frac{4}{3}\pi r^3$	$A = 4\pi r^2$

TABLE S
FORMULAS FROM TRIGONOMETRY

$$\sin A = \frac{\text{side opposite angle } A}{\text{hypotenuse}}$$

$$\cos A = \frac{\text{side adjacent angle } A}{\text{hypotenuse}}$$

$$\tan A = \frac{\text{side opposite angle } A}{\text{side adjacent angle } A}$$

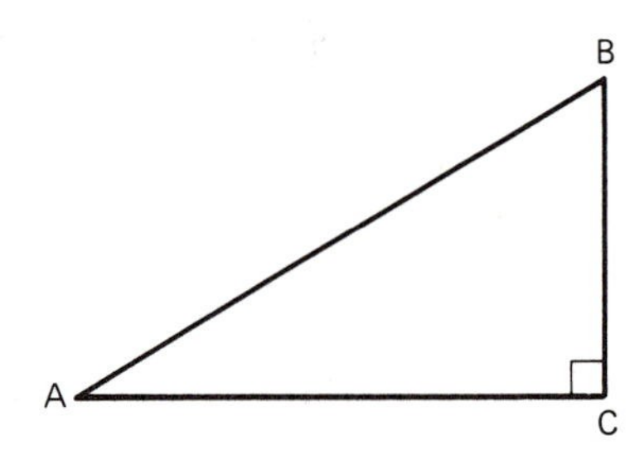

TABLE T
TRIGONOMETRY RATIOS

Angle	Sine	Cosine	Tangent	Angle	Sine	Cosine	Tangent
0	0.000	1.000	0.000	45	0.707	0.707	1.000
1	0.017	0.999	0.017	46	0.719	0.695	1.036
2	0.035	0.999	0.035	47	0.731	0.682	1.072
3	0.052	0.999	0.052	48	0.743	0.669	1.111
4	0.070	0.998	0.070	49	0.755	0.656	1.150
5	0.087	0.996	0.087	50	0.766	0.643	1.192
6	0.105	0.995	0.105	51	0.777	0.629	1.235
7	0.122	0.993	0.123	52	0.788	0.616	1.280
8	0.139	0.990	0.141	53	0.799	0.602	1.327
9	0.156	0.988	0.158	54	0.809	0.588	1.376
10	0.174	0.985	0.176	55	0.819	0.574	1.428
11	0.191	0.982	0.194	56	0.829	0.559	1.483
12	0.208	0.978	0.213	57	0.839	0.545	1.540
13	0.225	0.974	0.231	58	0.848	0.530	1.600
14	0.242	0.970	0.249	59	0.857	0.515	1.664
15	0.259	0.966	0.268	60	0.866	0.500	1.732
16	0.276	0.961	0.287	61	0.875	0.485	1.804
17	0.292	0.956	0.306	62	0.883	0.469	1.881
18	0.309	0.951	0.325	63	0.891	0.454	1.963
19	0.326	0.946	0.344	64	0.899	0.438	2.050
20	0.342	0.940	0.364	65	0.906	0.423	2.145
21	0.358	0.934	0.384	66	0.914	0.407	2.246
22	0.375	0.927	0.404	67	0.921	0.391	2.356
23	0.391	0.921	0.424	68	0.927	0.375	2.475
24	0.407	0.914	0.445	69	0.934	0.358	2.605
25	0.423	0.906	0.466	70	0.940	0.342	2.747
26	0.438	0.899	0.488	71	0.946	0.326	2.904
27	0.454	0.891	0.510	72	0.951	0.309	3.078
28	0.469	0.883	0.532	73	0.956	0.292	3.271
29	0.485	0.875	0.554	74	0.961	0.276	3.487
30	0.500	0.866	0.577	75	0.966	0.259	3.732
31	0.515	0.857	0.601	76	0.970	0.242	4.011
32	0.530	0.848	0.625	77	0.974	0.225	4.331
33	0.545	0.839	0.649	78	0.978	0.208	4.705
34	0.559	0.829	0.675	79	0.982	0.191	5.145
35	0.574	0.819	0.700	80	0.985	0.174	5.671
36	0.588	0.809	0.727	81	0.988	0.156	6.314
37	0.602	0.799	0.754	82	0.990	0.139	7.115
38	0.616	0.788	0.781	83	0.993	0.122	8.144
39	0.629	0.777	0.810	84	0.995	0.105	9.514
40	0.643	0.766	0.839	85	0.996	0.087	11.43
41	0.656	0.755	0.869	86	0.998	0.070	14.30
42	0.669	0.743	0.900	87	0.999	0.052	19.08
43	0.682	0.731	0.933	88	0.999	0.035	28.64
44	0.695	0.719	0.966	89	0.999	0.017	57.29
45	0.707	0.707	1.000	90	1.000	0.000	—

ANSWERS TO ODD-NUMBERED PROBLEMS

Note: Answers to even-numbered problems may be found in the Teacher's Guide.

CHAPTER 1

Pages 3–4

1. 3 yd, 9 ft, 274.32 cm
3. 412.16 km
5. (a) 19.48 ft
   (b) 2.795 in.
   (c) 3.048 cm
7. smaller
9. 9

Page 5

1. 720 s
3. 28,800 s
5. 26,280 hr
7. 106

Page 7

1. 15,575 N
3. 890 N
5. 418 N

Page 9

1. —18
3. —29
5. 3
7. —4
9. —18
11. —2
13. 35
15. —60
17. —4
19. 3

Page 11

1. 3.26×10^2
3. 8.264×10^2
5. 6.432×10^0
7. 2.24×10^{-3}
9. 2.99×10^{-4}
11. 7.32×10^{17}

Page 11

1. 8.62
3. 0.000631
5. 0.768
7. 777,000,000
9. 69.3
11. 96,100

Page 12

1. 10^{-1}
3. 10^{20}
5. 0.000030 or 3.0×10^{-5}
7. 1.04

Page 13

1. 10^1
3. 10^0 or 1
5. 0.04 or 4×10^{-2}
7. 70,000 or 7×10^4

Page 14

1. 3
3. 4
5. 2
7. 3
9. 5
11. 3

CHAPTER 2

Pages 19–21

1. 50.3 cm^2
3. 9.62 in^2
5. 101 ft^2
7. large lot = 14,300 ft^2
   lot 1 = 2100 ft^2
   lot 2 = 3800 ft^2
   lot 3 = 3850 ft^2
   lot 4 = 4550 ft^2
9. large end = 10.7 cm^2
   small end = 3.46 cm^2
11. 22.0 in^2

Page 22

1. 4.18 m^2
3. 1.39 m^2
5. 1940 ft^2
7. 0.687 ft^2
9. 7.91 in^2
11. 9.67 cm^2
13. 25.8 cm^2

Pages 26–27

1. lateral surface area = 36.0 cm^2
   total surface area = 76.0 cm^2
   volume = 40.0 cm^3
3. 284 cm^3
5. 2170 ft^2
7. 76.9 in^2
9. 143 in^3
11. 127 in^2

Page 28

1. 513 ft^3
3. 1.77 in^3
5. 1440 cm^3
7. 0.0486 ft^3
9. 255,000 cm^3
11. 623 cm^3

CHAPTER 3

Page 32

1. (a) 3
   (b) 4
   (c) 2
   (d) 1.4
   (e) 0.6
   (f) 3.6
   (g) 2.4

Page 35

1. 61 miles at 35°
3. 1300 miles at 181°
5. 36 miles at 5°
7. 38 miles at 296°
9. 3700 miles at 336°
11. 15 miles at 126°
13. 6.5 miles at 7°

Page 36

1. 80
3. 90
5. 50
7. 120
9. 33.3 mi/hr, north
11. 62 mi/hr, west

Page 38

1. 10 ft/s^2
3. 10 ft/s^2
5. 20 ft/s^2

CHAPTER 4

Pages 41–42

1. $x = \frac{4}{3}$
3. $x = 17$
5. $x = 0$
7. $x = \frac{35}{2}$ or 17.5
9. $x = 3$
11. $y = \dfrac{B}{A}$
13. $y = 7$
15. $y = 12.5$
17. $x = \dfrac{C + D}{AB}$
19. $x = 26.5$

Page 43

1. $y = 12$
3. $x = 55$
5. $m = -150$
7. $l = -15.5$
9. $w = 1\frac{5}{7}$ or $\frac{12}{7}$
11. $m = 99$
13. $k = 42$
15. $h = 2$
17. $x = -50$
19. $y = 6\frac{2}{17}$

Page 45

1. $s = vt$
3. $x_i = x - v_i t - \frac{1}{2}at^2$
5. $a = \dfrac{v^2 - v_i^2}{2(s - s_i)}$
7. $t = \dfrac{w}{p}$
9. $v = \dfrac{2KE}{m}$
11. $g = \dfrac{PE}{mh}$
13. $a = \dfrac{F}{m}$
15. $v_i = 2v_{ave} - v_f$

Page 46

1. $x = \pm 6$
3. $x = \pm 7$
5. $x = \pm 3$
7. $x = \pm 0.8$
9. $t = \sqrt{2d/a}$

CHAPTER 5

Pages 52–53

1. 5.05 ft/s
3. 128 ft
5. 2.13 ft/s^2

Page 53

1. 4.38 s
3. 1.55 s
5. 34.5 s
7. 4.40 s
9. 81.3 ft/s

CHAPTER 6

Pages 60–61

1. 30.0
3. 744
5. 252
7. 1.71
9. 11.7
11. 0.518
13. 5250 N
15. 1320 lb
17. 400 ft/s^2
19. (a) 14.0 ft/s^2
    (b) 10.6 ft/s^2

Pages 64–65

1. 15.0, left
3. 3.00, right
5. 4.00, left
7. 4.00 ft/s^2
9. 0.509 m/s^2

Page 66

1. 322
3. 1.73
5. 0.652
7. 77.1

Page 69

1. 80.0
3. 765

5. 902
7. (a) 12,600 slug ft/s
   (b) 107 mi/hr
   (c) 5800 lb, 2580 lb
9. (a) 55,200 kg m/s
   (b) 47.2 m/s

CHAPTER 7

Pages 74–76

1. 0.940
3. 0.384
5. 0.982
7. 0.656
9. 0.306
11. 27°
13. 4°
15. 46°
17. 67°
19. 49°
21. $B = 65°$
    $a = 8.46$
    $b = 18.1$
23. $A = 48°$
    $B = 42°$
    $a = 12.5$
25. $A = 24°$
    $B = 66°$
    $c = 24.1$
27. $B = 70°$
    $b = 23.9$
    $c = 25.4$
29. $A = 50°$
    $a = 18.2$
    $b = 15.2$
31. (a) 10°
    (b) 0.704 in.
    (c) 2.41 in.
33. $C = 2.72$ in.
    $D = 2.28$ in.

Pages 77–78

1. $b = 8.49$
3. $c = 21.6$
5. $a = 10.2$
7. $c = 24.8$
9. $a = 8.60$
11. $C = 16.1$ cm, $B = 8.0$ cm

Pages 83–85

1.

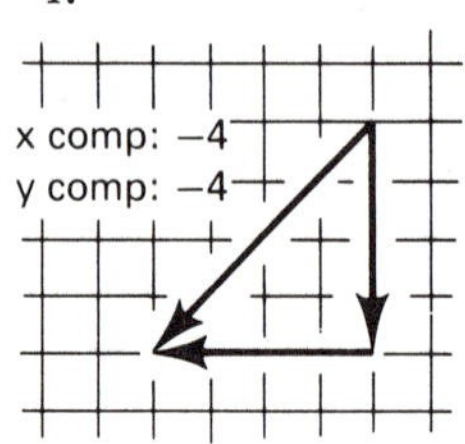

3.

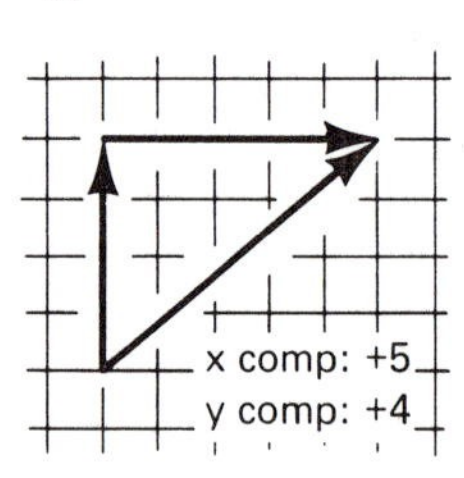

5.

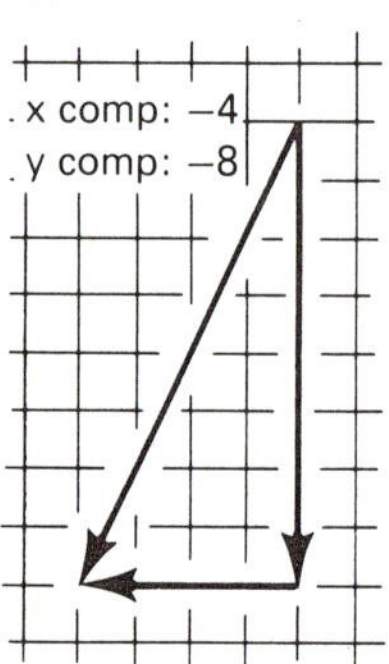

7.

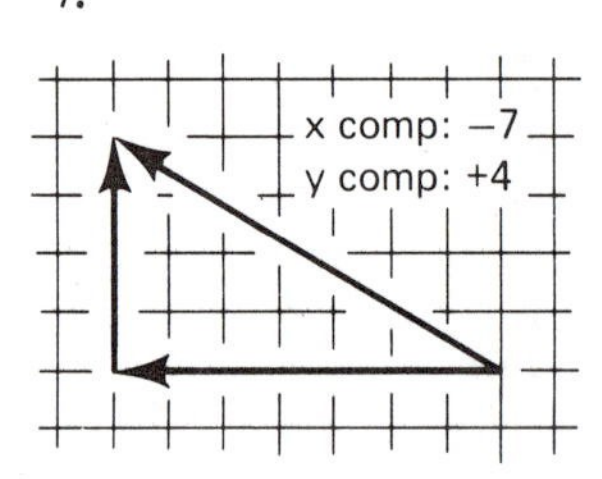

9.

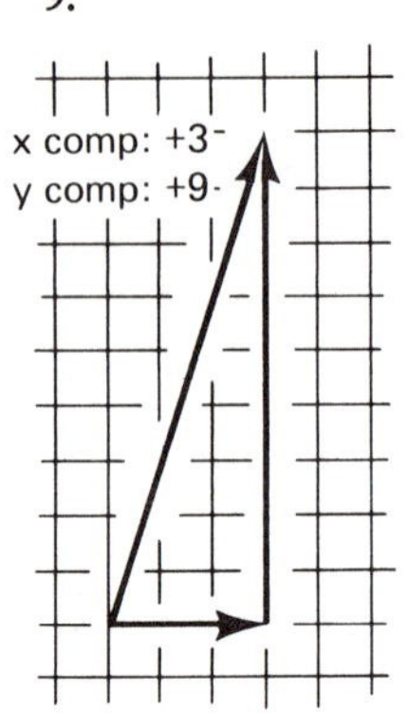

11.

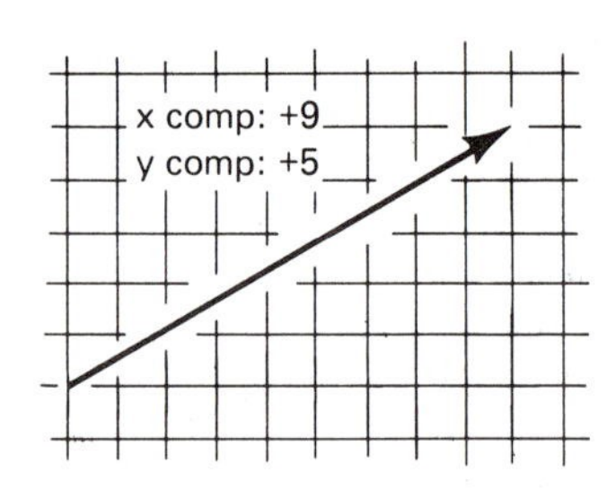

13.

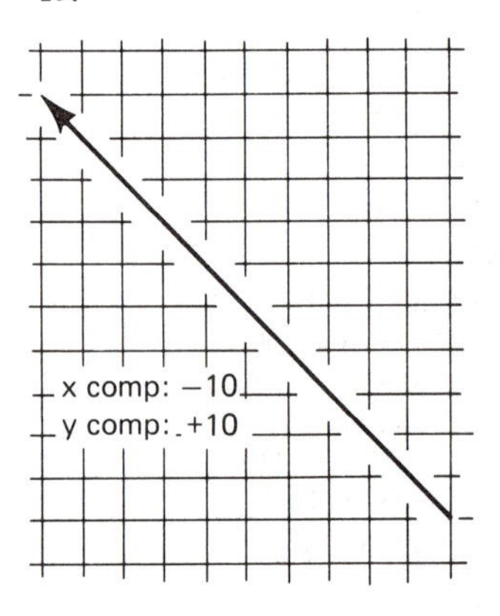

15.

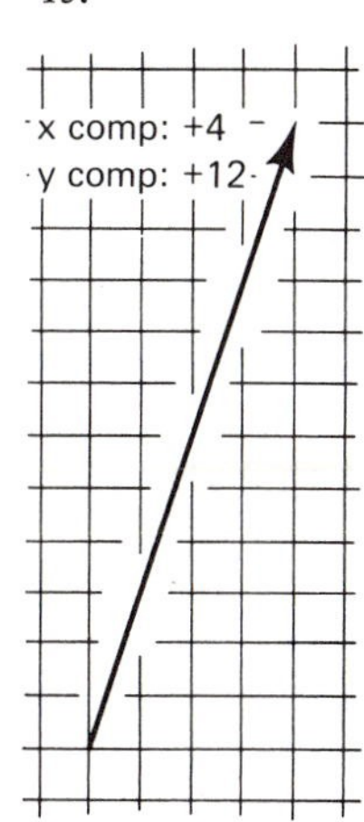

17.

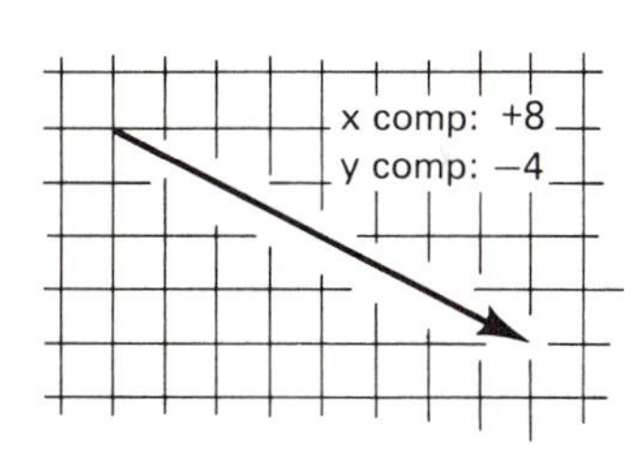

19.

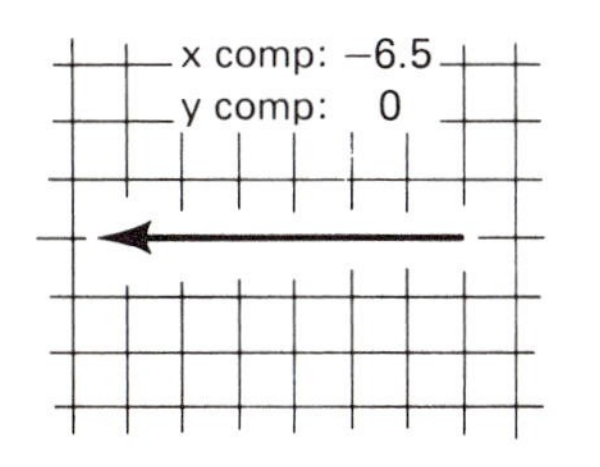

Pages 92–93

	x component	y component
1.	+5.67	+12.1
3.	−4.60	−4.60
5.	+10.4	−4.86
7.	−7.04	−26.3
9.	−5.36	+1.95

	Bearing	Magnitude
11.	32°	9.43
13.	248°	20.6
15.	137°	40.3
17.	251°	39.8
19.	117°	49.2

CHAPTER 8

Pages 96–97

1. 100 lb
3. 500 N
5. 200 lb
7. 400 N

Pages 101–102

1. $F_1 = 70.7$ N
   $F_2 = 70.7$ N
3. $F_1 = 823$ N
   $F_2 = 475$ N
5. $F_1 = 577$ lb
   $F_2 = 289$ lb
7. $F_1 = 433$ lb
   $F_2 = 500$ lb
9. $T_1 = 1440$ lb
   $T_2 = 1440$ lb
11. $C = 3000$ lb
    $T = 2600$ lb
13. 5540 N

CHAPTER 9

Page 107

1. 34.3 ft lb
3. 0.445 N
5. 21,600 ft lb
7. 4410 J
9. 0.300 N
11. 1.67×10^5 ft lb

Pages 111–112

1. 18.9 ft lb/s
3. 5 W
5. 12.4 ft lb/s
7. (a) 1.68 hp
   (b) 219 N
9. 59.6 s
11. 1490 W

Pages 116–117

1. 8080 ft lb
3. 217 J
5. (a) 80.7 ft/s
   (b) 3.09×10^6 ft lb
7. 4.48 ft/s
9. (a) 13.2 ft lb
   (b) 21.3 ft lb
11. 27.8 ft/s
13. 72.8 m/s

CHAPTER 10

Pages 125–126

1. 14.8
3. 36.3
5. 52.4
7. 2.39
9. 48.8
11. 1.55
13. 2.27
15. 41.1
17. 4.00
19. 2.50
21. 4.00
23. 0.500

Pages 128–129

1. 14.0
3. 27.1
5. 52.4
7. 48.8
9. 20.4
11. 438 lb
13. 429 N
15. 2010 N
17. 3.34

Pages 132–134

1. 1
3. 3
5. 6
7. 2

9.

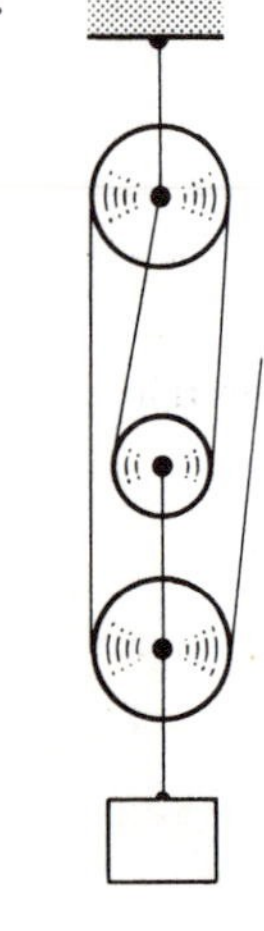

11.

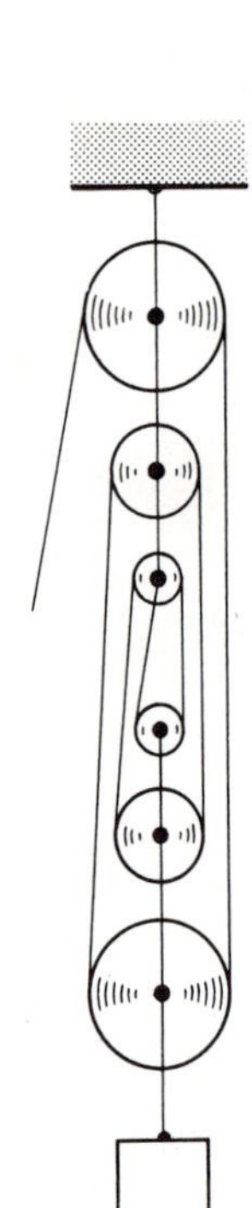

13.

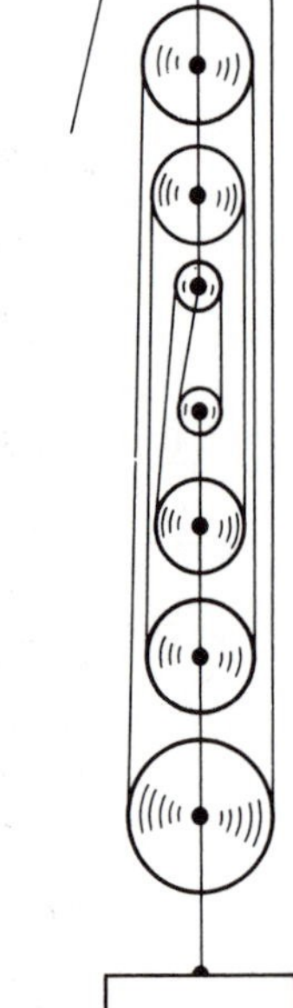

15. 2.00
17. 30 ft
19. 82.0 m
21. 100 lb
    120 ft

Pages 136–137

1. 12.8
3. 21.2
5. 36.3
7. 4.62
9. 5.61
11. 4.00
13. No
15. 3.84
17. 1.33 m

Pages 138–139

1. 2.30
3. 142
5. 2.28
7. 29.3
9. 35.2
11. 4.85 lb
13. 0.887 in.
15. 37.7
17. 189 lb
19. 33,500 N

Page 141

1. 15.0
3. 40.0
5. 75.0 lb

CHAPTER 11

Pages 146–147

1. 154 rev/min
3. 660 rev/min
5. 26.5 rev/s

Pages 148–149

1. 96.0 lb ft
3. 1490 lb
5. 84.3 lb ft
7. 3.36 lb ft

Page 151

1. 4350 N
3. 79.4 slugs
5. 5.53 m/s
7. 23.2 N
9. 135 ft

Pages 154–155

1. 18.7 hp
3. 7280 ft lb/s
5. 4.23 ft lb
7. 1.64 ft lb
9. 34.7 rev/min

CHAPTER 12

Pages 161–162

1. 73.5
3. 200
5. 858
7. 13
9. 167
11. 44.9 rpm
13. 75 teeth
15. 75 teeth
17. 42 teeth
19. 130 teeth
21. 69 teeth

Pages 166–170

1. counterclockwise
3. clockwise
5. clockwise
7. clockwise
9. clockwise
11. 1160 rpm
13. 576 rpm
15. 1480 rpm
17. 40 teeth
19. 20 teeth

Pages 172–173

1. 1930
3. 3600
5. 147
7. 71.4 rpm
9. 23.2 in.
11. clockwise
13. counter clockwise
15. counter clockwise

CHAPTER 13

Page 179

1. 62 lb; 103 lb
3. 27,600 N; 10,700 N

Pages 181–182

1. 22.6
3. 42.0
5. (a) 9850 N
   (b) 15,150 N
7. (a) 18,300 lb
   (b) 23,700 lb
9. (a) 20.5 lb
   (b) 22.5 lb
11. 101,000 N; 90,000 N

CHAPTER 14

Page 188

1. 50.0 in.
3. 49.0 N
5. 0.00977 in.
7. 90.0 cm

Pages 194–195

1. 2.80 g/cm^3
3. 1750 lb
5. 5420 cm^3
7. 1200 lb/ft^3
9. 2.70 g/cm^3
11. 0.680 g/cm^3
13. 55.1 gal
15. 0.0340 lb/in^3
17. 1,490,000 cm^3

CHAPTER 15

Page 201

1. 21.7 lb/in^2
3. 16.5 lb/ft^3
5. 4900 lb
7. 0.164 ft or 1.97 in.

Page 203

1. 60 lb
3. 12.5
5. 6.67
7. 33.3 N
9. 3600 lb

Page 206

1. 13 lb
3. 75.5 lb
5. 30,600 ft^3; 795 tons

CHAPTER 16

Page 212

1. 21.1°
3. 121°
5. 257°
7. —12.2°
9. 23°
11. 610°
13. 140°
15. 223°
17. 5727°
19. 899°C

Page 217

1. 173
3. 3800
5. 51,000
7. 278
9. 39,100 BTU
11. 130,000 cal

Page 219

1. 428°F
3. 0.051 cal/g°C
5. 95.4°F
7. 80.9°C

CHAPTER 17

Page 224

1. 0.304 ft
3. 0.102 m
5. 200.104 m
7. 0.540 ft
9. 0.75146 in.

Page 225

1. 0.329 in^2
3. 88.969 cm^2
5. 60.0461 cm^2

Page 227

1. 652.9 l
3. 299.2 ft^3
5. 3754.3 ft^3
7. 0.581 cm^3

CHAPTER 18

Page 231

1. 288°K
3. 44°C
5. 532°R
7. 90°F
9. 222 cm^3
11. −16°F
13. 380 cm^3
15. 1380 l

Page 233

1. 237 cm^3
3. 75.0 psi
5. 1.58 g/cm^3
7. 0.180 lb/ft^3
9. 62.5 cm
11. 2440 g/cm^3
13. 75.0 psi
15. 20.8 ft^3

Pages 234–235

1. 1280 in^3
3. 765 cm^3
5. 2010°R
7. 295°C
9. 2200 psi

CHAPTER 19

Page 242

1. 1120 cal
3. 10,700 BTU
5. 25,600 cal
7. 6070 BTU
9. 11,837.4 BTU

CHAPTER 21

Pages 253–254

1. 1.07 Ω
3. 0.0165 Ω/ft
5. 1.04 Ω
7. 0.684 Ω

CHAPTER 22

Page 257

1. 9.58 A
3. 220 V

Page 263

1. 14.0 Ω
3. 60.0 Ω
5. 0.750 A
7. 378 V
9. 23.0 Ω

Pages 268–269

1. 3.44 Ω
3. 2.00 A
5. 1.21 Ω
7. 1.67 A
9. 30.3 V

Pages 273–275

1. 3.00 Ω
3. 8.89 A
5. 2.23 A
7. 21.24 Ω
9. 44.6 V
11. 5.41 A
13. 10.0 Ω
15. 17.31 Ω
17. 30.0 V
19. 6.93 V
21. 14.44 Ω
23. 55.6 V
25. 0.120 A

Pages 276–279

1.

	V	I	R
Batt.	12.0 V	9.00 A	1.33 Ω
R_1	12.0 V	6.00 A	2.00 Ω
R_2	12.0 V	3.00 A	4.00 Ω

3.

	V	I	R
Batt.	36.0 V	6.00 A	6.00 Ω
R_1	36.0 V	2.00 A	18.0 Ω
R_2	36.0 V	3.00 A	12.0 Ω
R_3	36.0 V	1.00 A	36.0 Ω

5.

	V	I	R
Batt.	50.0 V	5.00 A	10.0 Ω
R_1	25.0 V	2.00 A	12.5 Ω
R_2	25.0 V	2.00 A	12.5 Ω
R_3	10.0 V	3.00 A	3.33 Ω
R_4	40.0 V	3.00 A	13.3 Ω

7.

	V	I	R
Batt.	36.0 V	9.00 A	4.00 Ω
R_1	12.0 V	6.00 A	2.00 Ω
R_2	12.0 V	3.00 A	4.00 Ω
R_3	24.0 V	6.00 A	4.00 Ω
R_4	24.0 V	3.00 A	8.00 Ω

9.

	V	I	R
Batt.	80.0 V	12.0 A	6.67 Ω
R_1	18.0 V	4.00 A	4.50 Ω
R_2	18.0 V	2.00 A	9.00 Ω
R_3	18.0 V	6.00 A	3.00 Ω
R_4	48.0 V	12.0 A	4.00 Ω
R_5	8.00 V	4.00 A	2.00 Ω
R_6	8.00 V	8.00 A	1.00 Ω
R_7	6.00 V	12.0 A	0.500 Ω

11.

	V	I	R
Batt.	65.0 V	5.00 A	13.0 Ω
R_1	10.0 V	0.500 A	20.0 Ω
R_2	10.0 V	1.00 A	10.0 Ω
R_3	10.0 V	2.50 A	4.00 Ω
R_4	10.0 V	1.00 A	10.0 Ω
R_5	25.0 V	5.00 A	5.00 Ω
R_6	30.0 V	5.00 A	6.00 Ω

CHAPTER 23

Pages 284–285

1. 1.4875 V
3. (a) 0.850 A
   (b) 4.50 V
   (c) 0.150 Ω
5. 0.405 A
7. 1.50 Ω

Page 288

1. 957 W
3. 0.455 A
5. 6.82 A
7. $.34

CHAPTER 27

Pages 312–313

1. 29.6 V
3. 113 V
5. 1560 V
7. 117 V
9. 12.0 A
11. 432 W
13. 26.9 Ω
15. 3.42 A
17. 12,600 kW

CHAPTER 28

Pages 320–321

1. 10 turns
3. 58.5 V
5. 350 turns
7. 4.67 A
9. 0.0733 A
11. 2200 V
13. 0.255 A

CHAPTER 29

Page 325

1. 1.88
3. 44.0
5. 251
7. 2.39
9. 0.154

Pages 327–328

1. 221
3. 3300
5. 304
7. 19°
9. 48°
11. 0.226 A
13. 0.00455 A
15. 0.0263 A

Page 329

1. 3.98
3. 2.65
5. 0.00199

Page 332

1. 1280
3. 8.64
5. 33.3
7. 14°
9. 8°
11. 0.0781 A
13. 1.74 A
15. 0.450 A

Page 334

1. $Z = 141\ \Omega$
   $I = 0.0355$ A
3. $Z = 636\ \Omega$
   $I = 0.0236$ A

Page 336

1. 2.25×10^4/s
3. 37.6/s

CHAPTER 30

Page 340

1. 1.50×10^9 m
3. 0.108 s
5. 500 s

Page 341

1. 7.48×10^{12} Hz
3. 3.08×10^{-4} m
5. 2,140 m

Page 342

1. 603
3. 43.2
5. 942

Page 344

1. 1.46 foot-candles
3. 3.51 foot-candles
5. 4750 lumens
7. 608 lumens

APPENDIX

Page 349

1. 5.40
3. 82.4
5. 3.58
7. 13.3
9. 75.1

Page 350

1. 598
3. 13,500
5. 1530
7. 0.000531
9. 0.306
11. 295 ft^2
13. 18.9 in^2
15. $286

Page 352

1. 0.0685
3. 0.0267
5. 5.81
7. 0.0270
9. 1.40
11. 0.292
13. 901
15. 0.000200

Page 355

1. 22.0
3. 0.661
5. 92.5
7. 15,500
9. 103
11. 0.199
13. 2860
15. 40,000
17. 51.9
19. 37,300
21. 70.0
23. 907

Page 356

1. 20.3
3. 70.6
5. 24,100
7. 8980
9. 6110
11. 224
13. 31.0
15. 4.70
17. 25,800
19. 1700

Page 357

1. 45.7
3. 96,700
5. 3,610,000,000
7. 0.00000250

Page 359

1. 0.818
3. 31,000
5. 4.95
7. 85.9
9. 17.0
11. 287
13. 1,420,000
15. 0.691
17. 23,800
19. 0.0331

Page 360

1. 3.72
3. 0.262
5. 6910
7. 0.00372
9. 538
11. 21,000
13. 0.320
15. 313
17. 0.498
19. 27,900,000

Page 362

1. 16.5
3. 0.267
5. 75.6
7. 0.303
9. 3.99
11. 397
13. 0.0247
15. 0.640
17. 2.72

19. 40.0
21. 0.687
23. 29.5
25. 0.000816

Page 362

1. 0.0299
3. 290
5. 0.0259
7. 6.26
9. 29,600
11. 0.691
13. 1,040,000
15. 67.2
17. 0.0412
19. 6810

Page 364

1. 256
3. 391,000
5. 0.00509
7. 261,000
9. 1.60

Page 366

1. 1.87
3. 7.49
5. 0.850
7. 0.192
9. 0.808

Page 366

1. 64.5
3. 40,700
5. 8.31
7. 0.0000507
9. 43.9
11. 1480
13. 6.89
15. 4.45
17. 85.7
19. 0.000358

Page 367

1. 24.1
3. 106
5. 710
7. 62,700
9. 22.8
11. 25.9
13. 36.7
15. 3.04

Page 368

1. 23.0
3. 740
5. 3.03
7. 10.1
9. 1.40
11. 6.85
13. 13,200
15. 2.09
17. 2620
19. 0.115
21. 0.460
23. 0.212
25. 182
27. 17.9
29. 15.8
31. 1.68
33. 1.17
35. 98.4
37. 7.46
39. 3.22
41. 12,800
43. 2.36
45. 7.12
47. 6040
49. 1.61

INDEX